MONITORING AND CONTROL OF INFORMATION-POOR SYSTEMS

MONITORING AND CONTROL OF INFORMATION-POOR SYSTEMS

AN APPROACH BASED ON FUZZY RELATIONAL MODELS

Arthur L. Dexter
Department of Engineering Science
University of Oxford
UK

A John Wiley & Sons, Ltd., Publication

Library of Congress Cataloging-in-Publication Data

Dexter, A. L. (Arthur L.)
  Monitoring and control of information-poor systems : an approach based on fuzzy relational models / Arthur L. Dexter.
    p. cm.
  Includes bibliographical references and index.
  ISBN 978-0-470-68869-4 (cloth)
  1. Engineering–Statistical methods.   2. Uncertainty–Mathematical models.   3. Fuzzy systems.
4. Control theory.   I. Title.
  TA340.D49 2012
  620.001′519542–dc23

                                                                              2011046728

A catalogue record for this book is available from the British Library.

Print ISBN: 978-0-470-68869-4

Typeset in 10/12pt Times by Aptara Inc., New Delhi, India
Printed and bound in Singapore by Markono Print Media Pte Ltd.

Contents

Preface

The monitoring and control of highly uncertain systems is an important and challenging practical problem. Methods of solution based on fuzzy techniques have generated considerable interest and the research literature is extensive. Perhaps surprisingly, most of the textbooks on fuzzy systems concentrate on the modelling and control of non-linear systems and very few consider explicit ways of taking uncertainties into account. The main exceptions are the seminal texts by Kacprzyk on multi-stage fuzzy control and by Mendel on type-2 fuzzy systems (although neither of these books deals with online control applications) and the book by Sousa and Kaymak that focuses on the use of fuzzy decision-making in modelling and control.

This book describes an approach to the monitoring and control of information-poor systems that is based on fuzzy relational models which generate fuzzy outputs. Suitable for adoption as a graduate course text, it should also be of interest to practising control engineers.

The book is divided into four parts. The aim of the first part of the book is to clarify why design decisions must take account of the uncertainty associated with optimal choices, and to explain how a fuzzy relational model (FRM) can be used to generate a fuzzy output which reflects the uncertainties associated with its predictions. The aim of the second part of the book is to give a brief introduction to fuzzy decision-making and to show how it can be used to design a predictive control scheme that is suitable for controlling information-poor systems using inaccurate measurements. The aim of the penultimate part of the book is to describe different ways in which fuzzy relational models generating fuzzy outputs can be identified online and to explain the practical issues associated with their online identification and application. The aim of the final part is to provide illustrative examples of the use of the previously described techniques in real applications.

Part I examines different approaches to analysing the behaviour of information-poor systems.

Chapter 1 introduces the reader to different types of information-poor systems, explains the basic sources of uncertainty and discusses the main considerations when designing or controlling an information-poor system.

Chapter 2 compares different ways of describing and propagating uncertainty. Its aim is to explain the limitations of using statistical approaches and to describe the advantages of the fuzzy approach in dealing with uncertainty in non-linear systems. Different methods of defining uncertainty (uncertainty intervals and probability distributions, fuzzy sets and fuzzy numbers) and propagating uncertainty (root-sum-squares, interval arithmetic, statistical methods, Monte Carlo methods and fuzzy arithmetic) are described, and a technique for aggregating fuzzy sets based on the extension principle is explained.

Chapter 3 explains how fuzzy random variables can be used to describe both the random and systematic errors associated with many measurements. A brief introduction to the theory of evidence and possibility theory is given and the definition of a fuzzy random variable is explained. A hybrid approach to the propagation of uncertainty is described. An approach to fuzzy sensor fusion based on the extension principle is described and the design of fuzzy sensors is explained.

Chapter 4 considers the most commonly used fuzzy models and explains why they do not give any indication of the uncertainty associated with their predictions. An introduction is given to linguistic fuzzy models, functional fuzzy models and fuzzy neural networks. Methods of identifying fuzzy models from data are described and a way of modifying expert rules to take account of uncertainty by using type-2 fuzzy sets is explained. The disadvantages of defuzzification are discussed.

Chapter 5 explains how a fuzzy relational model can be used to generate a fuzzy output which reflects the uncertainties associated with its predictions (the fuzzy FRM), and demonstrates that precise estimation of the model's parameters is inappropriate and unnecessary. The concept of the generic fuzzy model is also introduced. Methods of estimating rule confidences from data are described, and a probabilistic interpretation of one fuzzy identification scheme is suggested. The effect on the output of a fuzzy FRM of structural errors and estimation based on limited amounts of training data are considered. A method of reducing the time required to generate the training data needed to identify a generic fuzzy model is described.

Part II considers the problem of controlling information-poor systems.

Chapter 6 gives a brief introduction to fuzzy decision-making and shows how it can be used to solve multi-step optimization problems in information-poor systems. A fuzzy approach to risk assessment in information-poor systems is described. The ideas behind fuzzy optimization (fuzzy goals and fuzzy constraints, fuzzy aggregation operators) are discussed and the basic methods of fuzzy ranking are reviewed. Multi-stage fuzzy decision-making based on fuzzy dynamic programming is explained and two widely used search methods (branch and bound, genetic algorithms) are described. Fuzzy decision-making based on intuitionistic fuzzy sets is also discussed.

Chapter 7 shows how a fuzzy FRM and fuzzy decision-making can be used to design a model-based predictive control scheme that is suitable for controlling information-poor systems. Two fuzzy approaches to the model-based control of uncertain systems are described. The practical limitations of multi-step fuzzy decision-making (the accumulation of uncertainty, the excessive computational demands and infeasibility issues) are discussed. A simplified approach to fuzzy FRM-based predictive control is described and a method of conditional defuzzification is suggested.

Chapter 8 explains how inaccurate measurements can be incorporated into fuzzy control schemes. The use of fuzzy references and fuzzy measurements is discussed and different ways of finding fuzzy measures of the tracking error and its derivative are described. A fuzzy inference scheme is described that can deal with fuzzy inputs. Fuzzy neural networks that are capable of generating fuzzy outputs when presented with fuzzy inputs are explained. A method of modelling input uncertainty using a fuzzy FRM is proposed and examples of its use are presented.

Chapter 9 considers ways in which unmeasured and poorly measured disturbances can be rejected in information-poor systems. A brief introduction is given to robust fuzzy control and feedback linearization using a fuzzy disturbance observer. A method of rejecting unmeasured disturbances using an internal model control scheme that is based on a fuzzy FRM, and a fuzzy FRM-based predictive controller that rejects poorly measured disturbances using a fuzzy disturbance measurement, are described. Results are presented that demonstrate the effectiveness of the proposed control schemes.

Part III addresses the problems associated with on-line learning in information-poor systems.

Chapter 10 describes different ways in which fuzzy FRMs can be identified online and explains the practical issues that must be taken into account. Several online fuzzy identification schemes (recursive fuzzy least-squares and recursive forms of the RSK method) are described and ways of reducing the effect of poor-quality and incomplete training data (incorporating prior knowledge, improving the information content of the training data, taking data distribution into account) are explained. Two approaches to reducing the associated computational demands are discussed (evolving FRMs and hierarchical FRMs).

Chapter 11 considers the issues associated with applying an adaptive version of the fuzzy FRM-based control scheme to information-poor systems. A brief introduction is given to robust adaptive fuzzy control and an adaptive version of the fuzzy FRM-based predictive controller is proposed. Ways of incorporating prior knowledge are considered, and a scheme for initializing the adaptive controllers using a generic FRM is suggested. Three methods of generating an optimal control signal when the model is only partially trained (weighted fuzzy decision-making, secondary controller and multi-model schemes) are described. Results are presented that show how well a feedforward version of the adaptive controller can reject the effects of a disturbance when the disturbance measurement is inaccurate.

Chapter 12 explains how a model-free control scheme, which is suitable for use in information-poor systems, can be designed. A brief introduction to model-free adaptive control is given and a fuzzy FRM-based direct adaptive control scheme based on feedback error learning is described. Results are presented that show how the controller can account for noise on the measurement of the controlled variable and handle unmeasured disturbances. Disturbance rejection using a fuzzy FRM-based direct adaptive controller with feedforward is also considered, and a way of accounting for noise on the measurement of an input disturbance is proposed.

Chapter 13 show how FRMs can be used to diagnose faults in information-poor systems without generating false alarms. An introduction is given to fault detection and isolation (FDI) in non-linear uncertain systems and a fuzzy FRM-based approach to FDI is explained. Schemes are described for measuring the similarity of FRMs, accumulating evidence of fault-free or faulty operation and generating robust generic models of faulty operation. A multi-step method of fault diagnosis is proposed.

Part IV presents some examples of the use of the previously described techniques in real applications.

Chapter 14 considers the problem of controlling thermal comfort in a large building. The most important sources of uncertainty are identified and the design constraints are specified. Previous approaches to dealing with the uncertainty are reviewed and the design of a fuzzy FRM-based controller is described. Results are presented that demonstrate the advantages of the proposed fuzzy controller in comparison to conventional PI control.

Chapter 15 considers the problem of identifying faults in air-conditioning systems. The most important sources of uncertainty are identified and design constraints are specified.

Previous approaches to dealing with the uncertainty are reviewed and the design of a fuzzy FRM-based monitoring scheme is described. Results obtained from computer simulation and from field trials in a real office building are used to evaluate the scheme.

Chapter 16 considers the problem of controlling heat exchangers. The most important sources of uncertainty are identified and design constraints are specified. Previous approaches to dealing with the uncertainty are reviewed and the design of a fuzzy FRM-based predictive controller, capable of controlling the outlet air temperature from a water-to-air heat exchanger, is described. Results are presented that compare the performance of the predictive controller to conventional PI control. The design of an internal model control scheme based on a fuzzy FRM is also described. In this case, the control performance is compared to that of a more conventional internal model control scheme based on a fuzzy model whose output is defuzzified.

Chapter 17 considers the problem of measuring spatially distributed quantities. A brief introduction to the issues associated with measuring spatial averages is given, and approaches to dealing with sensor bias are reviewed. As an example, two ways of estimating the average temperature of air flowing down a large duct are described and an uncertainty analysis is undertaken. A method of using fuzzy data fusion to improve the measurement accuracy is described, and results are present that demonstrate the advantages of using this technique in an automated commissioning scheme.

About the Author

Arthur Dexter was educated at Rotherham Grammar School and St Catherine's College, Oxford. After completing his doctorate, he spent one year as the Gas Council Research Fellow in the Department of Engineering Science at Oxford before leaving in 1971 to accept the offer of a lectureship at Trinity College Dublin. He returned to Oxford in 1981 to take up a tutorial fellowship at Worcester College and a lectureship in the Department of Engineering Science. He was subsequently promoted, first to a Reader in Engineering Science in 1996 and then to a Professor of Engineering Science in 2002. He retired from this post in 2010 and is now an Emeritus Fellow of Worcester College. He is a member of the Institution of Engineering and Technology and a UK Chartered Engineer.

Throughout his career he has worked on a wide variety of projects involving the design of monitoring and control systems in the both the building services and process industries. His main areas of expertise are the implementation of embedded microprocessor systems, the design and simulation of computer-control schemes, the application of neurofuzzy techniques to highly uncertain systems and the monitoring and control of the thermal environment inside large buildings. He has co-authored two books on the design of microcomputers, co-edited a book on automated fault detection and diagnosis in buildings and published over 100 research papers in research journals and conference proceedings.

Acknowledgments

I should first like to acknowledge the debt I owe to those whose pioneering work on the design and application of fuzzy control systems first sparked my interest in the monitoring and control of information-poor systems over 30 years ago.

The important contribution made by the many graduate students I have supervised over the last 20 years, who worked tirelessly on debugging the software and performing the experiments needed to assess the performance of techniques we developed together, is also gratefully acknowledged. My particular thanks go to Woei Wan Tan, Darius Ngo, Richard Thompson, Amanda Lee, Yue Wu, Jean-Camille de Barros, Huiling Tan, Bilal Kadri and Andrew Wright, whose research formed the basis of many of the ideas and techniques described in this book.

I should also like to pay tribute to Nicky, my original project editor, whose totally unexpected and tragic, premature death came as a terrible shock to all of us who worked with her.

Finally, a special thank you goes to my wife, my three daughters and my grandchildren both for allowing me the time and personal space I needed to complete the book and for bringing so much happiness into my life.

Part I

Analysing the behaviour of information-poor systems

1

Characteristics of Information-Poor Systems

1.1 Introduction to Information-Poor Systems

The term *information poor* was first used by Howell in 1991 to describe systems in which the quality of the information about the system is poor and/or the quantity of the information about the system is low (Howell, 1991).

> Such plants may have a bare minimum of sensors available with which to operate the process, the sensors may output at frequencies which are low relative to the dynamics of the plant, ... and there may be considerable uncertainty surrounding any models that are available. (Howell, 1994)

The term was originally proposed as an antonym to describe systems that were not information rich. The main properties of information-poor systems are poor-quality measurement systems, relatively low-frequency data collection, susceptible to unknown, abnormal modes of operation and inaccurate models.

A wide variety of engineering, biological and economic systems could be described as information-poor.

1.1.1 Blast Furnaces

The process inside a blast furnace is highly complex, and both time varying and spatially varying (Martin *et al.*, 2007). The internal conditions are very difficult to monitor and the composition of the inputs (the fuel, the ore and the coke) varies in an unpredictable manner.

1.1.2 Container Cranes

A container crane is a complex, non-linear, multi-input multi-output system, which has significant measurement noise associated with its sensors and large external disturbances (e.g. changes in the magnitude and direction of the wind) (Mendonca *et al.*, 2006).

Monitoring and Control of Information-Poor Systems: An Approach based on Fuzzy Relational Models, First Edition.
Arthur L. Dexter.
© 2012 John Wiley & Sons, Ltd. Published 2012 by John Wiley & Sons, Ltd.

1.1.3   Cooperative Control Systems

Cooperative control schemes are used to control the movement of a group of cooperating autonomous guided vehicles (AGVs) (Innocenti *et al.*, 2007), autonomous air vehicles (AAVs) or unmanned underwater vehicles (UUVs) (Hou and Allen, 2008). A key issue is how to overcome the effects of the uncertainties arising from imperfections in communications between the controllers of individual vehicles (Gil *et al.*, 2008a), the uncertain dynamics of the robots (Dong and Farrell, 2009) and the uncertainties associated with imperfect sensor information and timing errors (Gil *et al.*, 2008a, b) so that the benefits of cooperation can still be realized. There can also be uncertainties associated with assessing the overall performance in an unmanned multivehicle environment (Li and Cassandras, 2006).

1.1.4   Distillation Columns

Distillation is a complex, non-linear, dynamic process with multiple inputs and outputs (Gormandy and Postlethwaite, 2001). There are significant process interactions and some unmeasured input disturbances (Molov *et al.*, 2004). As a result, many assumptions are necessary even to derive a simplified physical model of a distillation column (Mahfouf *et al.*, 2002).

1.1.5   Drug Administration

There is significant variability between patients in terms of their response to the administration of a particular drug. In addition, it is difficult to measure the effect of the drug (e.g. the depth of anaesthesia) and drug interaction mechanisms are complex (Nunes *et al.*, 2005). Previously developed phenomenological models have both structural and parametric uncertainties due to a lack of experimental data (Gueorguieva *et al.*, 2005). Indeed, some researchers believe that biological systems are so complex it is not possible to produce a precise model of the process (Mahfouf *et al.*, 2001).

1.1.6   Electrical Power Generation and Distribution

There are many sources of uncertainty associated with the generation and distribution of electrical power (future energy demands, reserve demands, market prices, probability of reserves actually being used) (Attaviriyanupap *et al.*, 2004). The increasing use of renewable energy systems to generate electrical power introduces further uncertainties as upcoming wind speeds, solar radiation levels and the availability of stored water cannot be predicted accurately (Liang and Liao, 2007).

1.1.7   Environmental Risk Assessment Systems

Information about environmental risk is usually incomplete and/or imprecise. The interpretability of results by non-technical decision-makers is another major cause of uncertainty (Darba *et al.*, 2008).

1.1.8 Financial Investment and Portfolio Selection

There is considerable uncertainty associated with predicting the behaviour of economic systems (e.g. predictions of future sales demands and future interest rates) (Schjaer-Jacobsen, 2002). For example, when assessing the return on investment in information technology, the relationship between investment and business success is unclear (Chou *et al.*, 2005). When taking decisions about investing in more sophisticated manufacturing systems, the available information (investment costs, expenses, lifetime of the product, depreciation, subjective judgements of managers, lack of objective data on which to base cash flow analysis) is vague and uncertain (Chan *et al.*, 2003). When determining the amount of credit required for (and the overall financial cost of) a construction project, the direct costs of the related activities depend on weather conditions, resource availability and productivity (Afshar and Fathi, 2009). When selecting from a number of possible industrial, building and service sector projects, the initial investment costs, profits, resource requirement and total available budget are all to some extent uncertain (Damghani *et al.*, 2011). When taking decisions about investing in production plant, there is always a lack of information, conflicting evidence, ambiguity and inaccurate data (Sakalli and Baykoc, 2010). When investing in real estate, both the market information (structural quality and location of and future financial returns from a property) and human factors (the merit of investing in the property and its architectural attractiveness) are imprecise and subjective (Hui, Lau and Lo, 2009). When considering investment in large software packages, key factors (e.g. adaptability and flexibility, service support in different countries) are often described linguistically (Erol and Ferrell Jr, 2003).

The uncertainty associated with choosing the optimal (in the sense of maximizing the overall return while minimizing the overall risk) makeup of a portfolio of financial assets (stocks, bonds etc.) (Magoc *et al.*, 2010) could be even greater as there are three highly unpredictable factors that can have a major influence on the decision: general economic factors (fiscal and monetary policy, inflation, devaluations, political events, social events), industrial factors (import and export quotas, excess supply, shortages, government regulations) and the companies themselves (dividend yields, cash flows, stock price) (Serguieva and Hunter, 2004). All have significant uncertainty associated with predicting their influence as they are usually based on expert knowledge involving linguistic vagueness (Huang, 2008).

1.1.9 Health Care Systems

Health risk is difficult to model because there is often a lack of data and only sparse, imperfect and heuristic information about the processes and the process parameters. For example, when assessing the risk associated with exposure to pollutants, intrinsic variability and extensive uncertainty exists in the physical, chemical and biological processes involved in the generation, transportation and deposition of the pollutants in the environment (Kentel and Aral, 2007).

1.1.10 Indoor Climate Control

Accurate mathematical models cannot be produced because most designs are unique and financial considerations restrict the amount of time and effort that can be put into deriving the models. Detailed design information is seldom available, and measured data from the actual plant are often a poor indicator of the overall behaviour because the associated thermodynamic and fluid mechanic processes are non-linear (Sousa *et al.*, 1997; Ghiaus *et al.*, 2007), buildings are subject to significant seasonal disturbances and test signals cannot usually be injected during normal operation (Dexter, 1999). In addition, the control objectives are poorly defined as thermal comfort is a highly subjective issue (Dexter and Trewhella, 1990) and key variables (spatially varying temperatures and flows) are difficult to measure accurately (Tan and Dexter, 2006).

1.1.11   NOx Emissions from Gas Turbines and Internal Combustion Engines

A gas turbine is a difficult-to-model non-linear process (Oh *et al.*, 2007). Uncertainty is introduced by the simplifying assumptions that are necessary when modelling a real thermodynamic process. For example, a multi-zone model of a spatially distributed system is often used to approximate the complex, non-linear combustion process and to estimate the thermodynamic variables required to calculate the NOx emissions (Kesgin and Heperkan, 2005).

1.1.12   Penicillin Production Plant

Biological processes have greater complexity and uncertainty than other types of systems. The nature of the media, cultures and raw material is poorly understood, causing inherent process variability and non-linearity of the process. There is a lack of good mathematical models, which often have time-varying parameters, and there is usually a need to incorporate uncertain qualitative knowledge. The effects of non-measurable information are also important. Some measurements must be made offline, and high uncertainty and low sampling rates are normally associated with laboratory measurements (Arauzo-Bravo *et al.*, 2004).

1.1.13   Polymerization Reactors

Polymerization is a non-linear multivariable process as complex interactions occur in the reactor. Potentially dangerous technological processes, such as polymerization, are characterized by a high level of uncertainty, large uncontrolled disturbances, frequently bad observability and very often no mathematical descriptions (Rusinov *et al.*, 2007). The controlled variables are difficult to measure (Wakabayashi *et al.*, 2009), and polymerization reactors are difficult to model because the heat transfer processes are highly non-linear and the viscosity of the fluid has time-varying characteristics (Altinten *et al.*, 2003).

1.1.14   Rotary Kilns

The operation of a rotary cement kiln involves complex non-linear processes which are difficult to model (Bavdaz and Kocijan, 2007). Most of the measurements are of poor quality, the variable chemical composition of the raw material cannot be measured online and there are also conflicting control objectives (Holmblad and Ostergaard, 1995). The process of manufacturing perlite in a rotary kiln has 'unwieldy' characteristics (complex interactions that vary depending on the type of clay), the weight of the finished product cannot be measured online (so infrequent sampling and laboratory analysis must be used) and there are significant disturbances in the feedstock of the kilns (Asayama *et al.*, 1994).

1.1.15 Solar Power Plant

The generation of solar power plant is a distributed process with significant, though measurable, disturbances (solar radiation, external air temperature, inlet oil temperature) (Flores *et al.*, 2005). The behaviour of solar power plant is therefore highly uncertain.

1.1.16 Wastewater Treatment Plant

The treatment of wastewater is a complex process with large load disturbances and other uncertainties (Huang *et al.*, 2009). The large amount of data collected online is information-poor (Ward *et al.*, 1986), many input disturbances cannot be measured directly (Muller *et al.*, 1997) and there are significant uncertainties associated with the values of the parameters of the complex mathematical models that have been proposed and with the online measurements (Tsai *et al.*, 1996). These result from an incomplete understanding of the biological mechanisms, the inaccuracies associated with the online monitored data, the non-linearity of the dynamic system, the presence of time delays and a lack of expert knowledge about how to control the process (Tsai *et al.*, 1994).

1.1.17 Wood Pulp Production Plant

The production of wood pulp is a complicated interactive process: pulp quality is difficult to measure, there are large input disturbances (e.g. the chip quality depends on the type of wood, the size distribution and the moisture content), no phenomenological models exist and the amount of available process data is limited (Qian and Tessier, 1995).

1.2 Main Causes of Uncertainty

It can be seen that there are three main reasons for the uncertainty associated with information-poor systems:

1. a poor understanding of the behaviour of the system, which is usually complex, non-linear and spatially distributed;
2. an inability to measure important variables accurately because of the restrictions imposed by economic costs and challenging operating environments;
3. incomplete or inconsistent specification of the design resulting from poorly understood or subjective design objectives and poor information management.

1.2.1   Sources of Modelling Errors

All engineering designs are based on some type of mathematical model of the behaviour of the system to be optimized or controlled. The accuracy of the mathematical model used for model-based design should always be taken into account in applications with significant uncertainties (Ning and Zaheeruddin, 2009).

There are two basic types of mathematical models: first-principles models and black-box models.

First-principles models are based on the laws of physics (in the case of engineering applications) and have parameters that depend solely on the properties of matter. The two basic causes of modelling errors are the use of inaccurate parameter values and the use of simplifying assumptions.

Black-box models (e.g. neural networks) rely entirely on training data obtained from the system to be modelled, and use no domain knowledge about the system. The main cause of modelling errors is the use of incomplete or inconsistent training data.

In practice, models are not first-principles or black-box but are based on a combination of domain knowledge and training data. Such models are called grey-box models. The two main causes of inaccuracy in these models are: the use of simplifying assumptions and incomplete or inaccurate calibration data.

Mathematical models of complex systems are often simplified to reduce computational demands (e.g. spatially distributed systems are described by lumped-parameter ordinary differential equations rather than partial differential equations; non-linear relationships are represented by linear approximations; high-frequency dynamics are neglected; the effects of unmeasured or less critical disturbances are ignored). The underlying structure of the resulting models is no longer correct and structural errors are introduced into their predictions.

It is usually impossible for the parameters of grey-box models to be determined from material properties and other physical constants, and the model must be calibrated using measured data from the actual system or design data supplied by the equipment manufacturer. If the system is information-poor, there will be uncertainty associated with measurements and design data will be incomplete or inaccurate. As a result, the model will not be calibrated correctly and parametric errors will be introduced into any model predictions.

1.2.2   Sources of Measurement Errors

It is unusual for electronic measurement noise and temperature drift to be a major problem when using modern electronic instrumentation, even in information-poor systems.

In practice, the main reason for measurement errors is the use of a limited number of measurements (in time or space) to generate an output that is representative of the time average and/or spatial average of the physical variable being measured. The output of the sensor is then merely an estimate of the quantity to be measured and is usually subject to both random and systematic errors.

Some quantities cannot be measured directly and they must be estimated or inferred from other measurements (Arauzo-Bravo *et al.*, 2004) using a mathematical model of the relationship between the measured variables and the variable to be estimated. In this case, estimation errors arise because of both modelling and measurement errors.

1.2.3 Reasons for Poorly Defined Objectives and Constraints

The design objectives are usually poorly defined because there is disagreement about the relative importance of different (and often conflicting) objectives (e.g. profit versus risk), because of the subjective nature of some objectives (e.g. human comfort, sensitivity to pain) and because the importance of failing to satisfy some of the design constraints is not fully understood (e.g. pollution levels).

1.3 Design in the Face of Uncertainty

Design, particularly engineering design, is often viewed as a constrained optimization problem. However, there are additional issues to be considered when the design specifications are uncertain (i.e. the design objectives, the constraints and the environment are not completely defined). In such cases, the importance of human insight and creativity cannot be overestimated and the final outcome of the design process is usually dependent on the power of human reasoning and expertise (Saridakis and Dentsoras, 2008).

The design process can be divided into three stages: conceptual system design, preliminary system design and detailed system design. The first stage, which is inevitably the one most characterized by imprecision and vagueness, has a great influence on the following stages including the important detailed design stage (Smith and Verma, 2004).

Multiple decision-making actions are at the heart of the design process and they are nearly always subject to significant uncertainty (Saridakis and Dentsoras, 2008). In practice, the decision-making involves both quantitative and qualitative criteria, and several different types of uncertainty (including human judgement) must be considered (Fu, 2008). In many engineering applications, the uncertainties arise because much of the information on which the design is based involves social and economic issues. Such factors are often vague and imprecise, and must be described linguistically. In industrial design problems, the uncertainty often arises because of a lack of data on which to base the design. Traditionally, statistical methods have been used to deal with the uncertainties (so-called reliability-based design optimization) but there is now growing interest in methods based on fuzzy techniques (so-called possibility-based design optimization), particularly in design problems involving both random and fuzzy variables (hybrid approaches) (Du *et al.*, 2006). Whatever approach is taken, it is most important that design decisions take account of the uncertainty associated with optimal choices when dealing with information-poor systems. It must also be accepted that any attempt at precise optimization is unlikely to yield meaningful results in practice.

References

Afshar, A. and Fathi, H. (2009) Fuzzy multi-objective optimization of finance-based scheduling for construction projects with uncertainties in costs. *Engineering Optimization*, **41**(11), 1063–1080.

Altinten, A., Erdogan, S., Hapoglu, H. and Alpbaz, M. (2003) Control of a polymerization reactor by fuzzy control method with genetic algorithm. *Computers and Chemical Engineering*, **27**(7), 1031–1040.

Arauzo-Bravo, M.J., Cano-Izquierdo, J.M., Gomez-Sanchez, E, *et al.* (2004) Automatization of a penicillin production process with soft sensors and an adaptive controller based on neuro fuzzy systems. *Control Engineering Practice*, **12**, 1073–1090.

Asayama, H., Burton, P., Gerstacker, J., *et al.* (1994) Fuzzy logic control of a perlite plant. *Computers and Control Engineering*, **5**(6), 293–298.

Attaviriyanupap, P., Kita, H., Tanaka, E., and Hasegawa, J. (2004) A fuzzy-optimization approach to dynamic economic dispatch considering uncertainties. *IEEE Transactions on Power Systems*, **19**(3), 1299–1307.

Bavdaz, G. and Kocijan, J. (2007) Fuzzy controller for cement raw material blending. *Transactions of Institute of Measurement & Control*, **29**(1), 17–34.

Chan, F.T.S, Chan, H.K. and Chan, M.H. (2003) An integrated fuzzy decision support system for multicriterion decision-making problems, *Proceedings of the Institution of Mechanical Engineers, Part B, Journal of Engineering Manufacture*, **217**(1), 11–27.

Chou, T-Y., Chou, S-C., and Tzeng, G-H. (2005) Evaluating IT/IS investments: a fuzzy multi-criteria decision model approach. *European Journal of Operational Research*, **170**(1), 1026–1046.

Damghani, K.K., Sadi-Nezhad, S., and Aryanezhad, M.B. (2011) A modular decision support system for optimum investment selection in presence of uncertainty: combination of fuzzy mathematical programming and fuzzy rule based system. *Expert Systems with Applications*, **38**(1), 824–834.

Darba, R.M., Eljarrat, E. and Barcelo, D. (2008) How to measure uncertainties in environmental risk assessment. *Trends in Analytical Chemistry*, **27**(4), 377–385.

Dexter, A.L. (1999) Control and fault detection in buildings. Proceedings of 3rd International Symposium on Heating, Ventilating and Air-conditioning, Shenzhen, PRC, **1**, 39–50.

Dexter, A.L. and Trewhella, D.W. (1990) Building control systems: fuzzy rule-based approach to performance assessment. *Building Services Engineering Research & Technology*, **11**(4), 115–124.

Dong, W. and Farrell, J.A. (2009) Decentralized cooperative control of multiple nonholonomic dynamic systems with uncertainty. *Automatica*, **45**, 706–710.

Du, L., Choi, K.K., Youn, B.D. and Gorsich, D. (2006) Possibility-based design optimization method for design problems with both statistical and fuzzy input data. *ASME Journal of Mechanical Design*, **128**(4), 928–935.

Erol, I. and Ferrell, W.G. Jr (2003) A methodology for selection problems with multiple, conflicting objectives and both qualitative and quantitative criteria. *International Journal of Production Economics*, **86**, 187–199.

Flores, A., Saez, D., Araya, J., *et al.* (2005) Fuzzy predictive control of a solar power plant. *IEEE Transactions on Fuzzy Systems*, **13**(1), 58–68.

Fu, G. (2008) A fuzzy optimisation method for multicriteria decision-making: an application to reservoir flood control operation. *Expert Systems with Application*, **34**, 145–149.

Ghiaus, C., Chicinas, A. and Inard, C. (2007) Grey-box identification of air-handling unit elements. *Control Engineering Practice*, **15**, 421–433.

Gil, A.E., Passino, K.M. and Cruz, J.B. (2008a) Stable cooperative surveillance with information flow constraints. *IEEE Transactions on Control Systems Technology*, **16**(5), 856–868.

Gil, A.E., Passino, K.M., Ganapathy, S. and Sparks, A. (2008b) Cooperative task scheduling for networked uninhabited air vehicles. *IEEE Transactions on Aerospace and Electronic Systems*, **44**(2), 561–581.

Gormandy, B.A. and Postlethwaite, B.E. (2001) Model-based control using fuzzy relational models. *IEEE International Conferences on Fuzzy Systems*, 1581–1584.

Gueorguieva, I., Nestorov, I.A., Aarons, L. and Rowland, M. (2005) Uncertainty analysis in pharmacokinetics and pharmacodynamics: application to naratriptan. *Pharmaceutical Research*, **22**(10), 1614–1626.

Holmblad, L.P. and Ostergaard, J-J. (1995) The FLS application of fuzzy logic. *Fuzzy Sets & Systems*, **70**, 135–146.

Hou, Y. and Allen, R. (2008) Intelligent behaviour-based team UUVs cooperation and navigation in a water flow environment. *Ocean Engineering*, **35**, 400–416.

Howell, J. (1991) Model-based fault diagnosis in information poor processes. PhD thesis, University of Glasgow, UK.

Howell, J. (1994) Model-based fault detection in information poor plants. *Automatica*, **30**(6), 929–943.

Huang, M., Wan, J., Ma, Y., *et al.* (2009) Control rules of aeration in a submerged biofilm wastewater treatment process using fuzzy neural networks. *Expert Systems with Applications*, **36**, 10428–10437.

Huang, X. (2008) Mean-entropy models for portfolio selection. *IEEE Transactions on Fuzzy Systems*, **16**(4), 1096–1101.

Hui, E.C.M., Lau, O.M.F. and Lo, T.K.K. (2009) Deciphering real estate investment decisions through fuzzy logic systems. *Property Management*, **27**(3), 163–177.

Innocenti, B., Lopez, B. and Salvi, J. (2007) A multi-agent architecture with cooperative fuzzy control for a mobile robot. *Robotics and Autonomous Systems*, **55**, 881–891.

Kentel, E. and Aral, M.M. (2007) Risk tolerance measure for decision-making in fuzzy analysis: a health risk assessment perspective. *Stochastic Environmental Research Risk Assessment*, **21**, 405–417.

Kesgin, U. and Heperkan, H. (2005) Simulation of thermodynamic systems using soft computing techniques. *International Journal of Energy Research*, **29**(7), 581–611.

Li, W. and Cassandras, C.G. (2006) A cooperative receding horizon controller for multivehicle uncertain environments. *IEEE Transactions on Automatic Control*, **51**(2), 242–257.

Liang, R-H. and Liao, J-H. (2007) A fuzzy-optimization approach for generation scheduling with wind and solar energy systems. *IEEE Transactions on Power Systems*, **22**(4), 1665–1674.

Magoc, T., Wang, X. and Modave, F. (2010) Application of fuzzy measures and interval computation to financial portfolio selection. *International Journal of Intelligent Systems*, **25**(7), 621–635.

Mahfouf, M., Abbod, M.F. and Linkens, D.A. (2001) A survey of fuzzy logic monitoring and control utilisation in medicine. *Artificial Intelligence in Medicine*, **21**, 27–42.

Mahfouf, M., Kandiah, S. and Linkens, D.A. (2002) Fuzzy model-based predictive control using an ARX structure with feedforward. *Fuzzy Sets and Systems*, **125**, 39–59.

Martin, R.D., Obeso, F., Mochon, J., *et al.* (2007) Hot metal temperature prediction in blast furnace using advanced model based on fuzzy logic tools. *Ironmaking & Steelmaking*, **34**(3), 241–247.

Mendonca, L.F., Sousa, J.M.C., Kaymak, U. and Sa da Costa, J.M.G. (2006) Weighting goals and constraints in fuzzy predictive control. *Journal of Intelligent & Fuzzy Systems*, **17**(5), 517–532.

Molov, S., Babuska, R., Abonyi, J. and Verbruggen, H.B. (2004) Effective optimization for fuzzy model predictive control. *IEEE Transactions on Fuzzy Systems*, **12**(5), 661–675.

Muller, A., Marsili-Libelli, S., Aivasidis, A., *et al.* (1997) Fuzzy control of disturbances in a wastewater treatment process. *Water Research*, **31**(12), 3157–3167.

Ning, M and Zaheeruddin, M. (2009) Fuzzy set-based uncertainty analysis of HVAC&R systems: simulation study. *Building Services Engineering Research and Technology*, **30**(3), 241–262.

Nunes, C.S., Mahfouf, M., Linkens, D.A. and Peacock, J.E. (2005) Modelling and multivariable control in anaesthesia using neural fuzzy paradigms: Part I. *Artificial Intelligence in Medicine*, **35**, 207–213.

Oh, S-K., Pedrycz, W. and Park, H-S. (2007) Fuzzy relation-based neural networks and their hybrid identification. *IEEE Transactions on Instrumentation and Measurement*, **56**(6), 2522–2537.

Qian, Y. and Tessier, P.J.C. (1995) Application of fuzzy relational modelling to industrial product quality control. *Chemical Engineering & Technology*, **18**(5), 330–336.

Rusinov, L.A., Rudakova, I.V. and Kurkina, V.V. (2007) Real-time diagnostics of technological processes and field equipment. *Chemometrics and Intelligent Laboratory Systems*, **88**, 18–25.

Sakalli, U.S. and Baykoc, O.F. (2010) An application of investment decision with random fuzzy outcomes. *Expert Systems with Applications*, **37**(4), 3405–3414.

Saridakis, K.M. and Dentsoras, A.J. (2008) Soft computing in engineering design – a review. *Advanced Engineering Informatics*, **22**, 202–221.

Schjaer-Jacobsen, H. (2002) Representation and calculation of economic uncertainties: intervals, fuzzy numbers, and probabilities. *International Journal of Production Economics*, **78**(1), 91–98.

Serguieva, A. and Hunter, J. (2004) Fuzzy interval methods in investment risk appraisal. *Fuzzy Sets and Systems*, **139**(2), 443–466.

Smith, C. and Verma, D. (2004) Conceptual system design evaluation: rating and ranking versus compliance analysis. *Systems Engineering*, **7**(4), 338–351.

Sousa, J.M., Babuska, R. and Verbruggen, H.B. (1997) Fuzzy predictive control applied to an air-conditioning system. *Control Engineering Practice*, **5**(10), 1395–1406.

Tan, H. and Dexter, A.L. (2006) Estimating airflow rates in air handling units from actuator control signals. *Building and Environment*, **41**(10), 1291–1298.

Tsai, Y.P., Ouyang, C.F., Chiang, W.L. and Wu, M.Y. (1994) Construction of an on-line fuzzy controller for the dynamic activated sludge process. *Water Research*, **28**(4), 913–921.

Tsai, Y-P., Ouyang, C-F., Wu, M-Y. and Chiang, W-L. (1996) Effluent suspended solid control of activated sludge process by fuzzy control approach. *Water Environment research*, **68**(6), 1045–1053.

Wakabayashi, C., Embirucu, M., Fontes, C. and Kalid, R. (2009) Fuzzy control of a nylon polymerization semi-batch reactor. *Fuzzy Sets & Systems*, **160**, 537–553.

Ward, R.C., Loftis, J.C. and McBride, G.B. (1986) The "data rich but information-poor" syndrome in water quality monitoring. *Environmental Management*, **10**(3), 291–297.

2

Describing and Propagating Uncertainty

2.1 Methods of Describing Uncertainty

Uncertainty in the true value of any quantity can be described in several different ways depending on the nature of the uncertainty and the depth of understanding of the underlying causes of the uncertainty (Ferson and Ginzburg, 1996).

2.1.1 Uncertainty Intervals and Probability Distributions

If the only available information is that the uncertainty is known to lie within a particular range, the only possible way of describing it is by specifying an uncertainty interval within which the true value must lie. For example, if the uncertainty interval is $[a, b]$, the standard uncertainty u_x associated with a quantity $x = (a + b)/2$ is given by $u_x = (b - a)/2$, where the true value of $x = x_T$ is given by $x_T = x \pm u_x$.

If the uncertainty is known to be a result of random effects, it may be possible for the probability of the quantity having a particular value to be found. In this case, the most probable value of the quantity x_m can found from its associated probability density function $p(x)$:

$$x_m = x | p(x_m) > p(x) \forall x \neq x_m \tag{2.1}$$

and, if the distribution is unimodal, a confidence interval $I_\lambda = [a, b]$ can be defined that specifies those values of the quantity whose probability is equal to a specified confidence level λ, where

$$\lambda = \Pr(a \leq x \leq b) = \int_a^b p(x)dx \tag{2.2}$$

If the probability distribution is (or can be assumed to be) normal, only two parameters (the mean μ_x and variance σ_x^2) are needed to describe the uncertainty, and the confidence

Monitoring and Control of Information-Poor Systems: An Approach based on Fuzzy Relational Models, First Edition.
Arthur L. Dexter.
© 2012 John Wiley & Sons, Ltd. Published 2012 by John Wiley & Sons, Ltd.

interval depends only on the confidence level and the standard deviation of the distribution (Papadopoulos and Yeung, 2001). For example, if $\lambda = 0.95$,

$$I_{0.95} \cong \mu_x \pm 2\sigma_x \tag{2.3}$$

where the mean of x is given by

$$\mu_x = \int_{-\infty}^{+\infty} xp(x)dx \tag{2.4}$$

and the variance of x is given by

$$\sigma_x^2 = \int_{-\infty}^{+\infty} (x - \mu_x)^2 p(x)dx \tag{2.5}$$

If the uncertainty is a result of both random and systematic effects, a fuzzy number may be a more appropriate way of describing the uncertainty.

2.1.2 Fuzzy Sets and Fuzzy Numbers

Fuzzy sets are an extension of Boolean or crisp sets to allow for partial membership of the set. For example, a particular value of the temperature is either a member, or not a member, of the Boolean set 'Temperature is between 8°C and 12°C':

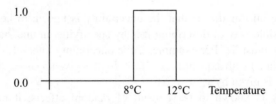

whereas the degree of membership of the temperature in the fuzzy set 'Temperature is about 10°C' can take any value in the range 0 (not a member of the set) to 1 (fully a member of the set). For example:

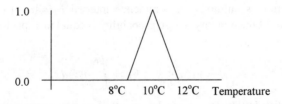

Each fuzzy set A has an associated membership function μ_A defined over a field of reference or Universe of Discourse (UoD) for x, where the degree of membership $\mu_A(x)$ might be interpreted as:

- the degree of truth that 'x is A'
- the likelihood that the average person would agree that 'x is A'

- the extent to which 'x is A'
- the degree of certainty that 'x is A'
- the possibility that 'x is A'.

A membership function that is defined over a discrete UoD is usually expressed as

$$\mu_A(x) = \mu_A(x_1)/x_1 + \mu_A(x_2)/x_2 + \mu_A(x_3)/x_3 \cdots$$

or specified as a vector of membership values

$$\mu_A(x) = [\mu_A(x_1), \mu_A(x_2), \mu_A(x_3), \cdots].$$

Note that the membership function of a crisp value is a delta function of unit height (a *fuzzy singleton*) positioned at that value on the universe of discourse.

It can be argued that randomness and fuzziness differ both conceptually and theoretically. Fuzziness describes the ambiguity associated with the definition of an event and measures the degree to which the event occurs, not whether it occurs. For example, there is fuzziness associated with the uncertainty arising when defining the event 'the weather is hot'. Randomness is associated with the uncertainty of the event occurring, for example, 'there is an 80% chance of the event occurring'. Events in the real world involve both types of uncertainty, for example, 'there is an 80% chance that tomorrow the weather will be hot' involves both types of uncertainty because it concerns the probability of a fuzzy event occurring. Fuzziness arises because of the ambiguity between an event and its opposite. It still exists even if the event has occurred, that is, when tomorrow's weather is observed. Note that, as fuzziness and randomness describe different types of uncertainty, there is no requirement for the area underneath a membership function to equal unity (as is the case with a probability density function).

2.2 Methods of Propagating Uncertainty

2.2.1 Interval Arithmetic

If the relationship between a quantity y and other quantities $(x_1, \ldots, x_N)$ (whose uncertainty intervals are known) can be expressed in terms of the four basic arithmetic operations (addition, subtraction, multiplication and division), the uncertainty interval for y can be calculated using interval arithmetic (Klir and Yuan, 1995), where

$$[a, b] + [c, d] = [(a + c), (b + d)],$$

$$[a, b] - [c, d] = [(a - d), (b - c)],$$

$$[a, b] . [c, d] = [\min(ac, ad, bc, bd), \max(ac, ad, bc, bd)]$$

and

$$[a, b] / [c, d] = [\min(a/c, a/d, b/c, b/d), \max(a/c, a/d, b/c, b/d)] \tag{2.6}$$

However, as interval arithmetic generates an upper bound on the results of each arithmetic operation, this approach will always result in an estimate of the uncertainty interval which is unrealistically large. For example, consider the case where $y = \sum_{i=1}^{N} x_i$ and each of the input quantities $x_i = [a, b]$. The uncertainty interval for y is given by $y = [Na, Nb]$ if interval arithmetic is used to propagate the uncertainties.

An alternative approach is to assume that the uncertainties in the quantities $(x_1, \ldots x_N)$ are small and that they arise from uncorrelated, random effects. For example, if a quantity y is a known function of the quantities $(x_1, \ldots x_N)$ and $y = f(x_1, \ldots x_N)$, the uncertainty interval associated with y can be calculated from the uncertainty intervals associated with $(x_1, \ldots x_N)$ using the root-sum-square method (Moffat, 1988):

$$u_y = \sqrt{\sum_{i=1}^{N} \left(\frac{\partial y}{\partial x_i} \right)^2 u^2(x_i)} \tag{2.7}$$

where $\frac{\partial y}{\partial x_i}$ are the sensitivity coefficients.

2.2.2   Statistical Methods

If $y = f(x_1, \ldots x_N)$ and the standard deviation of x_i is $\sigma(x_i)$, it can be shown (Papadopoulos and Yeung, 2001) that the standard deviation of y is given by

$$\sigma_y = \sqrt{\sum_{i=1}^{N} \left(\frac{\partial y}{\partial x_i} \right)^2 \sigma^2(x_i) + \sum_{i=1}^{N} \sum_{\substack{j=1 \\ j \neq i}}^{N} \left(\frac{\partial y}{\partial x_i} \right) \left(\frac{\partial y}{\partial x_j} \right) \sigma(x_i, x_j)} \tag{2.8}$$

where $\sigma(x_i, x_j)$ are the correlation coefficients and $\frac{\partial y}{\partial x_i}$ are the sensitivity coefficients.

It should be noted that this approach assumes that the uncertainties in the inputs are relatively small or that the relationship $y = f(x_1, \ldots x_N)$ is approximately linear. The same approach has also been used to propagate uncertainty intervals (Restivo and Sousa, 2008).

2.2.3   Monte Carlo Methods

One of the problems in using the previous approach is that the sensitivity coefficients may be difficult, if not impossible, to find in complex non-linear systems where the relationship between the inputs and the outputs is difficult to model. In such cases, an estimate of the probability distribution for y can be obtained from a large number of computer simulations of the relationship between y and $(x_1, \ldots x_N)$, in which the values of input variables are chosen randomly from their (assumed known) probability distributions. This method of propagating uncertainties can handle large uncertainties in the inputs and also takes care of input dependencies automatically (Papadopoulos and Yeung, 2001; Ferrero et al., 2002; Wubbeler et al., 2008). However, some concern has been expressed about the accuracy of the resulting probability distributions, even when they are estimated from a relatively large number of simulations (Willink, 2006).

Example 2.1

The output y of a simple non-linear static system is given by $y = (5.0 + n)^3$, where n is a normally distributed random variable with a mean of 0.0 and a standard deviation of 1.0. Figure 2.1 shows the probability density function of the output estimated from the results of 10^6 simulations of the system, using a histogram with 1001 equally spaced bins from 0.0 to 1000.0.

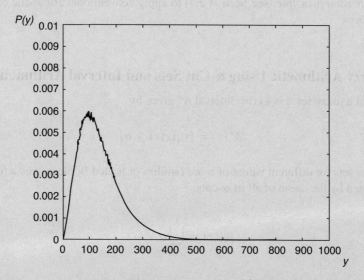

Figure 2.1 Probability density function estimated using the Monte Carlo method.

It can be seen that the distribution is still relatively noisy, even when it is based on a very large number of simulations.

The main disadvantages of this approach are the large number of simulations required to obtain an accurate estimate of the probability density function (Ferson, 1996) and the assumptions that the probability distributions of the inputs are known and, in practice, that the inputs are independent (Darba *et al.*, 2008).

2.2.4 Fuzzy Arithmetic

Fuzzy arithmetic can be used to propagate the uncertainties if they are described using fuzzy numbers. Fuzzy arithmetic is an extension of conventional arithmetic to fuzzy numbers. These are normal, convex, fuzzy sets with closed support, which have a quantitative meaning (e.g. 'close to 1.3'). Note that a fuzzy set A is normal if it has unit height where the height $h(A)$

of a fuzzy set A is equal to $\sup(\mu_A)$, has closed support if the crisp set $S(A)$ defined by $S(A) = \{x \in X | \mu_A > 0\}$ is a closed interval and is convex if

$$\forall x, y \in X \; \forall \lambda \in [0, 1] : \mu_A(\lambda x + (1 - \lambda)y) \geq \min[\mu_A(x), \mu_A(y)].$$

There are two ways of implementing fuzzy arithmetic (Klir and Yuan, 1995). The first method decomposes fuzzy numbers into a set of crisp intervals, uses interval arithmetic on the intervals and produces a fuzzy result from the resulting crisp intervals. The second method uses the *extension principle* (see Section 2.4) to apply conventional arithmetic operations to fuzzy numbers.

2.3   Fuzzy Arithmetic Using α-Cut Sets and Interval Arithmetic

The α-cut of a fuzzy set A is a crisp interval A^α given by

$$A^\alpha(x) = \{x | \mu_A(x) \geq \alpha\} \tag{2.9}$$

As the α-cut sets for different values of α are families of nested Boolean sets, a fuzzy set can be represented by the union of all its α-cuts:

$$\mu_A(x) = \bigcup_{\alpha=0}^{1} \alpha A^\alpha(x) \tag{2.10}$$

Interval arithmetic can therefore be used to perform arithmetic operations on fuzzy numbers.

Example 2.2

Consider two fuzzy numbers A and B with triangular membership functions defined by points $(-1, +1, +3)$ and $(+1, +3, +5)$.

The α-cut sets for A and B are $[(2\alpha - 1), (3 - 2\alpha)]$ and $[(2\alpha + 1), (5 - 2\alpha)]$, respectively, and the α-cut sets for the fuzzy number $A + B$ are therefore $[4\alpha, (8 - 4\alpha)]$.

The resulting fuzzy number has a triangular membership function defined by the points $(0, +4, +8)$.

Example 2.3

The results of fuzzy arithmetic based on the use of interval arithmetic and 11 α-cuts from 0.0 to 1.0 in increments of 0.1 are shown in Figures 2.2–2.5.

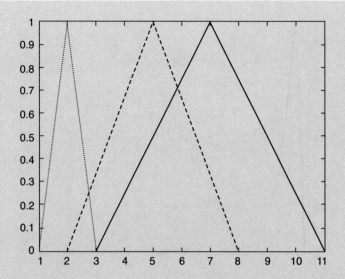

Figure 2.2 Calculation of $Z = X + Y$ (solid line) using α-cuts and interval arithmetic, where X (dotted line) and Y (dashed line) are two fuzzy numbers defined over a discrete UoD.

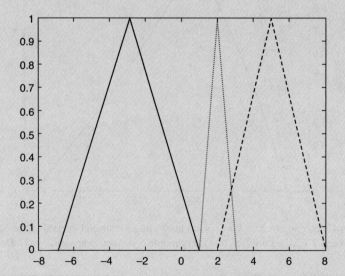

Figure 2.3 Calculation of $Z = X - Y$ (solid line) using α-cuts and interval arithmetic, where X (dotted line) and Y (dashed line) are two fuzzy numbers defined over a discrete UoD.

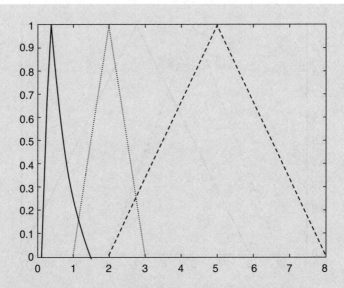

Figure 2.4 Calculation of $Z = X/Y$ (solid line) using α-cuts and interval arithmetic, where X (dotted line) and Y (dashed line) are two fuzzy numbers defined over a discrete UoD.

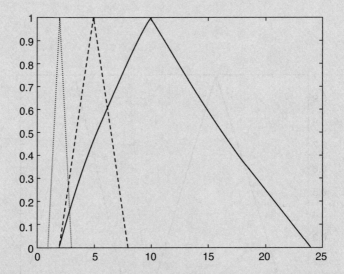

Figure 2.5 Calculation of $Z = X * Y$ (solid line) using α-cuts and interval arithmetic, where X (dotted line) and Y (dashed line) are two fuzzy numbers defined over a discrete UoD.

It can be seen that fuzzy multiplication and division do not generate fuzzy numbers with triangular membership functions, even when they are applied to fuzzy numbers with triangular membership functions. However, the computational demands of fuzzy arithmetic can be small in practice, as only two α-cuts at 0.0 and 1.0 need to be used to obtain satisfactory results in many applications.

2.4　Fuzzy Arithmetic Based on the Extension Principle

The extension principle allows any crisp function $f: x \to z$ to be fuzzified $f: X \to Z$ so that its use can be extended to the fuzzy sets X and Z:

$$\mu_Z(z) = \sup_{x\,|\,z=f(x)} \{\mu_X(x)\} \tag{2.11}$$

where sup{ } is the supremum or maximum value of any set of real numbers. The extension principle can also be applied to functions of more than one argument. For example, if the function has two arguments

$$\mu_Z(z) = \sup_{x,y\,|\,z=f(x,y)} \{T[\mu_X(x), \mu_Y(y)]\} \tag{2.12}$$

where T is a t-norm. Note that there are many arithmetic and logical functions that can be used to perform the basic fuzzy operations (fuzzy complement, fuzzy intersection and fuzzy union), all of which are generalizations of the classical set operations (Klir and Yuan, 1995). Operators that can be used to implement fuzzy intersection are called t-norms. Operators that can be used to implement fuzzy union are called t-conorms or s-norms.

Example 2.4

Figure 2.6 shows the results of using the extension principle to calculate $Z = (X + 0.3Y)^{0.5}$ where X and Y are fuzzy numbers with triangular membership functions. In this example, the product operator is used as the t-norm.

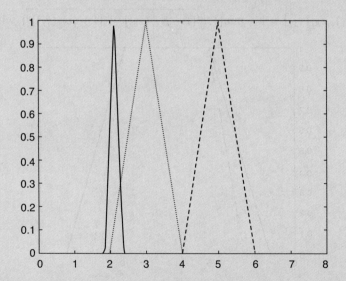

Figure 2.6　Calculation of $Z = (X + 0.3Y)^{0.5}$ (solid line) using the extension principle, where X (dotted line) and Y (dashed line) are two fuzzy numbers defined over a discrete UoD.

Example 2.5

Consider the problem of finding the arithmetic average of two numbers x_1 and x_2 whose values are known to lie in a particular range $[b_L, b_U]$, but the probability of them taking a particular value in this range is unknown and there is some uncertainty δ about the precise values of the upper and lower bounds of the range. One way of representing the uncertainty associated with each of the values is to model them as trapezoidal fuzzy intervals, as shown in Figure 2.7.

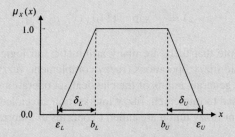

Figure 2.7   Representation of the uncertainty using a trapezoidal fuzzy interval.

The extension principle is to be used to find the average of the two estimated values (see Figures 2.8–2.10). Thus,

$$\mu_Y(y) = \sup_{y=\frac{1}{2}(x_1+x_2)} \left\{ t\left[\mu_{X_1}(x_1), \mu_{X_2}(x_2)\right] \right\}$$

where the t-norm is minimum, product or Yager's t-norm:

$$t\left[\mu_{X_1}(x_1), \mu_{X_2}(x_2)\right] = 1 - \min\{1, \sqrt[p]{[1 - \mu_{X_1}(x_1)]^p + [1 - \mu_{X_2}(x_2)]^p}\}, \text{ with } p = 2$$

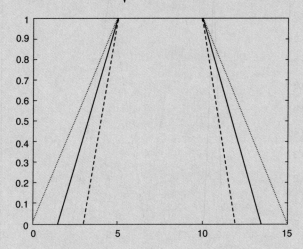

Figure 2.8   Result of averaging two fuzzy intervals using the extension principle with min as the t-norm, where X_1 (dotted line), X_2 (dashed line) and $Z = \dfrac{(X_1 + X_2)}{2}$ (solid line).

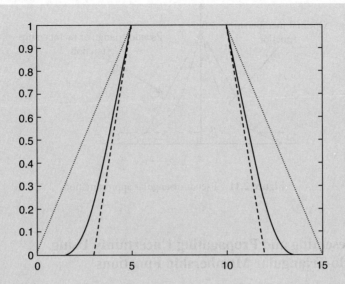

Figure 2.9   Result of averaging two fuzzy intervals using the extension principle with product as the t-norm, where X_1 (dotted line), X_2 (dashed line) and $Z = \dfrac{(X_1 + X_2)}{2}$ (solid line).

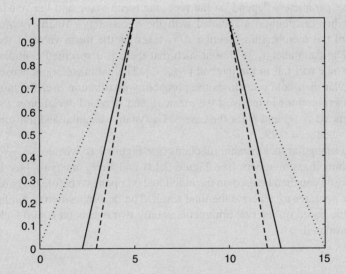

Figure 2.10   Result of averaging two fuzzy intervals using the extension principle with the Yager t-norm (p = 2), where X_1 (dotted line), X_2 (dashed line) and $Z = \dfrac{(X_1 + X_2)}{2}$ (solid line).

It can be seen that, in all three cases, the kernel of the fuzzy number remains the same after averaging and that the result is relatively insensitive to the choice of the t-norm.

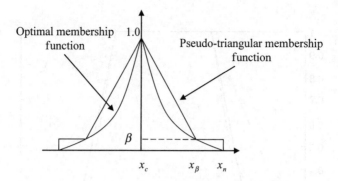

Figure 2.11 Pseudo-triangular approximation.

2.5   Representing and Propagating Uncertainty Using Pseudo-Triangular Membership Functions

The computational demands of fuzzy arithmetic can be greatly reduced by parameterizing the fuzzy numbers and propagating the uncertainty using a decomposition method (Mauris *et al.*, 2001). A simple worst-case approach results in pseudo-triangular approximations to the membership functions (see Figure 2.11), characterized by only four parameters. The values of these parameters depend on the type, the mean value and the standard deviation of the probability distribution associated with the uncertainty (Mauris *et al.*, 2000). The modal value of the membership function x_c is taken as the mean value of the probability distribution. The parameter x_n is chosen such that there is a specified confidence level (γ) that the value of x will fall in the interval $[-x_n, x_n]$. The parameter x_β is chosen so that the pseudo-triangular distribution encloses the probability distribution over the range $[-x_n, x_n]$. Table 2.1 lists expressions in terms of the mean x_m and standard deviation σ for each of the four parameters x_c, x_n, x_β and β for the case of Gaussian, triangular and uniform probability distributions, when $\gamma = 0.99$.

The pseudo-triangular membership functions (see Figure 2.12 where $x_i = \frac{x_\beta}{(1-\beta)}$) are easily decomposed into fuzzy numbers (see Figure 2.13) and fuzzy intervals (see Figure 2.14). Interval arithmetic can then be used on the individual components before they are recombined using the max operator to generate the final result. The decomposition simplifies the use of fuzzy arithmetic based on interval arithmetic as only two α-cuts (at $\alpha = 0$ and $\alpha = 1$) now need to be considered.

Table 2.1 Parameter values for three commonly used probability laws.

Probability law	x_c	x_n	x_β	β
Gaussian	x_m	$x_m + 2.58\sigma$	$x_m + 1.54\sigma$	0.12
Triangular	x_m	$x_m + 2.45\sigma$	$x_m + 1.63\sigma$	0.11
Uniform	x_m	$x_m + 1.73\sigma$	$x_m + 1.73\sigma$	0

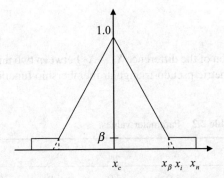

Figure 2.12 Pseudo-triangular membership function.

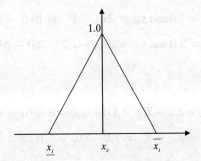

Figure 2.13 Fuzzy number generated by the decomposition scheme.

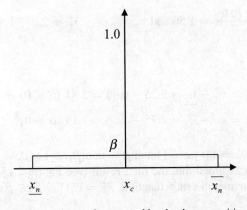

Figure 2.14 Fuzzy interval generated by the decomposition scheme.

Example 2.6

Consider the calculation of the difference $X_1 - X_2$ between two uncertain quantities represented by two symmetric pseudo-triangular membership functions whose parameters are given in Table 2.2.

Table 2.2   Parameter values.

	x_c	x_n	x_β	β
X_1	5.0	7.0	6.0	0.2
X_2	1.0	2.0	1.5	0.1

Therefore,

$$\overline{x_{n_1}} = x_{n_1} = 7.0 \text{ and } \underline{x_{n_1}} = 2x_{c_1} - \overline{x_{n_1}} = 10.0 - 7.0 = 3.0,$$

$$\overline{x_{n_2}} = x_{n_2} = 2.0 \text{ and } \underline{x_{n_2}} = 2x_{c_2} - \overline{x_{n_2}} = 2.0 - 2.0 = 0.0,$$

and

$$\underline{x_n} = \underline{x_{n_1}} - \overline{x_{n_2}} = 3.0 - 2.0 = 1.0 \ (\alpha = 0)$$

$$\overline{x_n} = \overline{x_{n_1}} - \underline{x_{n_2}} = 7.0 - 0.0 = 7.0 \ (\alpha = 0)$$

Also,

$$\overline{x_{i_1}} = \frac{x_{\beta_1} - \beta_1 x_{c_1}}{(1 - \beta_1)} = 6.25 \text{ and } \underline{x_{i_1}} = 2x_{c_1} - \overline{x_{i_1}} = 10.0 - 6.25 = 3.75,$$

$$\overline{x_{i_2}} = \frac{x_{\beta_2} - \beta_2 x_{c_2}}{(1 - \beta_2)} = 1.56 \text{ and } \underline{x_{i_2}} = 2x_{c_2} - \overline{x_{i_2}} = 2.0 - 1.56 = 0.44.$$

Therefore,

$$\overline{x_i} = \overline{x_{i_1}} - \underline{x_{i_2}} = 6.25 - 0.44 = 5.81 \ (\alpha = 0) \text{ and}$$

$$\underline{x_i} = \underline{x_{i_1}} - \overline{x_{i_2}} = 3.75 - 1.56 = 2.19 \ (\alpha = 0)$$

Also $x_c = x_{c_1} - x_{c_2} = 4.0 \ (\alpha = 1)$ and $\beta = \min[\beta_1, \beta_2] = 0.1$.

Therefore, if it is assumed that the final result (see Figure 2.15) is also a symmetric pseudo-triangular membership function, $\overline{x_\beta} = \overline{x_{i_1}}(1 - \beta) + \beta x_c = 5.63$ and $\underline{x_\beta} = 2x_c - \overline{x_\beta} = 2.37$.

It should be noted that this method only approximates the correct result if the fuzzy arithmetic involves product or division operations.

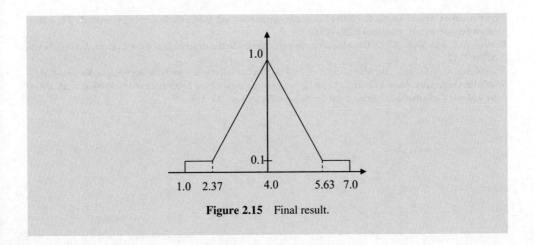

Figure 2.15   Final result.

2.6   Summary

Different ways of describing and propagating uncertainty have been described. The limitations of using statistical approaches have been explained and the advantages of the fuzzy approach to dealing with uncertainty in non-linear systems have been explained. Different methods of defining uncertainty (uncertainty intervals and probability distributions, fuzzy sets and fuzzy numbers) and propagating uncertainty (root-sum-squares, interval arithmetic, statistical methods, Monte Carlo methods and fuzzy arithmetic) have been described, and a technique for aggregating fuzzy sets based on the extension principle has been explained. A way of reducing the computational demands of fuzzy arithmetic by parameterizing the fuzzy numbers and using a decomposition method has been presented.

References

Darba, R.M., Eljarrat, E. and Barcelo, D. (2008) How to measure uncertainties in environmental risk assessment. *Trends in Analytical Chemistry*, **27**(4), 377–385.

Ferrero, A., Lazzaroni, M. and Salicone, S. (2002). A calibration procedure for a digital instrument for electric power quality measurement. *IEEE Transactions on Instrumentation and Measurement*, **51**(4), 716–722.

Ferson, S. (1996) What Monte Carlo methods cannot do. *International Journal of Human and Ecological Risk Assessment*, **2**(4), 990–1007.

Ferson, S. and Ginzburg, L.R. (1996) Different methods are needed to propagate ignorance and variability. *Reliability Engineering and System Safety*, **54**(2–3), 133–144.

Klir, G.J. and Yuan, B. (1995) *Fuzzy Sets and Fuzzy Logic: Theory and Applications*, Prentice Hall, New Jersey.

Mauris, G., Lasserre, V. and Foulloy, L. (2000) Fuzzy modelling of measurement data acquired from physical sensors. *IEEE Transactions on Instrumentation and Measurement*, **49**(6), 1201–1205.

Mauris, G., Lasserre, V. and Foulloy, L. (2001) A fuzzy approach for the expression of uncertainty in measurement. *Measurement*, **29**, 165–177.

Moffat, R.J. (1988) Describing the uncertainties in experimental results. *Experimental Thermal and Fluid Science*, **1**(1), 3–17.

Papadopoulos, C.E. and Yeung, H. (2001) Uncertainty estimation and Monte Carlo simulation methods. *Flow Measurement and Instrumentation*, **12**, 291–298.

Restivo, M.T. and Sousa, C. (2008) Measurement uncertainties in the experimental field. *Sensors & Transducers*, **95**(8), 1–12.

Willink, R. (2006) On using the Monte Carlo method to calculate uncertainty intervals. *Metrologia*, **43**, L39–L42.

Wubbeler, G., Kryatek, M. and Elster, C. (2008) Evaluation of measurement uncertainty and its numerical calculation by a Monte Carlo method. *Measurement Science and Technology*, **19**, 1–4.

3

Accounting for Measurement Uncertainty

3.1 Measurement Errors

The *de facto* standard for the evaluation of measurement uncertainty is the IEC-ISO Guide to the Expression of Uncertainty in Measurement (GUM) (Joint Committee for Guides in Metrology, 2008). The Guide defines measurement uncertainty as "a parameter, associated with the result of a measurement, which characterizes the dispersion of values that could reasonably be attributed to the measurand". It is recognized that the uncertainty results from two basic types of errors – random errors (Type A errors) and systematic errors (Type B errors) – and assumed that the measurement can be corrected for recognized systematic errors. Standard uncertainty is defined and the use of confidence intervals and levels of confidence to relate an expanded or overall uncertainty to the standard uncertainty is advocated. The guide goes on to specify how the expanded uncertainty can be calculated when a series of observations of the measurand are available (e.g. 95% confidence level for a coverage factor of two standard deviations) and when other kinds of information (e.g. manufacturers' specifications, calibration information) about the uncertainty must be utilized.

Until recently, the main mathematical tool for dealing with measurement uncertainty has been probability theory. The limitations of this approach have now been recognized (Ferrero and Salicone, 2005) and some of the assumptions underlying the proposed techniques for describing and propagating measurement uncertainties have now been relaxed (e.g. normal distribution) (Urbanski and Wasowski, 2003; Ferrero and Salicone, 2007).

3.2 Introduction to Fuzzy Random Variables

One way of dealing with the representation of the measurement uncertainty associated with single observations that results from both random and systematic errors is to describe them using fuzzy random variables (Ferrero and Salicone, 2004, 2006).

Monitoring and Control of Information-Poor Systems: An Approach based on Fuzzy Relational Models, First Edition. Arthur L. Dexter.
© 2012 John Wiley & Sons, Ltd. Published 2012 by John Wiley & Sons, Ltd.

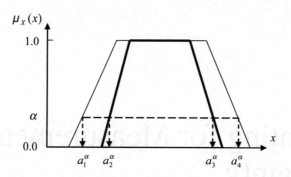

Figure 3.1   Membership function of a fuzzy random variable.

3.2.1   Definition of a Fuzzy Random Variable

The membership function of a fuzzy random variable is defined in terms of its α-cuts (see Figure 3.1), where each α-cut is represented by four values:

$$A^\alpha = \left[\alpha_1^\alpha, \alpha_2^\alpha, \alpha_3^\alpha, \alpha_4^\alpha\right] \tag{3.1}$$

The inner interval $\left[\alpha_2^\alpha, \alpha_3^\alpha\right]$ represents the part of the uncertainty which is associated with unknown or systematic errors. The outer intervals $\left[\alpha_1^\alpha, \alpha_2^\alpha\right]$ and $\left[\alpha_3^\alpha, \alpha_4^\alpha\right]$ represent the part of the uncertainty which is associated with random errors.

If $\alpha_1^\alpha = \alpha_2^\alpha$ and $\alpha_3^\alpha = \alpha_4^\alpha$, the fuzzy random variable can be used to represent situations where there is only random uncertainty or only systematic uncertainty, depending on whether a statistical approach or fuzzy arithmetic is used to propagate the measurement uncertainties.

3.2.2   Generating Fuzzy Random Variables from a Knowledge of the Random and Systematic Errors

As a fuzzy variable is by definition convex, the focal elements of its possibility distribution (i.e. the α-cut sets) are nested and it can be shown that there is a connection between the evidence theory and possibility theory (Mauris *et al.*, 2000).

The theory of evidence is based on two non-additive measures: belief and plausibility (Klir and Yuan, 1995). For example, let A be a crisp subset of the universal set X. The basic assignment $m(A)$ is a measure of the evidence supporting the claim that a given element of X belongs to set A. It should be noted that the value of $m(A)$ pertains solely to one set, set A, and not, for example, to any subsets of A.

$$\text{Bel}(A) = \sum_{B|B\subseteq A} m(B) \tag{3.2}$$

$$\text{Pl}(A) = \sum_{B|A\cap B\neq 0} m(B) \tag{3.3}$$

Now consider the α-cuts of a fuzzy variable A,

$$A^{1.0} \subset A^{0.9} \subset A^{0.8} \subset \ldots \subset A^{0.1} \subset A^{0.0} \text{ etc.}$$

In possibility theory, the equivalent of belief is necessity $Nec(A^\alpha)$ and the equivalent of plausibility is possibility $Pos(A^\alpha)$ where

$$Nec(A^\alpha) = \sum_{i=\alpha}^{1} m(A^i) \tag{3.4}$$

and

$$Pos(A^\alpha) = \sum_{i=0.0}^{1.0} m(A^i) = 1 \tag{3.5}$$

It can be shown (Klir and Yuan, 1995) that,

$$Nec(A^\alpha) = 1 - Pos(\overline{A^\alpha}) \tag{3.6}$$

$$Pos(A^\alpha) = \sup_{x \in A^\alpha} \{\mu_A(x)\} \tag{3.7}$$

Therefore

$$Pos(\overline{A^\alpha}) = \sup_{x \in \overline{A^\alpha}} \{\mu_A(x)\} = \alpha \tag{3.8}$$

and

$$Nec(A^\alpha) = 1 - \alpha \tag{3.9}$$

Because necessity is a measure of belief that a particular value x belongs to the set A^α, it must be equal to the degree of confidence λ that x lies within the confidence interval $I_\lambda = [\min\{A^\alpha\}, \max\{A^\alpha\}]$. Hence,

$$\lambda = 1 - \alpha \tag{3.10}$$

Fuzzy random variables can therefore be generated using the approach described by Ferrero and Salicone (2006). First, given an estimate of the probability density function of the random error obtained from multiple measurements, the confidence levels λ are calculated for a number of confidence intervals covering the full range. A fuzzy set can then be formed from the α-cuts based on the confidence intervals, where $\alpha = 1 - \lambda$. If the probability distribution approaches zero asymptotically, the confidence interval associated with a confidence level close to unity (e.g. 0.99) is used to approximate the α-cut set for $\alpha = 0$. Second, given that manufacturers' data or calibration data indicate that the systematic errors lie within a given range but with unknown probability distribution, the systematic errors can be described by a crisp set over this range. The possibility distribution of the fuzzy random variable can then be formed by dividing the fuzzy set describing the random errors into two at its median value and inserting the crisp set between the two parts.

It should be noted that the computational demands associated with propagating the measurement uncertainties can be reduced significantly if the probability density function of the random errors are, or can be assumed to be, normal. In this case, the α-cut set for $\alpha = 1$ can be approximated by the mean value of the probability distribution and the random errors can be represented by a fuzzy set with a triangular membership function, whose support is equal to 6σ.

3.3   A Hybrid Approach to the Propagation of Uncertainty

In many cases the direct measurement of a physical quantity is unfeasible and an estimate of its value must be obtained from two or more primary measurements. If there is uncertainty associated with the primary measurements, a method of propagating these uncertainties must be used to determine the uncertainty associated with the estimated value. The propagation of the systematic error is different from the propagation of random errors.

The most straightforward approach to error propagation is to assume that the random effects affecting the measurement process can be represented by a normal distribution (Ferrero and Salicone, 2004). Under this assumption, the outer intervals of the fuzzy random variable can be related to the standard deviation of the probability distribution.

Consider the case where a fuzzy random variable C is related to two fuzzy random variables A and B by

$$C = f(A, B) \tag{3.11}$$

where the function f is summation, difference, product or division. Let A, B and C be represented by the α-cuts $A^\alpha = \left[a_1^\alpha, a_2^\alpha, a_3^\alpha, a_4^\alpha\right]$, $B^\alpha = \left[b_1^\alpha, b_2^\alpha, b_3^\alpha, b_4^\alpha\right]$ and $C^\alpha = \left[c_1^\alpha, c_2^\alpha, c_3^\alpha, c_4^\alpha\right]$, respectively.

The internal interval $C^\alpha = \left[c_2^\alpha, c_3^\alpha\right]$ is calculated using interval arithmetic. For example, if f is a sum:

$$c_2^\alpha = a_2^\alpha + b_2^\alpha$$

and

$$c_3^\alpha = a_3^\alpha + b_3^\alpha$$

The outer values of the external intervals $C^\alpha = \left[c_1^\alpha, c_2^\alpha\right]$ and $C^\alpha = \left[c_3^\alpha, c_4^\alpha\right]$ are calculated from the appropriate standard deviation. Thus,

$$c_1^\alpha = c_2^\alpha - K_\lambda \sigma_C \tag{3.12}$$

or, if $K_\lambda \sigma_C = 3\sigma_C^\alpha$,

$$c_1^\alpha = c_2^\alpha - 3\sigma_C^\alpha \tag{3.13}$$

and

$$c_4^\alpha = c_3^\alpha + 3\sigma_C^\alpha \tag{3.14}$$

where

$$\sigma_C^\alpha = \sqrt{\left(\frac{\partial f}{\partial A}\right)^2 (\sigma_A^\alpha)^2 + \left(\frac{\partial f}{\partial B}\right)^2 (\sigma_B^\alpha)^2 + 2\left(\frac{\partial f}{\partial A}\right)\left(\frac{\partial f}{\partial B}\right)\sigma^\alpha(A, B)} \quad \text{etc.} \tag{3.15}$$

and

$$\sigma_A^\alpha = \frac{K_\lambda \sigma_A}{3} = \frac{(a_2^\alpha - a_1^\alpha)}{3} \quad \text{etc.} \tag{3.16}$$

The sensitivity coefficients are evaluated at the centres of the α-cuts when $\alpha = 1$. In this case, the sensitivity coefficients are both equal to unity and $\sigma^\alpha(A, B) = \sigma_A^\alpha \cdot \sigma_B^\alpha \cdot r(A, B)$, where $r(A, B)$ is the correlation factor between the two distributions.

Example 3.1

Figures 3.2 and 3.3 depict the results of estimating the electrical resistance ($r = v/i$) and power ($p = v\,i$), respectively, from uncertain measurements of the voltage (v) and current (i) when they are represented by the fuzzy random variables shown in the figures.

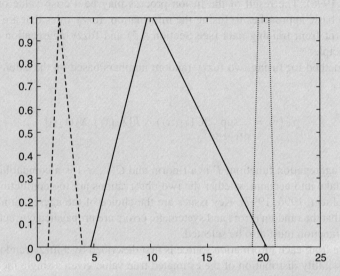

Figure 3.2 Estimated electrical resistance (solid line) given the fuzzy random variables representing the voltage (dotted line) and current (dashed line).

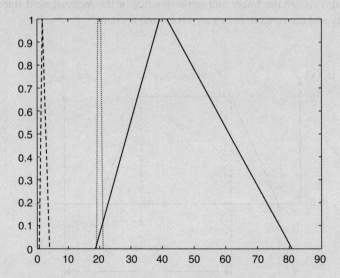

Figure 3.3 Estimated electrical power (solid line) given the fuzzy random variables representing the voltage (dotted line) and current (dashed line).

3.4 Fuzzy Sensor Fusion Based on the Extension Principle

Fuzzy sensor fusion is a method of combining uncertain information about the same measurand from two or more sources. The information may be in the form of measurements from conventional sensors (Jassar *et al.*, 2009) or it may come from a mixture of both quantitative and qualitative sources, including fuzzy sensors (see Section 3.5) and conventional sensors (Perrot *et al.*, 1996). The result of the fusion process may be a crisp value or a fuzzy set. There are two basic approaches to fusing the information: fuzzy rules that are elicited from experts or learnt from training data (see Section 4.5) and fuzzy aggregation based on the extension principle.

A general method for fusing two fuzzy random numbers based on the extension principle is given by

$$\mu_Y(y) = \sup_{g(x_1, x_2) = y} \{C(x_1, x_2) \wedge T[X_1(x_1), X_2(x_2)]\} \qquad (3.17)$$

where g is an aggregation function, T is a t-norm and $C(x_1, x_2)$ is a compatibility function, introduced to take into account whether the two observations are too conflicting to combine (Yager and Kelman, 1996, 1997). Key issues are the choice of the aggregation function and the t-norm so that the random errors and systematic errors are propagated correctly. A suitable compatibility function must also be selected.

The estimate from each information source is first described by a fuzzy random number. A plot of the possibility distribution of the estimated true value given a single measurement x_m and the estimated measurement uncertainty $\hat{\varepsilon}$ is depicted by Figure 3.4. The values of b_L and b_U are determined from *a priori* knowledge about the lower and upper bound of the systematic error, and ε_L and ε_U are the lower and upper bounds of the measurement uncertainty when both systematic and the random errors are taken into account.

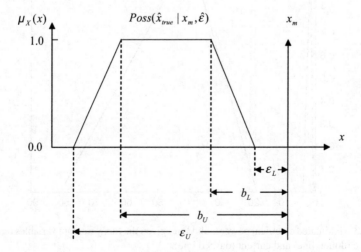

Figure 3.4 Possibility distribution of a fuzzy estimate.

Example 3.2

Consider fuzzy fusion based on weighted average to combine fuzzy random variables describing two uncertain but independent observations taken at a single sample time.

A crisp weighted average has been shown to be an efficient way of reducing the estimation errors if there are only random errors. For example, if there are two uncertain observations (x_1 and x_2) of the same variable, an overall estimate of the variable is given by:

$$\hat{x} = \alpha x_1 + (1 - \alpha)x_2 \qquad (3.18)$$

where α determines the weight given to each of the uncertain information sources. In statistical data fusion schemes, the measurement uncertainty is often assumed to be random error with zero mean and normal distribution, so α is inversely related to the variances (σ_i^2) of the random errors associated with each measurement. We therefore have:

$$\alpha = \frac{\sigma_2^2}{\sigma_1^2 + \sigma_2^2} \qquad (3.19)$$

In this example, fuzzy sensor fusion is based on Equation (3.17) with:

1. the fusion operator $g(x_1, x_2) = \alpha x_1 + (1 - \alpha)x_2$, where the weight α is inversely related to the size of the supports of the fuzzy random numbers, that is,

$$\alpha = \frac{\left(\frac{\varepsilon_{U_2} - \varepsilon_{L_2}}{2}\right)^2}{\left[\left(\frac{\varepsilon_{U_1} - \varepsilon_{L_1}}{2}\right)^2 + \left(\frac{\varepsilon_{U_2} - \varepsilon_{L_2}}{2}\right)^2\right]} = \frac{(\varepsilon_{U_2} - \varepsilon_{L_2})^2}{[(\varepsilon_{U_1} - \varepsilon_{L_1})^2 + (\varepsilon_{U_2} - \varepsilon_{L_2})^2]};$$

2. the compatibility function

$$C(x_1, x_2) = \begin{cases} 0 & \text{if } |x_1 - x_2| \geq D_{max} \\ 1 - \frac{|x_1 - x_2|}{D_{max}} & \text{if } |x_1 - x_2| < D_{max}; \end{cases}$$

3. the Yager t-norm with $p = 2$.

Three cases are considered in which the two observations are described by the fuzzy random numbers X_1 (dotted line) and X_2 (dashed line) shown in Figures 3.5–3.7. The results of fuzzy fusion with (solid line) and without (dashdot line) the use of a compatibility function, and the crisp values of the estimates after height defuzzification (*and o, respectively), are also indicated.

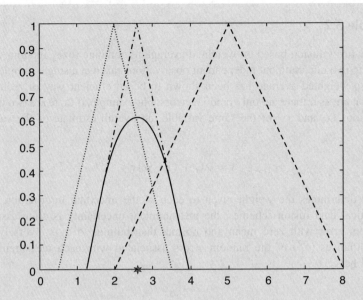

Figure 3.5 Results of fuzzy fusion when measurement uncertainties arise from random errors only.

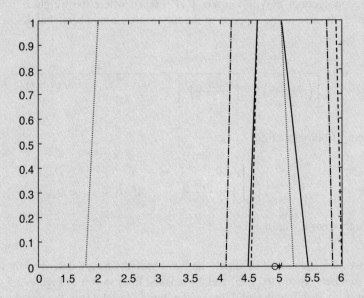

Figure 3.6 Results of fuzzy fusion when uncertainties associated with systematic errors are much larger than random errors.

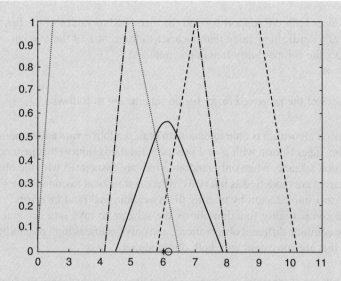

Figure 3.7   Results of fuzzy fusion when there are both systematic and random errors.

In the first case, the two uncertain observations are corrupted by random error only. The results of the fuzzy fusion are shown in Figure 3.5. It can be seen that with the inclusion of the compatibility function, the fuzzy fusion scheme reduces the membership values of the fuzzy estimate. However, the crisp values obtained after height defuzzification are the same for fuzzy weighted average with or without compatibility function. When the two fuzzy numbers to be combined are symmetric, the results after fuzzy weighted average and height defuzzification are equal to the crisp weighted average of the centres of the two fuzzy numbers.

In the second case, the uncertainties associated with the *a priori* knowledge about the systematic error are much larger than random errors. The results of fuzzy fusion are shown in Figure 3.6. It can be seen that, when the uncertainties associated with the two measurements are mostly systematic errors, the fuzzy fusion scheme reduces the width of the interval in the final estimation with the inclusion of compatibility function. This is because there is a small random error associated with both measurements in this case, and the values of the two uncertain measurements are therefore considered to be compatible only when the difference between them is small. It can be seen that the fuzzy estimate generated by the fuzzy fusion scheme incorporating the compatibility function is almost equal to the fuzzy intersection of the fuzzy random variables. It should also be noted that the crisp values after height defuzzification are different for fuzzy fusion with and without considering compatibility, and the defuzzified value of the result with compatibility is closer to the centre of the overlap area.

In the third case, there are both systematic and random errors associated with the uncertain observations. The results of fuzzy fusion are shown in Figure 3.7. It can be seen that, as in the second case, the fuzzy fusion scheme incorporating the compatibility function reduces both the support for and the membership values of the fuzzy estimate and produces a defuzzified value that is closer to the centre of the overlap area than is the case when the compatibility function is not used.

The advantages of the proposed fuzzy fusion scheme are as follows.

- The same fusion algorithm is able to deal with both random errors and systematic errors.
- The use of the Yager t-norm with $p = 2$ produces similar results to that generated by a crisp statistical fusion scheme, when only random errors are associated with the observations.
- Systematic errors are modelled as intervals, whereas statistical fusion schemes can only take systematic errors into account by treating them as additional random errors.
- The use of a compatibility function allows the scheme to take into account whether it is reasonable to combine different observations, and only generates high possibilities for values of the result that are consistent with both observations.

The main weak point of fuzzy fusion is that it is not known in what sense the fuzzy estimate is optimal when there are systematic errors associated with the measurements. There is therefore no universally accepted approach to selecting the aggregation operators and values of the parameters used in the fuzzy fusion scheme.

3.5 Fuzzy Sensors

A soft sensor (also known as an intelligent or smart sensor) uses software running on dedicated hardware to process available raw measurements and generate an estimate of the variable to be sensed (Schodel, 1994; Benoit and Foulloy, 2001). Such sensors must be used when direct measurement of the variable is impossible, uneconomic or too inaccurate.

Inferential sensors are soft sensors that estimate the primary variable from (usually crisp) measurements of one or more secondary variables that can be measured easily online (Angelov and Kordon, 2010). For example, inferential sensors have been used to estimate important process variables in a penicillin production process such as biomass concentration or the viscosity of the broth, which can only be measured directly using laboratory analysis (Arauzo-Bravo *et al.*, 2004), and to estimate the average temperature in a large multi-storey building from measurements defining the external environment (Jassar *et al.*, 2009).

A fuzzy sensor is the name given to a soft sensor that uses fuzzy rules to relate the raw measurements to the estimated variable, or a soft sensor that generates a fuzzy estimate of the variable. The fuzzy output might provide a linguistic description of a qualitative concept (Mauris *et al.*, 1994) such as colour, thermal comfort or air quality, or give an indication of the uncertainty associated with the estimated variable (Lee and Dexter, 2005).

In the former case, the rule-base generates the degrees of membership of a limited number of labelled fuzzy output sets from the measured data (Benoit and Foulloy, 2001; Mauris and Foulloy, 2002). For example, in the case of the air quality sensor (Wide *et al.*, 1997), the fuzzy output sets are *polluted air quality*, *medium air quality* and *clean air quality*. In the case of the thermal comfort sensor (Mauris and Foulloy, 2002), the fuzzy output sets are *uncomfortable*, *acceptable* and *comfortable*. The fuzzy rules, which relate the crisp measured inputs to the fuzzy output sets, are either based on expert knowledge (Bombardier *et al.*, 2009) or identified from training data (Jassar *et al.*, 2009) using a fuzzy identification scheme (see Section 4.5.2). The training data are obtained from detailed computer simulations (Lee and Dexter, 2005) or from online measurements of the input variable(s) and offline laboratory analysis of the output variable (Arauzo-Bravo *et al.*, 2004).

3.6 Summary

The difference between the random and systematic errors associated with many measurements has been discussed. A brief introduction to the theory of evidence and possibility theory has been given and a method of describing both types of measurement errors using fuzzy random variables has been explained. A hybrid approach to the propagation of measurement uncertainty has been described. An approach to fuzzy sensor fusion based on the extension principle has also been described and a brief introduction to fuzzy sensors has been given.

References

Angelov, P. and Kordon, A. (2010) Adaptive inferential sensors based on evolving fuzzy models. *IEEE Transactions on Systems, Man, and Cybernetics, Part B*, **40**(2), 529–539.

Arauzo-Bravo, M.J., Cano-Izquierdo, J.M., Gomez-Sanchez, E., *et al.* (2004) Automatization of a penicillin production process with soft sensors and an adaptive controller based on neuro fuzzy systems. *Control Engineering Practice*, **12**, 1073–1090.

Benoit, E. and Foulloy, L. (2001) High functionalities for intelligent sensors, application to fuzzy colour sensor. *Measurement*, **30**, 161–170.

Bombardier, V., Schmitt, E. and Charpentier, P. (2009) A fuzzy sensor for color matching vision system. *Measurement*, **42**, 189–201.

Ferrero, A. and Salicone, S. (2004) The random-fuzzy variables: a new approach to the expression of uncertainty in measurement. *IEEE Transactions on Instrumentation and Measurement*, **53**(5), 1370–1377.

Ferrero, A. and Salicone, S. (2005) A comparative analysis of the statistical and random-fuzzy approaches in the expression of uncertainty in measurement. *IEEE Transactions on Instrumentation and Measurement*, **54**(4), 1475–1481.

Ferrero, A. and Salicone, S. (2006) Fully comprehensive mathematical approach to expression of uncertainty in measurement. *IEEE Transactions on Instrumentation and Measurement*, **55**(3), 706–712.

Ferrero, A. and Salicone, S. (2007) Modeling and processing measurement uncertainty within the theory of evidence: mathematics of random-fuzzy variables. *IEEE Transactions on Instrumentation and Measurement*, **56**(3), 704–716.

Jassar, S., Liao, Z. and Zhao, L. (2009) Adaptive neuro-fuzzy based inferential sensor model for estimating the average air temperature in space heating systems. *Building & Environment*, **44**(8), 1609–1616.

Joint Committee for Guides in Metrology (2008) *Evaluation of measurement data – guide to the expression of uncertainty in measurement*, JCGM 100.

Klir, G.J. and Yuan, B. (1995) *Fuzzy Sets and Fuzzy Logic: Theory and Applications*, Prentice Hall, New Jersey.

Lee, P.S. and Dexter, A.L. (2005) A fuzzy sensor for measuring the mixed air-temperature in air-handling units. *Measurement*, **37**(1), 83–93.

Mauris, G. and Foulloy, L. (2002) A fuzzy symbolic approach to formalize sensory measurements: an application to a comfort sensor. *IEEE Transactions on Instrumentation and Measurement*, **51**(4), 712–715.

Mauris, G., Benoit, E. and Foulloy, L. (1994) Fuzzy symbolic sensors – from concept to applications. *Measurement*, **12**, 357–384.

Mauris, G., Berrah, L., Foulloy, L. and Haurat, A. (2000) Fuzzy handling of measurement errors in instrumentation. *IEEE Transactions on Instrumentation and Measurement*, **49**(1), 89–93.

Perrot, N., Trystram, G., Le Guennec, D. and Guely, F. (1996) Sensor fusion for real-time quality evaluation of biscuit during baking. Comparison between Bayesian and fuzzy approaches. *Journal of Food Engineering*, **29**, 301–315.

Schodel, H. (1994) Utilization of fuzzy techniques in intelligent sensors. *Fuzzy Sets and Systems*, **63**, 271–292.

Urbanski, M.K. and Wasowski, J. (2003) Fuzzy approach to the theory of measurement inexactness. *Measurement*, **34**, 67–74.

Wide, P., Winquist, F. and Driankov, D. (1997) An air quality sensor system with fuzzy classification. *Measurement Science Technology*, **8**, 138–146

Yager, R. R. and Kelman, A. (1996) Fusion of fuzzy information with considerations for compatibility, partial aggregation and reinforcement. *International Journal of Approximate Reasoning*, **15**, 93–122.

Yager, R. R., and Kelman, A. (1997) A general approach to the fusion of imprecise information. *International Journal of Intelligent Systems*, **12**, 1–29.

4

Accounting for Modelling Errors in Fuzzy Models

4.1 An Introduction to Rule-Based Models

There are four basic types of fuzzy models: linguistic or Mamdani models, functional or T-S models, neurofuzzy models and fuzzy relational models. The first three types of models are described in this chapter. Fuzzy relational models are considered in Chapter 5.

4.2 Linguistic Fuzzy Models

This type of fuzzy model consists of IF-THEN rules where both the antecedent and the consequent are fuzzy propositions that can be interpreted linguistically.

The output of the model is derived using a fuzzy inferencing scheme. Fuzzy inferencing may generate a fuzzy output but, if necessary, a defuzzification scheme can be used to convert this into a crisp numerical value (see Section 4.6).

4.2.1 Fuzzy Rules

Fuzzy IF-THEN rules are of the form:

```
              Antecedent                    Conclusion or Consequent

IF(        )AND/OR(        )AND/OR .......... THEN(        )
   clause            clause
   or premise        or premise
```

where each clause in the antecedent is a description of an input variable in terms of a fuzzy set and the consequent is a description of the output variable in terms of a fuzzy set.

Monitoring and Control of Information-Poor Systems: An Approach based on Fuzzy Relational Models, First Edition. Arthur L. Dexter.
© 2012 John Wiley & Sons, Ltd. Published 2012 by John Wiley & Sons, Ltd.

4.2.2   Fuzzy Inferencing

Fuzzy inference is usually based on the use of generalized *modus ponens*. For example.

Assertion: X is A_1
Rule: IF X is A_2 THEN Z is B_2
Conclusion: Z is B_1

where B_1 is similar to B_2 if A_1 is similar to A_2.

Example 4.1

Consider fuzzy inference using the single rule,

$$\text{IF } X \text{ is } A \text{ THEN } Z \text{ is } C$$

where the fuzzy sets A and C have the following membership functions

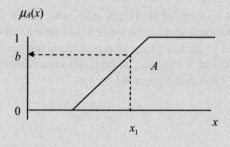

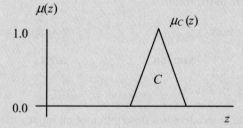

When $x = x_1$, the degree of satisfaction of the rule is the degree of membership of the fuzzy set "X is A" is equal to b.

If the min operator is used to represent fuzzy inference, the possibility of the output having the numerical value z is given by $\mu(z) = \min[\mu_C(z), b]$.

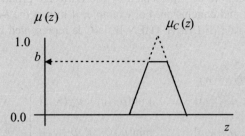

The simplest way of combining fuzzy outputs inferred by the different rules is to use the max operator to represent fuzzy-OR. For example, the result of partially firing two rules with different conclusions ("U is C" and "U is D") and different degrees of satisfaction is

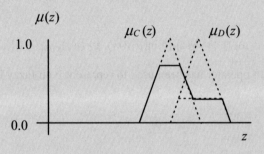

It should be noted that, although a linguistic fuzzy model produces a fuzzy output, the output may be difficult to interpret linguistically if more than one rule is fired.

4.2.3 Compositional Rules of Inference

The Cartesian product allows operators to be applied to fuzzy sets that are defined on more than one universe of discourse (e.g. if A and B are fuzzy sets defined on different, discrete universes of discourse). The fuzzy set

$$\text{"}U \text{ is } A\text{" AND "}V \text{ is } B\text{"}$$

is a Cartesian product $A \times B$, and its membership function can be described by a matrix R whose elements are given by

$$R_{i,j} = \mu_{A \times B} = T\left\{\mu_A(u_i), \mu_B(v_j)\right\}$$

where T is a t-norm.

A fuzzy rule can therefore be represented by the Cartesian product of the fuzzy sets describing its antecedent and conclusion. For example, if min is the t-norm used to represent fuzzy implication, the rule IF U is A THEN W is C is represented by the relational matrix $R = A \times C =$

$$
\text{CONCLUSION} \to w_j
$$

$$
\begin{bmatrix}
\min[\mu_A(u_1), \mu_C(w_1)] & \cdots & \min[\mu_A(u_1), \mu_C(w_n)] \\
\min[\mu_A(u_m), \mu_C(w_1)] & \cdots & \min[\mu_A(u_m), \mu_C(w_n)]
\end{bmatrix}
\quad
\begin{array}{c}
\text{ANTECEDENT} \\
\downarrow \\
u_i
\end{array}
$$

and the output can be inferred from $W = U \circ R$, where $\circ$ indicates compositional inference.

Consider the case where it is asserted that "U is D"; the membership functions are defined over discrete universes of discourse and sup-min compositional inference is used. The output membership function $\mu_W(w_j)$ is given by

$$
\mu_W(w_j) = \sup_{u_i}\{\min[\mu_D(u_i), R_{ij}]\} \tag{4.1}
$$

Thus,

$$
\mu_W(w_j) = \sup_{u_i}\{\min[\mu_D(u_i), \mu_C(w_j), \mu_A(u_i)]\} \tag{4.2}
$$

Note that here the min operator has been used to represent both fuzzy implication and fuzzy inference.

Example 4.2

Let the membership function of the antecedent be

$$
\mu_A(u_i) = \begin{bmatrix} 0.0 & 0.2 & 0.5 & 0.8 & 1.0 \end{bmatrix}
$$

and the membership function of the consequent be

$$
\mu_C(w_j) = \begin{bmatrix} 1.0 & 0.8 & 0.4 & 0.1 & 0.0 \end{bmatrix}
$$

The relational matrix is therefore

$$
\xrightarrow{\mu_C(w_j)}
$$

$$
R = \begin{bmatrix}
0.0 & 0.0 & 0.0 & 0.0 & 0.0 \\
0.2 & 0.2 & 0.2 & 0.1 & 0.0 \\
0.5 & 0.5 & 0.4 & 0.1 & 0.0 \\
0.8 & 0.8 & 0.4 & 0.1 & 0.0 \\
1.0 & 0.8 & 0.4 & 0.1 & 0.0
\end{bmatrix}
$$

If the assertion is crisp, for example if $\mu_D(u_i) = 1.0\delta(u_2)$, the compositional rule of inference gives

$$
\mu_W(w_j) = \sup_{u_i}\{\min[\mu_D(u_i), \mu_C(w_j), \mu_A(u_i)]\} = \min[\mu_C(w_j), \mu_A(u_2)]
$$

where $\mu_A(u_2) = 0.2$. Therefore

$$\mu_W(w_j) = \begin{bmatrix} 0.2 & 0.2 & 0.2 & 0.1 & 0.0 \end{bmatrix}$$

If the assertion "U is D" is fuzzy, for example if

$$\mu_D(u_i) = \begin{bmatrix} 0.1 & 0.6 & 1.0 & 0.6 & 0.1 \end{bmatrix} \neq \mu_A(u_i)$$

the compositional rule of inference is

$$\mu_W(w_j) = \sup_{u_i}\{\min[\mu_D(u_i), \mu_C(w_j), \mu_A(u_i)]\} = \sup_{u_i}\{\min[\mu_D(u_i), R_{ij}]\}$$

where $\min[\mu_D(u_i), R_{ij}]$ is

$$\begin{bmatrix} 0.0 & 0.0 & 0.0 & 0.0 & 0.0 \\ 0.2 & 0.2 & 0.2 & 0.1 & 0.0 \\ 0.5 & 0.5 & 0.4 & 0.1 & 0.0 \\ 0.6 & 0.6 & 0.4 & 0.1 & 0.0 \\ 0.1 & 0.1 & 0.1 & 0.1 & 0.0 \end{bmatrix}$$

Therefore $\sup_{u_i}\{\min[\mu_D(u_i), R_{ij}]\}$ is

$$\begin{bmatrix} \sup \begin{bmatrix} 0.0 \\ 0.2 \\ 0.5 \\ 0.6 \\ 0.1 \end{bmatrix} & \sup \begin{bmatrix} 0.0 \\ 0.2 \\ 0.5 \\ 0.6 \\ 0.1 \end{bmatrix} & \cdots \end{bmatrix}$$

and

$$\mu_W(w_j) = \begin{bmatrix} 0.6 & 0.6 & 0.4 & 0.1 & 0.0 \end{bmatrix}$$

4.2.3.1 An Alternative Interpretation

From Equation (4.2),

$$\mu_W(w_j) = \sup_{u_i}\{\min[\mu_D(u_i), \mu_C(w_j), \mu_A(u_i)]\} = \sup_{u_i}\{\min[\mu_D(u_i), \mu_A(u_i), \mu_C(w_j)]\}$$

Therefore,

$$\mu_W(w_j) = \sup_{u_i}\{\min[\min[\mu_D(u_i), \mu_A(u_i)], \mu_C(w_j)]\}$$

or

$$\mu_W(w_j) = \min[\sup_{u_i}\{[\min[\mu_D(u_i), \mu_A(u_i)]\}, \mu_C(w_j)]$$

but the possibility of A given D or $Poss(A|D)$ is given by

$$Poss(A|D) = \sup_{u_i}\{[\min[\mu_D(u_i), \mu_A(u_i)]\}$$

Therefore,

$$\mu_W(w_j) = \min[Poss(A|D), \mu_C(w_j)] \qquad (4.3)$$

Note that the degree of satisfaction of the rule is equal to $Poss(A|D)$ when sup-min inferencing is used.

Example 4.3

Consider again the case when $\mu_A(u_i) = \begin{bmatrix} 0.0 & 0.2 & 0.5 & 0.8 & 1.0 \end{bmatrix}$ and $\mu_C(w_j) = \begin{bmatrix} 1.0 & 0.8 & 0.4 & 0.1 & 0.0 \end{bmatrix}$.

1. If the assertion is crisp $\mu_D(u_i) = 1.0\delta(u_2)$,

$$Poss(A|D) = \mu_A(u_2) = 0.2$$

and

$$\mu_W(w_j) = \min[Poss(A|D), \mu_C(w_j)] = \begin{bmatrix} 0.2 & 0.2 & 0.2 & 0.1 & 0.0 \end{bmatrix}.$$

2. If the assertion is fuzzy $\mu_D(u_i) = \begin{bmatrix} 0.1 & 0.6 & 1.0 & 0.6 & 0.1 \end{bmatrix}$,

$$Poss(A|D) = 0.6$$

and

$$\mu_W(w_j) = \min[Poss(A|D), \mu_C(w_j)] = \begin{bmatrix} 0.6 & 0.6 & 0.4 & 0.1 & 0.0 \end{bmatrix}.$$

4.2.3.2 Relational Matrices for Rules Having More than One Clause in Their Antecedents

For example, consider the rule

IF U is A AND V is B THEN W is C

The rule can be represented by a three-dimensional relational array R whose elements are given by

$$R_{ijk} = \mu_{A \times B \times C}(u_i, v_j, w_k) = \min[\mu_A(u_i), \mu_B(v_j), \mu_C(w_k)] \qquad (4.4)$$

Hence $W = U \circ V \circ R$ or

$$\mu_W(w_k) = \sup_{u_i, v_j}\{\min[\mu_D(u_i), \mu_E(v_j), R_{ijk}]\} \qquad (4.5)$$

when it is asserted that "U is D" and "V is E".

A relational matrix can also be used to describe the entire fuzzy rule-base. Consider a set of rules:

$$\text{IF } X \text{ is } X_I \text{ THEN } U \text{ is } U_I \text{ for } I = 1, 2 \ldots N$$

The fuzzy conclusion obtained from the rule-base is given by:

$$U = X \circ R$$

where the relational matrix R satisfies the equations:

$$U = X_I \circ R \quad \text{for} \quad I = 1, 2, \ldots N$$

A possible estimate of the relational matrix is given by

$$\hat{R} = \bigcup_{I=1}^{N} R_I = \max_{I=1}^{N}[R_I]$$

where the relational matrix representing the jth rule is given by

$$R_I = \mu_{X_I x U_I} = \min[\mu_{X_I}, \mu_{U_I}]$$

Note however that, unless the rules are fully consistent and there is no interaction between them,

$$U = X_I \circ \hat{R} \neq X_I \circ R_I.$$

4.3 Functional Fuzzy Models

The rules in a functional fuzzy model have antecedents of the same form as those in a linguistic fuzzy model. However, the consequents of the rules are crisp functions of the input variables. For example, the affine form of a Takagi-Sugeno or T-S fuzzy model consists of rules R_i with the structure:

$$R_i : \text{IF } \mathbf{x} \text{ is } \mathbf{A_i} \text{ THEN } y_i = \mathbf{a_i}^T \mathbf{x} + b_i, \quad i = 1, 2, \ldots .K \tag{4.6}$$

where $\mathbf{x}$ is a crisp input vector, $\mathbf{A_i}$ is a multi-dimensional fuzzy set, y_i is the crisp scalar output of the ith rule, $\mathbf{a_i}$ is a parameter vector, b_i is a scalar constant and K is the number of rules in the rule-base. The output of the T-S model is given by:

$$y = \frac{\sum_{i=1}^{K} \gamma_i(\mathbf{x}) y_i}{\sum_{i=1}^{K} \gamma_i(\mathbf{x})} \tag{4.7}$$

where $\gamma_i(\mathbf{x})$ is the degree of satisfaction of the ith rule (the degree of fulfilment of the ith rule's antecedent). It can be seen that each rule is a local linear approximation of the non-linear relationship between y and $\mathbf{x}$.

In this case, $\gamma_i = \mu_{A_i}(\mathbf{x})$ is the degree of membership of $\mathbf{x}$ in the multi-dimensional fuzzy set A_i.

Alternatively, the conjunctive form of the antecedent can be used:

$$R_i : \text{IF } x_1 \text{ is } A_{i,1} \text{ AND } x_2 \text{ is } A_{i,2} \text{ AND } \ldots \text{ AND } x_p \text{ is } A_{i,p} \text{ THEN } y_i = \mathbf{a_i}^T \mathbf{x} + b_i \tag{4.8}$$

where $\mathbf{x}^T = \begin{bmatrix} x_1 x_2 \ldots x_p \end{bmatrix}$. The degree of satisfaction of the ith rule is then given by:

$$\gamma_i = \mu_{A_{i,1}} \wedge \mu_{A_{i,2}} \wedge \ldots \wedge \mu_{A_{i,p}} \tag{4.9}$$

It should be noted that a functional model gives no indication of the uncertainty associated with the outputs it produces.

4.4 Fuzzy Neural Networks

Fuzzy neural networks (FNNs) are a synthesis of artificial neural networks (ANNs) and fuzzy models. FNNs are capable of incorporating expert knowledge and can be learnt from training data. There are several different types of FNN (Figueiredo and Gomide, 1999; Mitra and Hayashi, 2000).

In some FNNs (see Figure 4.1), the network is structured so that it is directly equivalent to a set of fuzzy rules, a fuzzy inferencing scheme and a defuzzification scheme (Brown and

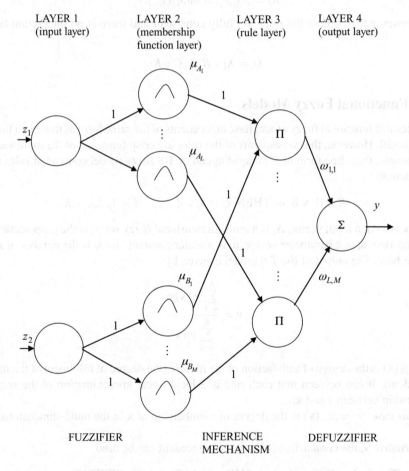

Figure 4.1 Four-layer two-input fuzzy neural network.

Harris, 1994; Lee *et al.*, 1996; Gao and Er, 2005). The main strength of this type of FNN is that the equivalent fuzzy rules can be easily found after the network has been trained (Lee *et al.*, 1996).

Consider the following set of rules for $i = 1, 2 \ldots L$, $j = 1, 2 \ldots M$ and $k = 1, 2 \ldots N$:

IF Z_1 is A_i AND Z_2 is B_j THEN Y is C_k ($R_{i,j,k}$)

Let $R_{i,j,k} = 1$ if the rule exists and $R_{i,j,k} = 0$ if the rule does not exist. Note that $\sum_{k=1}^{N} R_{i,j,k} = 1 \forall i, j$ if the set of rules is consistent.

Using sum-product inferencing and height defuzzification, the value of the output is given by

$$y = \frac{\sum_{i=1}^{L} \sum_{j=1}^{M} \sum_{k=1}^{N} \prod(\mu_{A_i}, \mu_{B_j}, \bar{y}_k, R_{i,j,k})}{\sum_{i=1}^{L} \sum_{j=1}^{M} \sum_{k=1}^{N} \prod(\mu_{A_i}, \mu_{B_j}, R_{i,j,k})} \tag{4.10}$$

where $\mu_{A_i}(z_1)$ and $\mu_{B_j}(z_2)$ are the membership functions of the input sets A_i and B_j, respectively, and $\bar{y}_k$ is the centroid of the output set C_k. Hence,

$$
y = \frac{\sum_{i=1}^{L} \sum_{j=1}^{M} \prod[\mu_{A_i}, \mu_{B_j} \sum_{k=1}^{N} \prod(\bar{y}_k, R_{i,j,k})]}{\sum_{i=1}^{L} \sum_{j=1}^{M} \prod(\mu_{A_i}, \mu_{B_j}) \sum_{k=1}^{N} R_{i,j,k}}
$$

$$
= \frac{\sum_{i=1}^{L} \sum_{j=1}^{M} \prod[\mu_{A_i}, \mu_{B_j}, \omega_{i,j}]}{\sum_{i=1}^{L} \sum_{j=1}^{M} \prod(\mu_{A_i}, \mu_{B_j})} = \frac{\sum_{i=1}^{L} \sum_{j=1}^{M} \prod[\mu_{A_i}, \mu_{B_j}, \omega_{i,j}]}{\prod\left(\sum_{i=1}^{L} \mu_{A_i}, \sum_{j=1}^{M} \mu_{B_j}\right)} \tag{4.11}
$$

where $\omega_{i,j} = \sum_{k=1}^{N} \prod(\bar{y}_k, R_{i,j,k})$. Therefore, assuming the input fuzzy sets have a partition of unity (i.e. $\sum_{i=1}^{L} \mu_{A_i} = 1$ and $\sum_{j=1}^{M} \mu_{B_j} = 1$),

$$y = \sum_{i=1}^{L} \sum_{j=1}^{M} \prod[\mu_{A_i}, \mu_{B_j}, \omega_{i,j}] = \sum_{i=1}^{L} \sum_{j=1}^{M} \omega_{i,j} \prod[\mu_{A_i}, \mu_{B_j}] \tag{4.12}$$

which is identical to the output of the FNN shown in Figure 4.1.

In other FNNs, the outputs of the neurons are calculated using t-norms and t-conorms (see Section 2.4) rather than conventional arithmetic (Pedrycz and Rocha, 1993; Manmohan *et al.*, 2003). The main weakness of fuzzy neural networks that generate a crisp output is that they give no indication of the uncertainty associated with the outputs they produce.

The third type of FNN is one that is capable of handling fuzzy inputs and generating fuzzy outputs (Ishibuchi *et al.*, 1993). FNNs of this type are described in more detail in Section 8.4. The main problem with fuzzy neural networks, which produce a fuzzy output, is that they are computationally demanding.

It should also be noted that any fuzzy model identified from training data is sometimes called a fuzzy-neural model (Barada and Singh, 1998).

4.5 Methods of Generating Fuzzy Models

There are two basic approaches to generating fuzzy rules:

1. elicitation of expert knowledge; and
2. estimation based on either measured data collected from the system to be modelled or data obtained by computer simulation of a similar system.

4.5.1 Modifying Expert Rules to Take Account of Uncertainty

There are two main sources of uncertainty associated with expert rules: uncertainty about the validity of the rules themselves (e.g. there may be apparent differences in the rules proposed by different experts) and uncertainty about the linguistic meaning of the fuzzy sets used in the rules (e.g. there is no universally agreed interpretation of the word 'large'). The fuzzy sets we have considered so far (so-called type-1 fuzzy sets) have membership functions whose membership values are crisp numerical values. One way of dealing with the uncertainty associated with expert rules is to use type-2 fuzzy sets, for which the membership values are themselves fuzzy sets characterized by secondary membership functions (Mendel, 2001).

4.5.1.1 Type-2 Fuzzy Sets

A type-2 fuzzy set, denoted as $\tilde{A}$, is characterized by a type-2 membership function $\mu_{\tilde{A}}(x, u)$ where

$$\tilde{A} = \left\{ (x, u), \mu_{\tilde{A}}(x, u) | \forall x \in X, \forall u \in J_x \subseteq [0, 1] \right\} \tag{4.13}$$

X is the universe of discourse over which the primary membership function is defined and J_x is the universe of discourse over which the secondary membership functions are defined. The secondary membership function for a particular value of x is a type-1 membership function.

The uncertainty associated with the primary memberships of a type-2 fuzzy set (see Figure 4.2) is a bounded region called the footprint of uncertainty (FOU).

The upper and lower membership functions, $\overline{\mu}_{\tilde{A}}(x)$ and $\underline{\mu}_{\tilde{A}}(x)$, are the two type-1 primary membership functions that define the bounds of the FOU of a type-2 fuzzy set. The primary

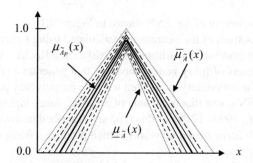

Figure 4.2   Example of a Footprint of Uncertainty.

membership function for which all grades of secondary membership are one, the so-called principal membership function $\mu_{\tilde{A}_P}(x)$, is indicated by the heavy solid line.

Consider a type-2 fuzzy set defined over discrete universes of discourse for x and u.

The type-2 fuzzy set is given by

$$\tilde{A} = \sum_{i=1}^{N}\left[\sum_{u\in J_{x_i}} f_{x_i}(u)/u\right]/x_i \tag{4.14}$$

where N is the number of discrete points defined over the universe of discourse.

A type-2 fuzzy set can be thought of as a collection of embedded type-2 fuzzy sets $\tilde{A}_e$ given by

$$\tilde{A}_e = \sum_{i=1}^{N}\left[f_{x_i}(\theta_i)/\theta_i\right]/x_i \tag{4.15}$$

where θ_i is a particular value of u when $x = x_i$ and $f(\theta_i)$ is the associated value of secondary membership. The associated embedded type-1 set is given by

$$A_e = \sum_{i=1}^{N}\theta_i/x_i. \tag{4.16}$$

Example 4.4

Consider the discrete type-2 fuzzy set shown in Figure 4.3. Figure 4.4 is an example of one of the embedded type-2 fuzzy sets. Its associated embedded type-1 fuzzy set is shown in Figure 4.5.

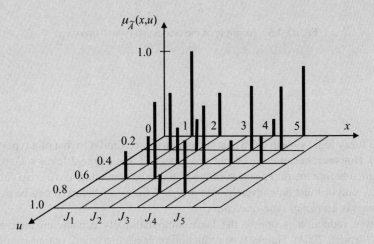

Figure 4.3   Example of a type-2 fuzzy set.

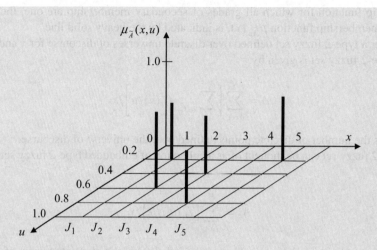

Figure 4.4   Example of an embedded type-2 fuzzy set.

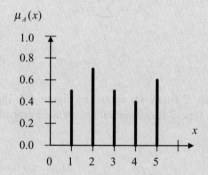

Figure 4.5   Example of the associated type-1 fuzzy set.

A type-2 fuzzy logic system (FLS) has a basic structure similar to that of a type-1 FLS (see Figure 4.6). However, because the output of a type-2 FLS is a type-2 fuzzy set, an additional step is required before the fuzzy output of the FLS can be converted into a crisp value (i.e. the type-2 fuzzy output must be converted into a type-1 fuzzy set before it can be defuzzified). This operation is known as *type-reduction*.

Height type reduction is one of the least computationally demanding methods of type reduction (Mendel, 2001). Height type reduction replaces the type-2 output set of each rule by a type-2 fuzzy singleton at the point $\bar{y}_l$ having the highest primary membership of the principal membership function of that output set. The type-2 fuzzy singleton has a secondary membership function $f_{\bar{y}_l}(\theta)$ that is a type-1 fuzzy set.

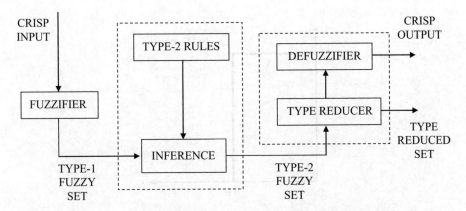

Figure 4.6 Type-2 fuzzy logic system.

The height type-reduced set can then be obtained by height defuzzification using the extension principle. We therefore have

$$Y_h(x) = \int_{\theta_1} \cdots \int_{\theta_M} \left[f_{\bar{y}_1}(\theta_1) * \cdots f_{\bar{y}_M}(\theta_M) \right] \Bigg/ \frac{\sum_{l=1}^{M} \bar{y}_l \theta_l}{\sum_{l=1}^{M} \theta_l} \tag{4.17}$$

where the integral sign denotes union over all possible values of θ and M is the number of rules rather than the number of points used on the discrete primary universe of discourse. It should be noted that, when only a single rule is fired, the result is a single point and all of the uncertainty associated with the antecedent and consequent sets is lost.

It can be seen that there are significant computational demands associated with manipulating type-2 fuzzy sets. The calculations are greatly simplified if the secondary membership functions used to define the type-2 fuzzy sets are fuzzy intervals.

Example 4.5

Consider the simple example shown in Figures 4.7–4.9.

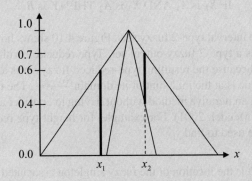

Figure 4.7 Example of an interval type-2 fuzzy set.

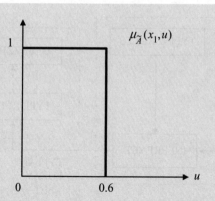

Figure 4.8   Associated secondary membership function at point x_1.

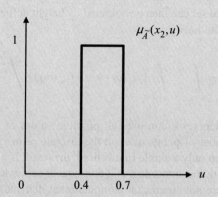

Figure 4.9   Associated secondary membership function at point x_2.

Fuzzy inference based on interval type-2 fuzzy rules is relatively straightforward. For example, consider inference based on a single fuzzy rule of the form

$$\text{IF } X_1 \text{ is } \tilde{A}_1 \text{ AND } X_2 \text{ is } \tilde{A}_2 \text{ THEN } Y \text{ is } \tilde{B},$$

where $\tilde{A}_1$, $\tilde{A}_2$ and $\tilde{B}$ are interval type-2 fuzzy sets. Figure 4.10 shows how inference using min as the t-norm generates a type-2 fuzzy output set. Type reduction with interval type-2 fuzzy sets is also simplified because the resulting type-reduced fuzzy set is a crisp interval $[y_L, y_U]$ and the defuzzified value is at the midpoint of its domain $\frac{(y_L + y_U)}{2}$. The type-reduced fuzzy set can be computed using an iterative method without having to consider all combinations of the embedded fuzzy sets (Mendel, 2001). For example, for height type reduction, the following iterative method can be used to find y_U.

1. Set $z_l = \bar{y}_l$, where $\bar{y}_l$ is the location of the fuzzy singleton associated with the output of the lth rule, and place the values z_l in ascending order.

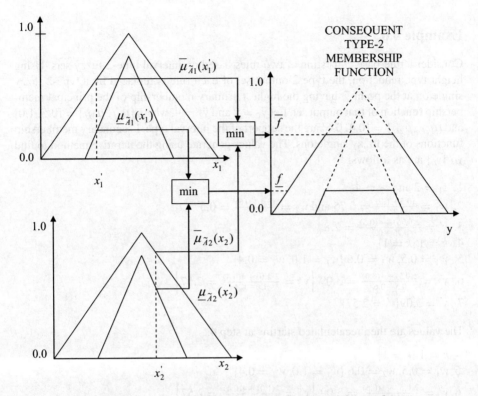

Figure 4.10 Fuzzy inference based on interval type-2 fuzzy rules.

2. Set $w_l = \frac{(\mu_{lL} + \mu_{lU})}{2} \forall l = 1, \ldots M$, where $[\mu_{lL}, \mu_{lU}]$ is the support of the secondary membership function of the fuzzy singleton associated with the output of the lth rule and M is the number of rules.

3. Compute $y' = \frac{\sum\limits_{l=1}^{M} z_l w_l}{\sum\limits_{l=1}^{M} w_l}$.

4. Find the values of k where $1 \le k \le (M-1)$ such that $z_k \le y' \le z_{k+1}$.

5. Set $w_l = \mu_{lL}$ for $l \le k$ and $w_l = \mu_{lU}$ for $l \ge (k+1)$.

6. Compute $y'' = \frac{\sum\limits_{l=1}^{M} z_l w_l}{\sum\limits_{l=1}^{M} w_l}$.

7. If $y'' = y'$ then $y_U = y''$; otherwise set $y' = y''$ and repeat the procedure from step 4.

The same method can be used to find y_L by making the following change to step 5.

5. Set $w_l = \mu_{lU}$ for $l \le k$ and $w_l = \mu_{lL}$ for $l \ge (k+1)$.

It has been shown that the iterative procedure converges in at most M iterations, where M is the number of rules (Karnik *et al.*, 1999).

Example 4.6

Consider a rule-base consisting of two rules based on interval type-2 fuzzy sets. Using height type reduction, the type-2 output set of each rule is replaced by a type-2 fuzzy singleton at the point $\bar{y}_l$ having the highest primary membership of the principal membership function of that output set. Let $\bar{y}_1 = 2$ and $\bar{y}_2 = 4$, where $[\mu_{1L}, \mu_{1U}] = [0.5, 1.0]$ and $[\mu_{2L}, \mu_{2U}] = [0.4, 0.6]$ are the support of the interval type-1 secondary membership functions of the fuzzy singletons. The values generate using the iterative method to find y_U $\{y_L\}$ are as follows:

1. $z_1 = 2$ and $z_2 = 4$
2. $w_1 = \frac{(0.5 + 1.0)}{2} = 0.75$ and $w_2 = \frac{(0.4 + 0.6)}{2} = 0.5$
3. $y' = \frac{2(0.75) + 4(0.5)}{0.75 + 0.5} = 2.8$
4. $k = 1 \{k = 1\}$
5. $w_1 = 0.5, w_2 = 0.6 \{w_1 = 1.0, w_2 = 0.4\}$
6. $y'' = \frac{2(0.5) + 4(0.6)}{0.5 + 0.6} = 3.09 \left\{ y'' = \frac{2(1.0) + 4(0.4)}{1.0 + 0.4} = 2.57 \right\}$
7. $y' = 3.09 \{y' = 2.57\}$

The values are then recalculated starting at step 4.

4 $k = 1 \{k = 1\}$
5 $w_1 = 0.5, w_2 = 0.6 \{w_1 = 1.0, w_2 = 0.4\}$
6 $y'' = \frac{2(0.5) + 4(0.6)}{0.5 + 0.6} = 3.09 \left\{ y'' = \frac{2(1.0) + 4(0.4)}{1.0 + 0.4} = 2.57 \right\}$
7 $y_U = 3.09 \{y_L = 2.57\}$

The type-reduced interval type-1 set is therefore $[2.57, 3.09]$ and the defuzzified value is $\frac{(2.57 + 3.09)}{2} = 2.83$.

Further simplifications have been suggested such as the use of mixed type-1 and type-2 fuzzy sets (Wu and Tan, 2006) and performing type reduction before the outputs of the rules are aggregated (Chaoui and Gueaieb, 2008).

The main advantage of using type-2 fuzzy rules is that they can take account of linguistic uncertainties (Karnik *et al.*, 1999). It is also claimed that they can produce better control performance compared to type-1 fuzzy logic controllers (Galluzzo *et al.*, 2008) and a smoother control surface (Melin and Castillo, 2003; Wu and Tan, 2006). The main weakness is the associated computational complexity and the loss of all of the information about the uncertainties following the processing of the output set to generate a crisp numerical value (Melin and Castillo, 2003).

4.5.2   Identifying Fuzzy Rules from Data

There are a number of steps in this process (Sugeno and Yasukawa, 1993; Barada and Singh, 1998; Chen and Linkens, 2004):

1. selection of the input and output variables;
2. definition of the antecedent and consequent structure;

3. determination of the number of rules that are required;
4. choice of membership functions; and
5. choice of other parameters of the model.

Cluster analysis can be used to generate rules from input/output data (so-called *product space clustering*).

4.5.2.1 Fuzzy c-Means Clustering

The problem of clustering is to find a number of cluster centres (prototypes) that can properly characterise the relevant classes of the data, given a finite data set. Classical cluster analysis uses crisp partitions to classify the data. Fuzzy clustering replaces the crisp partitions by fuzzy partitions. The fuzzy partitions are often called fuzzy-c partitions where c is the number of fuzzy classes in the partition.

The problem of fuzzy clustering is to find a fuzzy partition, and the associated cluster centres, by which the structure of a set of data is best represented.

Given a set of data $Z = [z_1 \quad z_2 \cdots z_k \cdots z_n]$, where z_k is the vector $[z_{k1} \quad z_{k2} \cdots \quad z_{kp}]^T = [x_k \mid y_k]^T$, y_k is the kth sample of the output of the system, x_k is the kth sample of the input vector of the system, and a fuzzy partition $P = [A_1 \quad A_2 \cdots A_i \cdots A_c]$, which satisfies $\sum_{i=1}^{c} \beta_{A_i}(z_k) = 1$, where $\beta_{A_i}(z_k)$ is the degree of membership of z_k in the fuzzy partition A_i, the c cluster centres $[v_1 v_2 \ldots v_i \ldots v_c]$ associated with the partition are given by:

$$v_i = \frac{\sum_{k=1}^{n} \beta_{A_i}(z_k)^m z_k}{\sum_{k=1}^{n} \beta_{A_i}(z_k)^m} \tag{4.18}$$

where $m > 1$ is a real number that governs the influence of the membership grades.

If $\|z_k - v_i\|^2 > 0$ for all values of i, the degree of membership of z_k in the fuzzy partition A_i is given by:

$$\beta_{A_i}(z_k) = \frac{1}{\sum_{j=1}^{c} \left(\frac{\|z_k - v_i\|^2}{\|z_k - v_j\|^2} \right)^{\frac{1}{m-1}}} \tag{4.19}$$

where $\|.\|$ is a vector norm and $\|z_k - v_i\|$ represents the distance between z_k and v_i. If $\|z_k - v_i\|^2 = 0$ for some values of $i \in I$, define $\beta_{A_i}(z_k)$ for $i \in I$ by any non-negative real numbers that satisfy $\sum_{i \in I} \beta_{A_i}(z_k) = 1$ and define $\beta_{A_i}(z_k) = 0$ for $i \notin I$. It should be noted that the cluster centre v_i is a weighted average of the data set Z in A_i and that the partition becomes fuzzier as the value of m increases.

The goal of fuzzy c-means clustering method is to find a fuzzy partition P that minimises a performance index J_m, which measures how well the clusters represent the data. The performance index is given by

$$J_m = \sum_{k=1}^{n} \sum_{i=1}^{c} \beta_{A_i}(z_k)^m \|z_k - v_i\| \tag{4.20}$$

Clearly the smaller the value of J_m, the better is the fuzzy partition P. Note that the clustering problem is a non-linear constrained optimization problem.

The fuzzy c-means algorithm is described as follows:

1. choose c, m and a small positive number ε (the stopping criterion);
2. set $t = 0$ and select an initial fuzzy partition $P^{(0)}$;
3. use Equation (4.18) to calculate the c cluster centres;
4. update $\beta_{A_i}(\mathbf{z}_k)$ using Equation (4.19); and
5. stop if

$$\left| P^{(t+1)} - P^{(t)} \right| = Max_{c,k} \left| \beta_{A_i}(\mathbf{z}_k)^{(t+1)} - \beta_{A_i}(\mathbf{z}_k)^{(t)} \right| \leq \varepsilon; \tag{4.21}$$

otherwise return to step 3.

It should be noted that there is no theoretical basis for choosing m but it has been established that the algorithm converges for any value of m, although $m = 2$ and $\varepsilon = 0.01$ are good choices for most problems.

4.5.2.2   Generating a Multi-Dimensional Membership Function

One possibility is to use the following expression for the multi-dimensional membership function of the antecedent of the ith rule:

$$\mu_i(\mathbf{x}_k) = \frac{1}{1 + \frac{\|\mathbf{x}_k - \mathbf{v}_i^*\|^2}{r^2}} \tag{4.22}$$

where v_i^* is the position of the centre of the ith cluster in the input space of the system and the value of r is chosen to vary the region of influence of each rule. Note that the membership function is convex, symmetrical about its centre and tends to zero far away from the centre.

Two practical considerations must be taken into account when generating rules using fuzzy clustering: model complexity and robustness to poor-quality training data. Prior knowledge of the system to be modelled is required to decide on the structure of the model. Techniques based on model cross-validation or rule merging (using a measure of fuzzy similarity to determine which antecedents have fuzzy sets that are similar to one another) can then be used to determine the most appropriate number of fuzzy clusters (rules). The quality of a fuzzy model produced by product-space clustering is very dependent on the information content of the data and it is more sensitive to inconsistent and incomplete data than other (more heuristic) approaches to generating the rules. It should also be noted that clustering will be computationally demanding for large datasets.

4.6   Defuzzification

In many applications, the fuzzy output from a linguistic model must be translated into a crisp value. Traditionally, this process is called defuzzification. Two commonly used methods are height defuzzification (see Figure 4.11) and mean of maxima defuzzification (see Figure 4.12).

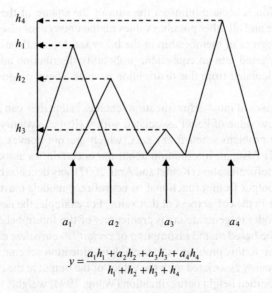

$$y = \frac{a_1 h_1 + a_2 h_2 + a_3 h_3 + a_4 h_4}{h_1 + h_2 + h_3 + h_4}$$

Figure 4.11 Example of height defuzzification.

It has however been argued (Martin and Klir, 2007) that the use of the term defuzzification is actually a misnomer because a fuzzy set involves two types of uncertainty: fuzziness (the boundaries of the set are imprecise) and ambiguity (the actual value of the variable described by the fuzzy set is only known to lie within a certain range). The term defuzzification should therefore imply a conversion of the fuzzy set to a crisp set, rather than a crisp set to a crisp value.

Most of the traditional defuzzification schemes (e.g. mean of maxima) are *ad hoc* in nature, although it has been shown (Roychowdhury and Pedrycz, 2001) that the well-known centre

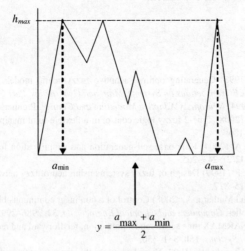

$$y = \frac{a_{max} + a_{min}}{2}$$

Figure 4.12 Example of mean of maxima defuzzification.

of area defuzzification scheme minimizes the sum of the square of the differences between the defuzzified value and all other possible values on the universe of discourse of the variable, weighted by their degrees of membership in the fuzzy set. Defuzzification schemes in which the fuzzy set is converted into an equivalent probability distribution and the expected value of the variable is calculated from this distribution have also been proposed (Yager and Filev, 1993a, b).

The main weaknesses of most defuzzification schemes is that they can produce a defuzzified value which has a low value of belief associated with it (Roychowdhury and Pedrycz, 2001). This is a particular problem when fuzzy sets, which are not convex, must be defuzzified (Pfluger *et al.*, 1992) Also, all information about the uncertainties associated with the fuzzy variable is lost after defuzzification (Kentel and Aral, 2007); any decision based on a defuzzified value (or the crisp output from a functional or neurofuzzy model) must assume that there is complete confidence in the correctness of that value. For example, the design of any controller that uses a fuzzy model to generate crisp predictions of the future behaviour of the process under control must be based on the assumption of *certainty equivalence*.

A possible solution to this problem is to use a defuzzification scheme that takes account of the amount of uncertainty associated with the values of the output at the centres of each output set. For example, modified height defuzzification (Wang, 1994) weights the value at the centre of each output set according to the support of the output set.

4.7 Summary

An introduction has been given to linguistic fuzzy models, functional fuzzy models and fuzzy neural networks. A way of modifying expert rules to take account of uncertainty by using type-2 fuzzy sets has been explained and a method of identifying fuzzy rules from data has been described. The disadvantages of defuzzification have been discussed and the reason why the most commonly used fuzzy models do not give any indication of the uncertainty associated with their predictions has been explained..

References

Barada, S. and Singh, H. (1998) Generating optimal adaptive fuzzy-neural models of dynamical systems with applications to control. *IEEE Transactions on Systems, Man, and Cybernetics, Part C*, **28**(3), 371–391.

Brown, M. and Harris, C. (1994) *Neurofuzzy Adaptive Modelling and Control*, Prentice-Hall, Hertfordshire, UK.

Chaoui, H. and Gueaieb, W. (2008) Type-2 fuzzy logic control of a flexible-joint manipulator. *Journal of Intelligent Robotic Systems*, **51**, 159–186.

Chen, M-Y. and Linkens, D.A. (2004) Rule-base self-generation and simplification for data-driven fuzzy models. *Fuzzy Sets and Systems*, **142**(2), 243–265.

Figueiredo, M. and Gomide, F. (1999) Design of fuzzy systems using neurofuzzy networks. *IEEE Transactions on Neural Networks*, **10**(4), 815–827.

Galluzzo, M., Cosenza, B. and Matharu, A. (2008) Control of a nonlinear continuous bioreactor with bifurcation by a type-2 fuzzy logic controller. *Computers and Chemical Engineering*, **32**, 2986–2993.

Gao, Y. and Er, M.J. (2005) NARMAX time series model prediction: feedforward and recurrent fuzzy neural network approaches. *Fuzzy Sets and Systems*, **150**, 331–350.

Ishibuchi, H., Fujioka, R. and Tanaka, H. (1993) Neural networks that learn from fuzzy If-Then rules. *IEEE Transactions on Fuzzy Systems*, **1**(1), 85–97.

Karnik, N.N., Mendel, J.M. and Liang, Q. (1999) Type-2 fuzzy logic systems. *IEEE Transactions on Fuzzy Systems*, **7**(6), 643–658.

Kentel, E. and Aral, M.M. (2007) Risk tolerance measure for decision-making in fuzzy analysis: a health risk assessment perspective. *Stochastic Environmental Research Risk Assessment*, **21**, 405–417.

Lee, K-M., Kwak, D-H. and Lee-Kwang, H. (1996) Fuzzy inference neural network for fuzzy model tuning. *IEEE Transactions on Systems, Man, and Cybernetics, Part B*, **26**(4), 637–645.

Manmohan, D.K., Chaturvedi, P.S. and Kalra, P.K. (2003) Neuro-fuzzy approach for development of new neuron model. *Soft Computing*, **8**, 19–27.

Martin, O. and Klir, G.J. (2007) Defuzzification as a special way of dealing with retranslation. *International Journal of General Systems*, **36**(6), 683–701.

Melin, P. and Castillo, O. (2003) A new method for adaptive model-based control of non-linear plants using type-2 fuzzy logic and neural networks. IEEE International Conference on Fuzzy Systems, 420–425.

Mendel, J.M. (2001) *Uncertain Rule-Based Fuzzy Logic Systems*, Prentice-Hall, New Jersey.

Mitra, S. and Hayashi, Y. (2000) Neuro-fuzzy rule generation: survey in soft computing framework. *IEEE Transactions on Neural Networks*, **11**(3), 748–768.

Pedrycz, W. and Rocha, A.F. (1993) Fuzzy set based models of neurons and knowledge-based networks. *IEEE Transactions on Fuzzy Systems*, **1**(4), 254–266.

Pfluger, N., Yen, J. and Langari, R. (1992) A defuzzification strategy for a fuzzy logic controller employing prohibitive information in command formulation. Proceedings of IEEE Conference on Fuzzy Systems, 717–723.

Roychowdhury, S. and Pedrycz, W. (2001) A survey of defuzzification strategies. *International Journal of Intelligent Systems*, **16**(6), 679–695.

Sugeno, M. and Yasukawa, T. (1993) A fuzzy-logic-based approach to qualitative modelling. *IEEE Transactions on Fuzzy Systems*, **1**(1), 7–31.

Wang, L-X (1994) *Adaptive Fuzzy Systems and Control – Design and Stability*, Prentice-Hall, New Jersey.

Wu, D. and Tan, W-W. (2006) A simplified type-2 fuzzy logic controller for real-time control. *Transactions of Instrumentation, Systems, and Automation Society*, **45**(4), 503–516.

Yager, R.R. and Filev, D. (1993a) On the issue of defuzzification and selection based on a fuzzy set. *Fuzzy Sets and Systems*, **55**, 255–271.

Yager, R. and Filev, D.P. (1993b) SLIDE: a simple adaptive defuzzification method. *IEEE Transactions on Fuzzy Systems*, **1**(1), 69–78.

5

Fuzzy Relational Models

5.1 Introduction to Fuzzy Relations and Fuzzy Relational Models

A crisp relation indicates the association (or lack of) between the elements of two or more sets. It is an alternative way of representing a set of linguistic rules. For example, consider the following rule-base relating two fuzzy variables X and Y, where A_i are the fuzzy sets used to describe X and C_k are the fuzzy sets used to describe Y:

$$R_{ij}: \text{IF } X \text{ is } A_i \text{ THEN } Y \text{ is } C_k \text{ for } i = 1 \ldots N \text{ and } k = 1 \ldots K$$

The crisp relation between Y and X can be defined by a characteristic function R which assigns a value of unity to each rule which relates Y to X and a value of zero to all other rules. Hence, $R_{A_i,C_k} = 1$ if one of the rules is

$$\text{IF } X \text{ is } A_i \text{ THEN } Y \text{ is } C_k,$$

and $R_{A_i,C_k} = 0$ otherwise. A relational array is a convenient way of specifying relations on finite universes, particularly when a computer is to be used to implement relations (Klir and Yuan, 1995). The dimensions of the array will depend on the number of clauses used in the antecedents of the rules and the size of the array will depend on the number of sets used to describe the fuzzy variables in the antecedents and conclusions. For example, let $X =$ valve position $= \{$nearly closed, half open, fully open$\}$, $Y =$ flow rate $= \{$low, medium, high$\}$ and the rule-base be:

- IF valve position is nearly closed THEN flow rate is low;
- IF valve position is half open THEN flow rate is medium;
- IF valve position is fully open THEN flow rate is high.

Then the crisp relational between valve position and flow can be represented by the two-dimensional relational array shown in Table 5.1.

A fuzzy relation is a generalization of a crisp relation in the same way that a fuzzy set is a generalization of a crisp or Boolean set. Fuzzy relations allow for degrees of association

Monitoring and Control of Information-Poor Systems: An Approach based on Fuzzy Relational Models, First Edition.
Arthur L. Dexter.
© 2012 John Wiley & Sons, Ltd. Published 2012 by John Wiley & Sons, Ltd.

Table 5.1   Example of a crisp relation.

	Low	Medium	High
Nearly closed	1	0	0
Half open	0	1	0
Fully open	0	0	1

between the sets. In this case, the elements of the relational array (known as a fuzzy relational array) have values in the range [0, 1]. An example is given in Table 5.2.

A fuzzy relation between fuzzy sets can be interpreted as a set of IF-THEN rules based on the fuzzy sets where the value of the element of the fuzzy relational array, which specifies the degree of association of the elements of the sets used in the antecedent and consequent of the rule, is the degree to which the rule is believed to be correct (the rule confidence) (Brown and Harris, 1994). For example, the third row of the fuzzy relational array in the simple example is equivalent to the following set of fuzzy rules:

- IF valve position is fully open THEN flow rate is low (0% belief);
- IF valve position is fully open THEN flow rate is medium (40% belief);
- IF valve position is fully open THEN flow rate is high (60% belief).

Fuzzy relational equations provide a convenient notation for representing fuzzy relations when performing mathematical analysis (Pedrycz and Gomide, 2007). There is a similarity between the notation used in fuzzy relational equations and that is used to represent matrix operations. The main difference is that the operators are t-norms and t-conorms rather than the usual algebraic operators. For example, the fuzzy equation

$$Y = X \circ R \tag{5.1}$$

denotes the inference of Y from a given X and associated fuzzy relational array R, where $\circ$ denotes compositional rule of inference based on a particular t-norm and t-conorm.

A fuzzy relational model (FRM) has a predefined set of linguistic rules, each of which has an associated rule confidence specifying to what extent the associated fuzzy rule is believed to describe the relationship between the inputs and the output of the system being modelled.

The values of the rule confidences can be obtained from expert or domain knowledge, but are more usually estimated from input and output data using a fuzzy identification scheme. It should be noted that, because the fuzzy identification scheme is able to change the value of the rule confidences, the number of rules actually used in the model can be varied.

Table 5.2   Example of a fuzzy relation.

	Low	Medium	High
Nearly closed	0.9	0.1	0.0
Half open	0.2	0.6	0.2
Fully open	0.0	0.4	0.6

For example, a first-order dynamic fuzzy relational model might be based on the following set of rules:

$$\textbf{IF } Y(n-1) \text{ is } A_i \textbf{ AND } U(n-1) \text{ is } B_j \textbf{ THEN } Y(n) \text{ is } C_k \quad (R_{A_i,B_j,C_k})$$

where A_i $(i = 1, 2, \ldots N)$, B_j $(j = 1, 2, \ldots M)$ and C_k $(k = 1, 2, \ldots .K)$ are fuzzy reference sets, and R_{A_i,B_j,C_k} are the rule confidences.

The choice of the reference sets is arbitrary but they must be:

- normal (the maximum value of its membership function is unity);
- convex (its membership function has a unique apex); and
- complete (there is a non-zero grade of membership of at least one reference set for every point on the universe of discourse).

In some applications, it may be helpful if the labels associated with the input reference sets are meaningful in terms of the particular system to be modelled.

All rule-based models suffer from the curse of dimensionality, no more so than the FRM. Although the problem has become less of a concern now that the cost of increased processing power and larger memories has decreased significantly, the computational demands can be significantly reduced by decomposing the overall model into a hierarchy of FRMs with fewer inputs (See Section 10.3.2).

5.2   Fuzzy FRMs

There are two basic types of fuzzy relational models: a fuzzy relational model, which uses a defuzzification scheme to produce a numeric output (known as a crisp FRM), and fuzzy relational model, which produces a fuzzy output (known as a fuzzy FRM).

A crisp FRM usually has a relatively small number of fuzzy reference sets and its output before defuzzification is normally difficult to interpret in a physically meaningful way.

Consider the case of a single-input single-output crisp FRM in which the input reference sets have triangular membership functions with a partition of unity (see Figure 5.1) and the rule confidences satisfy the condition $\sum_{k=1}^{K} R_{A_i,C_k} = 1.0$.

If sum-product inferencing and height defuzzification are used, the value of the numeric output y is given by

$$y = \frac{\sum_{i=1}^{N} \sum_{k=1}^{K} \mu_{A_i}(x) R_{A_i,C_k} \bar{y}_k}{\sum_{i=1}^{N} \sum_{k=1}^{K} \mu_{A_i}(x) R_{A_i,C_k}} = \sum_{i=1}^{N} \sum_{k=1}^{K} \mu_{A_i}(x) R_{A_i,C_k} \bar{y}_k \qquad (5.2)$$

where $\mu_{A_i}(x)$ is the membership function of the input set A_i and $\bar{y}_k$ is the centroid of the output set C_k. Alternatively, the expression for the output of the crisp FRM can be rewritten as

$$y = \sum_{i=1}^{N} \mu_{A_i} w_i \qquad (5.3)$$

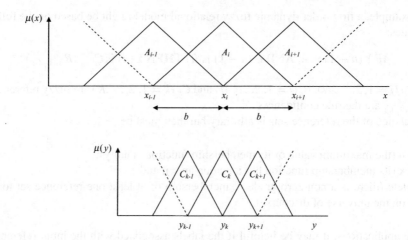

Figure 5.1   Fuzzy reference sets used to describe the input and output of a single input FRM.

where $w_i = \sum\limits_{k=1}^{K} R_{A_i,C_k} \bar{y}_k$. It can be seen that the expression for y is identical to that derived in Section 4.4 for the output of a fuzzy neural network (Brown and Harris, 1994).

A fuzzy-output FRM has a relatively large number of equally spaced output sets and the elements of the fuzzy relational array are the possibilities of particular values of the output given that rules with a particular antecedent are fully fired.

If sum-product inferencing is used:

$$Poss(C_k|x) = \sum_{i=1}^{N} \mu_{A_i}(x) R_{A_i,C_k} \tag{5.4}$$

If $x_i \leq x < x_{i+1}$, where x_i and x_{i+1} are the positions of the apexes of the fuzzy reference sets A_i and A_{i+1} on the universe of discourse,

$$Poss(C_k|x) = \sum_{i=1}^{N} \mu_{A_i}(x) R_{A_i,C_k} = [1 - \mu_{A_{i+1}}(x)] R_{A_i,C_k} + \mu_{A_{i+1}}(x) R_{A_{i+1},C_k} \tag{5.5}$$

and

$$\mu_{A_{i+1}}(x) = \frac{x - x_i}{x_{i+1} - x_i} = \frac{x - x_i}{b} \tag{5.6}$$

We therefore have

$$Poss(C_k|x) = \left[1 - \left(\frac{x - x_i}{b}\right)\right] R_{A_i,C_k} + \left[\left(\frac{x - x_i}{b}\right)\right] R_{A_{i+1},C_k}$$

$$\tag{5.7}$$

$$= R_{A_i,C_k} + \left(\frac{x - x_i}{b}\right) \left[R_{A_{i+1},C_k} - R_{A_i,C_k}\right]$$

It can be seen that the value of $Poss(C_k|x)$ varies linearly with x from R_{A_i,C_k} to R_{A_{i+1},C_k}.

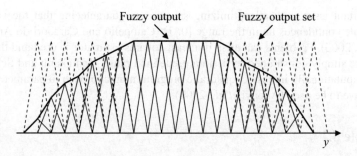

Figure 5.2 Fuzzy output produced by a fuzzy FRM of a highly uncertain system.

The possibility distribution of the output generated by a fuzzy FRM depends on:

- the method used to estimate the elements of its fuzzy relational array (see Section 5.3);
- inconsistencies in the training data used to identify the model (such inconsistencies will occur if there are un-modelled disturbances acting on the system from which the data are collected or measurement noise on the sensors used to collect the data; see Section 5.4.1); and
- the number of reference sets that are used to describe the inputs (see Section 5.4.2).

An example of the fuzzy output generated by the fuzzy FRM when the prediction is very uncertain is shown in Figure 5.2. The output generated by the fuzzy FRM approximates the possibility distribution indicated by the continuous solid line. It can be seen that the numeric value produced by any type of defuzzification scheme would be unrepresentative, as the information about the modelling uncertainty is then lost. It should be noted that a large number of the fuzzy output sets must be used if the output is to provide a good representation of the uncertainty.

Fuzzy FRMs can be used to described the uncertainties associated with the behaviour of an information-poor system, including the measurement noise, unmeasured (and therefore un-modelled) disturbances and uncertainty associated with modelling the input-output relationship, arising from incorrect structure and inaccurate estimation of its parameters.

5.3 Methods of Estimating Rule Confidences from Data

Many different fuzzy identification schemes have been proposed that can be used to estimate the elements of the FRM's fuzzy relational array (the rule confidences) from input/output data (Pedrycz, 1984; Postlethwaite, 1991; Chen *et al.*, 1994; Postlethwaite *et al.*, 1997; Bourke and Fisher, 2000a, b; Sing and Postlethwaite, 2000). There are two basic types: optimizing schemes and non-optimizing schemes.

Optimizing schemes find the values of the rule confidences that minimize a performance index. The sum of the squares of the defuzzified prediction errors is commonly used as a performance index when training crisp FRMs (*cf.* the training of neurofuzzy models), but some optimizing schemes minimize the difference between the possibility values defining the fuzzy predictions of an FRM and those of the numeric, or even fuzzy, training data (Wong *et al.*, 2000; Campello and Caradori do Amaral, 2006).

An important issue with all optimizing schemes is guaranteeing that the values of the estimated rule confidences lie in the range [0, 1] (Campello and Caradori do Amaral, 1998; Wong *et al.*, 2000). Problems can arise if a least-squares estimator is used and the parameter estimates are simply constrained to lie in a predefined range (Marakami and Seborg, 2000). A more computationally intensive, quadratic programming approach to parameter estimation may then have to be used (Abonyi *et al.*, 2000).

Example 5.1

Consider the following first-order plus time-delay fuzzy FRM:

$$Y(n) = Y(n-1) \circ U(n-d-1) \circ R$$

where the time delay is assumed to be an integer number d of sampling intervals.

The elements of the fuzzy relational array R can be estimated from the training data using quadratic programming to minimize the fuzzy performance index,

$$J = \sum_{n=1}^{L} \sum_{k=1}^{K} \left[Poss_{y_k}(n) - \hat{P}oss_{C_k}(n) \right]^2$$

subject to the constraints $0 \le \hat{R}_{A_i,B_j,C_k} \le 1 \; \forall i \forall j \forall k$ where

$$\hat{P}oss_{C_k}(n) = \underset{i=1}{\overset{N}{Max}} \, \underset{j=1}{\overset{M}{Max}} \{ \mu_{A_i}(y[n-1]) \mu_{B_j}(u[n-d-1]) \hat{R}_{A_i,B_j,C_k} \}$$

and $Poss_{y_k}(n)$ is obtained by finding the simplest fuzzy representation of the numerical value of the output $y(n)$. For example, if $y_k \le y[n] \le y_{k+1}$, $Poss_{y_k}(n) = h$ and $Poss_{y_{k+1}}(n) = (1-h)$ where $h = \frac{y_{k+1} - y[n]}{y_{k+1} - y_k}$ and $Poss_{y_s}(n) = 0 \; \forall s \ne k \, or \, k+1$.

Non-optimizing schemes match the rules to the training data by calculating the possibilities of the fuzzy reference sets (used in the antecedents and conclusions of the rules) given the training data, and using them to estimate the rule confidences (Xu, 1989).

A major criticism of these fuzzy identification schemes has been that it is unclear in what sense the resulting model fits the training data. Although theoretical analysis of the performance of FRMs is very difficult when they are identified using a non-optimizing fuzzy identification scheme, analysis is possible if the FRM is based on fuzzy sets with triangular membership functions. For example, it has been shown that a crisp FRM can be produced that has no prediction errors when the values of its inputs coincide with the apexes of the fuzzy input sets (Wu and Dexter, 2003).

The main factors which must be taken into account when evaluating the various fuzzy identification schemes (Postlethwaite, 1991) are as follows.

- *The modelling capabilities of the resulting FRM*: It is difficult to compare the performance of an FRM (which has a fuzzy output) with other types of fuzzy models whose outputs are numerical values when there are significant uncertainties associated with the predicted behaviour and the model must generate a satisfactory estimate of the output distribution, not just a good estimate of the mean value.

- *Computational demands of the fuzzy identification scheme*: The computational demands of non-optimizing schemes are far less than those of optimizing schemes. This issue will be particularly important if the fuzzy identification has to be performed online (see Chapter 10).
- *The sensitivity of the estimates to the distribution of the training data*: Non-optimizing schemes are generally more sensitive to the distribution of the training data (Sing and Postlethwaite, 2000) but the performance of all of the identification schemes will be poor if they are used to identify models of non-linear systems and the training data are not uniformly distributed over the input space (Laukonen and Passino, 1995; Kelkar and Postlethwaite, 1998).

Example 5.2

Consider the use of the following simple fuzzy identification (Xu and Lu, 1987) to identify an FRM, which has rules of the form:

IF X_1 is A_i AND X_2 is B_j THEN Y is C_k

where the reference sets for X_1, X_2 and Y are A_1 to A_N, B_2 to B_M and C_1 to C_K, respectively.

If there are L crisp datasets available – $\{x_1[1], x_2[1], y[1]\}$ to $\{x_1[L], x_2[L], y[L]\}$ – the rule confidences may be estimated as follows:

$$\hat{R}_{A_i,B_j,C_k} = \underset{n=1}{\overset{L}{Max}}\{\hat{R}_{A_i,B_j,C_k}(n)\}$$

where

$$\hat{R}_{A_i,B_j,C_k}(n) = \prod \{Poss(A_i|x_1[n]), Poss(B_j|x_2[n]), Poss(C_k|y[n])\}$$

and $Poss(A_i|x_1[n]) = \mu_{A_i}(x_1[n])$, $Poss(B_j|x_2[n]) = \mu_{B_j}(x_2[n])$ and $Poss(C_k|y[n]) = \mu_{C_k}(y[n])$.

The quality of the resulting fuzzy model will depend on both the choice of the reference sets and the consistency of the training data used to estimate the rule confidences (Shaw and Kruger, 1992).

A more robust fuzzy identification scheme must be used if the training data are corrupted by noise. For example, consider the estimation of the rule confidences of the two-input single-output FRM using the following fuzzy identification scheme first proposed by Ridley, Shaw and Kruger (Ridley *et al.*, 1988):

$$\hat{R}_{A_i,B_j,C_k} = \frac{\sum_{n=1}^{L} \prod \{Poss(A_i|X_1[n]), Poss(B_j|X_2[n]), Poss(C_k|Y[n])\}}{\sum_{n=1}^{L} \prod \{Poss(A_i|X_1[n]), Poss(B_j|X_2[n])\}}$$

$$= \frac{\sum_{n=1}^{L} \prod \{\mu_{A_i}(x_1[n]), \mu_{B_j}(x_2[n]), \mu_{C_k}(y[n])\}}{\sum_{n=1}^{L} \prod \{\mu_{A_i}(x_1[n]), \mu_{B_j}(x_2[n])\}} \tag{5.8}$$

in which the training data are weighted according to the grade of membership of the input set, that is, data that are closer to the centres of the input fuzzy sets have more influence on the estimated rule confidence than data at the edges of the sets. Note that the estimated rule confidence is a weighted average over L datasets; this so-called RSK fuzzy identification scheme is therefore less sensitive to noise than the previous scheme.

5.4   Estimating Probability Density Functions from Data

There are two general approaches to estimating probability density functions from data: parametric methods and non-parametric methods (Bishop, 1995). Parametric methods assume a particular functional form for the distribution (e.g. a Gaussian distribution) and use optimization to find the values of the parameters of the function that best fit the function to the training data. Non-parametric methods (e.g. kernel-based methods) do not require the form of the distribution to be predefined in advance, but generate an estimate of the probability density functions in terms of the training data themselves. Mixture models (Herzallah and Lowe, 2003) combine the advantages of both approaches: they can be used to estimate probability density functions of any form and all of the training data points do not have to be stored. A general approach to estimating probability density functions using a mixture model is shown in Figure 5.3a. Here, a set of kernel functions is used to estimate the probability density functions. The data are used to train the neural network, which generates the values of the mixing coefficients and the parameters of the kernels from the input vector. One way of reducing the computational demands of this approach is to use kernel functions with predefined values for their parameters, a fuzzy FRM to generate the mixing coefficients (see Figure 5.3b) and a non-optimizing fuzzy identification scheme to estimate the rule confidences of the fuzzy FRM.

For example, if both the kernel functions and the membership functions of the input reference sets of the fuzzy FRM are equally spaced isosceles triangles with a partition of

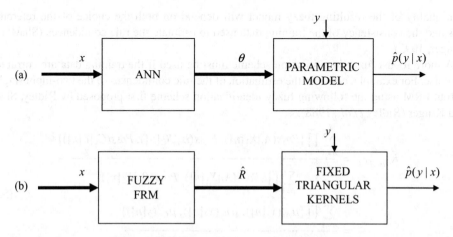

Figure 5.3 Estimation of a probability density function using a mixture model based on (a) an ANN (b) a fuzzy FRM.

unity, the estimated value of the conditional probability density function for $y_k \leq y \leq y_{k+1}$ is given by:

$$\hat{p}(y|x) = M_k \mu_{C_k}(y) + M_{k+1} \mu_{C_{k+1}}(y)$$ (5.9)

where the mixing coefficients are:

$$M_k = \sum_{i=1}^{N} \sum_{j=1}^{M} \mu_{A_i}(x) \mu_{B_j}(x) \hat{R}_{A_i, B_j, C_k}$$ (5.10)

and

$$M_{k+1} = \sum_{i=1}^{N} \sum_{j=1}^{M} \mu_{A_i}(x) \mu_{B_j}(x) \hat{R}_{A_i, B_j, C_{k+1}}$$ (5.11)

5.4.1 Probabilistic Interpretation of RSK Fuzzy Identification

If the FRM of an information-poor system has a large number of output reference sets and the RSK fuzzy identification scheme is used to identify the model, it can be shown that there is a direct relation between the possibility distribution of the fuzzy output and the probability distributions of the un-modelled disturbances and measurement noise. The fuzzy output therefore provides some indication of the uncertainties associated with the predictions of the model.

Case 1

First consider the case where the fuzzy FRM is defined over discrete universes of discourse, has only one input and each point on the universes of discourse coincides with the apex of a reference set. Then $x[n] = x_i$, where x_i is the position of the apex of one of the input fuzzy sets, and $y[n] = y_k$ where y_k is the position of the apex of one of the output fuzzy sets. In this case,

$$\mu_{A_i}(x[n]) = 1 \text{ and } \mu_{A_s}(x[n]) = 0, \forall s \neq i$$ (5.12)

and

$$\mu_{C_k}(y[n]) = 1 \text{ and } \mu_{C_p}(y[n]) = 0, \forall p \neq k$$ (5.13)

The estimated rule confidences are then given by

$$\hat{R}_{A_i, C_k} = \frac{\sum_{n=1}^{L} \mu_{A_i}(x[n]) \mu_{C_k}(y[n])}{\sum_{n=1}^{L} \mu_{A_i}(x[n])} = \frac{\sum_{n=1}^{M_{ik}} 1}{\sum_{n=1}^{M_i} 1} = \frac{M_{ik}}{M_i}$$ (5.14)

where M_{ik} is the number of times that $x[n] = x_i$ and $y[n] = y_k$ (i.e. the number of times the rule "IF X is A_i THEN Y is C_k" is fired by the training data) and M_i is the number of times $x[n] = x_i$ (i.e. the number of times rules which have the antecedent "X is A_i" are fired by the training data). If M_{ik} and M_i are very large,

$$\hat{R}_{A_i,C_k} = \frac{M_{ik}}{M_i} = \frac{M_{ik}/L}{M_i/L} \Rightarrow \frac{\Pr(y = y_k, x = x_i)}{\Pr(x = x_i)} = \Pr(y = y_k | x = x_i) \qquad (5.15)$$

Case 2

Consider a single-input fuzzy FRM whose (1) input is defined over a discrete universe of discourse, where each point on the universe of discourse coincides with the apex of a reference set and (2) output is described by a finite number of output reference sets over a continuous output universe of discourse.

$$\hat{R}_{A_i,C_k} = \frac{\sum\limits_{n=1}^{L} \mu_{A_i}(x[n])\mu_{C_k}(y[n])}{\sum\limits_{n=1}^{L} \mu_{A_i}(x[n])} = \frac{\sum\limits_{n=1}^{M_{ik}} \mu_{C_k}(y[n])}{\sum\limits_{n=1}^{M_i} 1} = \frac{\sum\limits_{n=1}^{M_{ik}} \mu_{C_k}(y[n])}{M_i} \qquad (5.16)$$

The expression for the estimated rule confidence is the same as that proposed for estimating probability density functions using fuzzy histograms (Runkler, 2004; Chmielewski, 2006). We therefore have

$$\hat{R}_{A_i,C_k} = \frac{\sum\limits_{n=1}^{M_{ik}} \mu_{C_k}(y[n])}{M_i} \cong \Pr(y = y_k | x = x_i) \qquad (5.17)$$

if the number of data points and the number of output sets is very large (Loquin and Strauss, 2008).

Case 3

Consider a fuzzy FRM whose input and output are defined over continuous universes of discourse and the number of input and output reference sets is finite.

If the noise on the training may be assumed to be ergodic and a very large number of datasets are used to estimate the rule confidences, we have

$$\hat{R}_{A_i,C_k} = \frac{\sum\limits_{n=1}^{L} \mu_{A_i}(x[n])\mu_{C_k}(y[n])}{\sum\limits_{n=1}^{L} \mu_{A_i}(x[n])} \simeq \frac{E\{\mu_{A_i}(x[n])\mu_{C_k}(y[n])\}}{E\{\mu_{A_i}(x[n]\}} \qquad (5.18)$$

and hence

$$\hat{R}_{A_i,C_k} \cong \frac{\int_{-\infty}^{+\infty} \int_{-\infty}^{+\infty} \mu_{A_i}(x)\mu_{C_k}(y)p(x,y)\,dx\,dy}{\int_{-\infty}^{+\infty} \mu_{A_i}(x)p(x)\,dx} = \frac{\int_{-\infty}^{+\infty} \int_{-\infty}^{+\infty} \mu_{A_i}(x)\mu_{C_k}(y)p(y|x)p(x)\,dx\,dy}{\int_{-\infty}^{+\infty} \mu_{A_i}(x)p(x)\,dx}$$

(5.19)

When the input and output fuzzy sets have equally spaced triangular membership functions with a partition of unity (see Figure 5.1),

$$\hat{R}_{A_i,C_k} \cong \frac{\int_{y_{k-1}}^{y_{k+1}} \int_{x_{i-1}}^{x_{i+1}} \mu_{A_i}(x)\mu_{C_k}(y)p(y|x)p(x)\,dx\,dy}{\int_{x_{i-1}}^{x_{i+1}} \mu_{A_i}(x)p(x)\,dx}$$

(5.20)

where x_{i-1} and x_{i+1} are the positions of the apexes of the membership functions of the $(i-1)$th and $(i+1)$th fuzzy input sets and y_{k-1} and y_{k+1} are the positions of the apexes the membership functions of the $(k-1)$th and $(k+1)$th fuzzy output sets (see Figure 5.4).

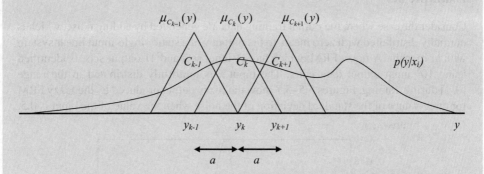

Figure 5.4   Using the output sets of a FRM to estimate the conditional PDF of the output.

If there are a very large number of fuzzy input and output sets, the width of each set will be very narrow and it may be assumed that the conditional probability density function $p(y|x) = p(y|x_i)$ is approximately constant when y is in the range $[y_{k-1}, y_{k+1}]$ and x is in the range $[x_{i-1}, x_{i+1}]$. Note that same simplifying assumption is made when proving the convergence of probability density function estimators based on kernel functions (Johnston and Kramer, 1994). We therefore have

$$\hat{R}_{A_i,C_k} \cong \frac{p(y|x_i) \int_{y_{k-1}}^{y_{k+1}} \mu_{C_k}(y) \int_{x_{i-1}}^{x_{i+1}} \mu_{A_i}(x)p(x)\,dx\,dy}{\int_{x_{i-1}}^{x_{i+1}} \mu_{A_i}(x)p(x)\,dx} = p(y|x_i) \int_{y_{k-1}}^{y_{k+1}} \mu_{C_k}(y)\,dy$$

(5.21)

Because the fuzzy output sets have triangular membership functions,

$$\hat{R}_{A_i,C_k} \cong p(y|x_i) \int_{y_{k-1}}^{y_k} \left(\frac{y - y_{k-1}}{y_k - y_{k-1}}\right) dy + p(y|x_i) \int_{y_k}^{y_{k+1}} \left(\frac{y_{k+1} - y}{y_{k+1} - y_k}\right) dy$$

$$= p(y|x_i) \left[\int_{y_{k-1}}^{y_k} \left(\frac{y - y_{k-1}}{a}\right) dy + \int_{y_k}^{y_{k+1}} \left(\frac{y_{k+1} - y}{a}\right) dy \right]$$

$$= \frac{p(y|x_i)}{a} \left[\left| \frac{(y - y_{k-1})^2}{2} \right|_{y_{k-1}}^{y_k} - \left| \frac{(y_{k+1} - y)^2}{2} \right|_{y_k}^{y_{k+1}} \right]$$

$$= \frac{p(y|x_i)}{2a} \left[a^2 + a^2\right] = p(y|x_i)\,[a] = \Pr(y_k < y \le y_{k+1}|x_i) \qquad (5.22)$$

Thus, the estimated rule confidence $\hat{R}_{A_i,C_k}$ of the fuzzy relational model is proportional to the probability that the output is in the range $[y_k, y_{k+1}]$ when the input is in the range $[x_{i-1}, x_{i+1}]$.

Example 5.3

Consider the case where the output training data are generated by adding noise, which is normally distributed with zero mean, to the output y of a static single-input linear system with unity gain. A fuzzy FRM, which has 21 input sets and 41 output sets, is identified using 10^5 input-output data pairs. The input x is uniformly distributed in the range $[0, 1]$ during training. Figures 5.5–5.7 show the fuzzy output produced by the fuzzy FRM for three values of the standard deviation of the noise when the value of the input is 0.5.

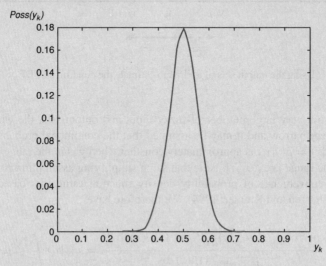

Figure 5.5 Fuzzy output when the input is 0.5 and the standard deviation of the noise is 0.05.

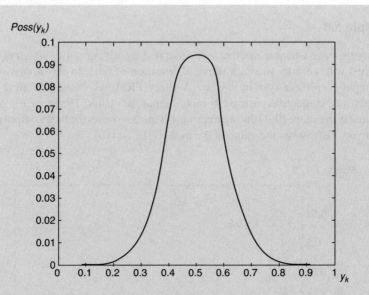

Figure 5.6 Fuzzy output when the input is 0.5 and the standard deviation of the noise is 0.1.

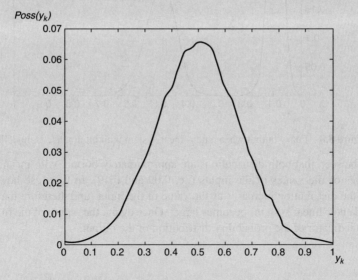

Figure 5.7 Fuzzy output when the input is 0.5 and the standard deviation of the noise is 0.15.

It can be seen that each of the distributions is roughly symmetric and normally distributed with a mean close to 0.5 (the value of the input); the widths of the distributions are approximately six times the standard deviations of the output noise. The output of the fuzzy FRM is therefore a good indicator of the probability distribution of the noisy output.

Example 5.4

Consider the case where the output y is generated by adding noise, which is normally distributed with zero mean and a standard deviation of 0.03, to the input x of a static single-input non-linear system $y = x^2$. A fuzzy FRM, which has 21 input sets and 41 output sets, is identified using 10^5 input-output data pairs. The input x is uniformly distributed in the range [0, 1] during training. Figure 5.8 shows the fuzzy output produced by the fuzzy FRM when the value of the input is 0.3 and 0.7.

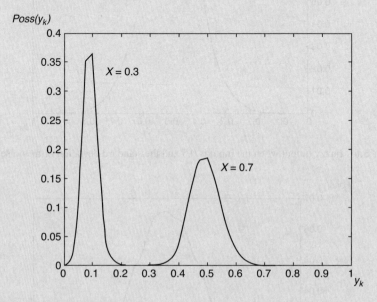

Figure 5.8   Fuzzy output when a non-linear system with input noise is modelled.

It can be seen that both distributions are approximately normal with means close to the squares of the values of the inputs (i.e. 0.09 and 0.49). In this case however, the width of the distribution increases as the value of the input (and therefore the effective gain of the non-linear system) becomes larger. Once again, the output of the fuzzy FRM is a good indicator of the probability distribution of the output.

Example 5.5

Consider the case where the output y is generated by adding noise, which is normally distributed with zero mean and a standard deviation of 10.0, to the input x of a first-order discrete-time linear dynamic system with unity gain, a time constant of 99.5 s and a sample time of 1 s. A fuzzy FRM, which has 11 input sets and 41 output sets, is identified using 10^5 input-output data pairs. During training, the input x is a uniformly distributed random variable in the range $[-9.5, +10.5]$. Figure 5.9 shows the fuzzy output produced by the fuzzy FRM after the value of the input x is stepped from 0.0 to 0.5.

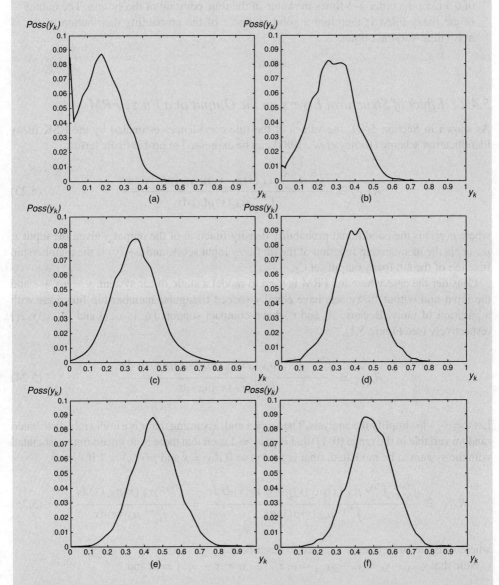

Figure 5.9 Time evolution of the fuzzy output when the input to a dynamic system is stepped from 0.0 to 0.5 where (a) $t = 42$ s; (b) $t = 82$ s; (c) $t = 122$ s; (d) $t = 162$ s; (e) $t = 242$ s; and (f) $t = 362$ s.

If edge effects are ignored, it can be seen that the output of the fuzzy FRM is an approximately symmetric distribution whose width is approximately constant and whose mean increases with time in an exponential manner, converging on a final value of 0.5 in of the order 4–5 times the value of the time constant of the system. The output of the fuzzy FRM is therefore a good indicator of the probability distribution of the noisy time-varying output.

5.4.2   Effect of Structural Errors on the Output of a Fuzzy FRM

As shown in Section 5.4.1, the values of the rule confidences estimated by the RSK fuzzy identification scheme (Ridley *et al.*, 1988) can be expressed in probabilistic terms:

$$\hat{R}_{A_i,C_k} \cong \frac{\int_{-\infty}^{+\infty} \int_{-\infty}^{+\infty} \mu_{A_i}(x)\mu_{C_k}(y)p(y|x)p(x)\,dx\,dy}{\int_{-\infty}^{+\infty} \mu_{A_i}(x)p(x)\,dx} \tag{5.23}$$

where $p(y|x)$ is the conditional probability density function of the output y given the input x, $\mu_{A_i}(x)$ is the membership function of the ith fuzzy input set A_i and $\mu_{C_k}(y)$ is the membership function of the kth fuzzy output set C_k.

Consider the case where the FRM is used to model a static linear system: $y = mx + c$ and the input and output fuzzy sets have equally spaced triangular membership functions with a partition of unity. Because A_i and C_k have compact support $[x_{i-1}, x_{i+1}]$ and $[y_{k-1}, y_{k+1}]$ respectively (see Figure 5.1),

$$\hat{R}_{A_i,C_k} \cong \frac{\int_{y_{k-1}}^{y_{k+1}} \int_{x_{i-1}}^{x_{i+1}} \mu_{A_i}(x)\mu_{C_k}(y)p(y|x)p(x)\,dx\,dy}{\int_{x_{i-1}}^{x_{i+1}} \mu_{A_i}(x)p(x)\,dx} \tag{5.24}$$

Let $v = \frac{y-c}{m}$ to simplify the analysis. Then $v = x$ and, assuming that x is a uniformly distributed random variable in the range $[0, 1]$ (that is $p(x) = 1$) and that there is no uncertainty associated with the system to be modelled, (that is $p(v|x) = 0$ if $v \neq x$ and $p(v|x) = 1$ if $v = x$).

$$\hat{R}_{A_i,V_k} \cong \frac{\int_{v_{k-1}}^{v_{k+1}} \int_{x_{i-1}}^{x_{i+1}} \mu_{A_i}(x)\mu_{V_k}(v)p(v|x)p(x)dxdv}{\int_{x_{i-1}}^{x_{i+1}} \mu_{A_i}(x)p(x)dx} = \frac{\int_{v_{k-1}}^{v_{k+1}} \mu_{A_i}(x)\mu_{V_k}(x)dx}{\int_{x_{i-1}}^{x_{i+1}} \mu_{A_i}(x)dx}, \tag{5.25}$$

where V_k is the kth reference set used to describe v.

Note that $y_{k+1} - y_k = y_k - y_{k-1} = a$, $x_{i+1} - x_i = x_i - x_{i-1} = b$, and

$$v_{k+1} - v_k = v_k - v_{k-1} = \frac{a}{m} \tag{5.26}$$

If it is assumed that $v_1 = x_1$ and $b = s\left(\frac{a}{m}\right)$, where s is an integer and $s \geq 2$, there are four cases that must be considered when evaluating equation (5.25).

Case 1: $x_{i-1} + \frac{a}{m} \leq v_k \leq x_i - \frac{a}{m}$ or $mx_{i-1} + c + a \leq y_k \leq mx_i + c - a$

$$\int_{v_{k-1}}^{v_{k+1}} \mu_{A_i}(x)\mu_{V_k}(x)dx = \int_{v_{k-1}}^{v_k} \left(\frac{x - x_{i-1}}{x_i - x_{i-1}}\right)\left(\frac{x - v_{k-1}}{v_k - v_{k-1}}\right)dx$$

$$+ \int_{v_k}^{v_{k+1}} \left(\frac{x - x_{i-1}}{x_i - x_{i-1}}\right)\left(\frac{v_{k+1} - x}{v_{k+1} - v_k}\right)dx$$

$$= \frac{m}{ab}\left[\int_{v_{k-1}}^{v_k} (x - x_{i-1})(x - v_{k-1})dx\right.$$

$$\left. + \int_{v_k}^{v_{k+1}} (x - x_{i-1})(v_{k+1} - x)dx\right] = \frac{a}{mb}(v_k - x_{i-1}) \quad (5.27)$$

Also,

$$\int_{x_{i-1}}^{x_{i+1}} \mu_{A_i}(x)dx = \frac{1}{b}\left[\int_{x_{i-1}}^{x_i} (x - x_{i-1})dx + \int_{x_i}^{x_{i+1}} (x_{i+1} - x)dx\right] = b \quad (5.28)$$

Therefore,

$$\hat{R}_{A_i,V_k} \cong \frac{a}{mb^2}(v_k - x_{i-1}) \quad (5.29)$$

and

$$\hat{R}_{A_i,C_k} \cong \frac{a}{mb^2}\left[\left(\frac{y_k - c}{m}\right) - x_{i-1}\right] \quad (5.30)$$

Case 2: $x_i + \frac{a}{m} \leq v_k \leq x_{i+1} - \frac{a}{m}$ or $mx_i + c + a \leq y_k \leq mx_{i+1} + c - a$

$$\int_{v_{k-1}}^{v_{k+1}} \mu_{A_i}(x)\mu_{V_k}(x)dx = \int_{v_{k-1}}^{v_k} \left(\frac{x_{i+1} - x}{x_{i+1} - x_i}\right)\left(\frac{x - v_{k-1}}{v_k - v_{k-1}}\right)dx$$

$$+ \int_{v_k}^{v_{k+1}} \left(\frac{x_{i+1} - x}{x_{i+1} - x_i}\right)\left(\frac{v_{k+1} - x}{v_{k+1} - v_k}\right)dx$$

$$= \frac{m}{ab}\left[\int_{v_{k-1}}^{v_k} (x_{i+1} - x)(x - v_{k-1})dx\right.$$

$$\left. + \int_{v_k}^{v_{k+1}} (x_{i+1} - x)(v_{k+1} - x)dx\right] = \frac{a}{mb}(x_{i+1} - v_k) \quad (5.31)$$

Therefore,

$$\hat{R}_{A_i, V_k} \cong \frac{a}{mb^2}(x_{i+1} - v_k) \tag{5.32}$$

and

$$\hat{R}_{A_i, C_k} \cong \frac{a}{mb^2}\left[x_{i+1} - \left(\frac{y_k - c}{m}\right)\right] \tag{5.33}$$

Case 3: $v_k = x_i$ or $y_k = mx_i + c$

$$\int_{v_{k-1}}^{v_{k+1}} \mu_{A_i}(x)\mu_{V_k}(x)\mathrm{d}x$$

$$= \int_{v_{k-1}}^{v_k} \left(\frac{x - x_{i-1}}{x_i - x_{i-1}}\right)\left(\frac{x - v_{k-1}}{v_k - v_{k-1}}\right)\mathrm{d}x + \int_{v_k}^{v_{k+1}} \left(\frac{x_{i+1} - x}{x_{i+1} - x_i}\right)\left(\frac{v_{k+1} - x}{v_{k+1} - v_k}\right)\mathrm{d}x$$

$$= \frac{a}{m} - \frac{a^2}{3m^2 b} \tag{5.34}$$

Therefore,

$$\hat{R}_{A_i, V_k} = \hat{R}_{A_i, C_k} \cong \frac{a}{mb} - \frac{a^2}{3m^2 b^2} \tag{5.35}$$

Case 4: $v_k = x_{i-1}$(or $v_k = x_{i+1}$) or $y_k = mx_{i-1} + c$ (or $y_k = mx_{i+1} + c$)

$$\int_{v_{k-1}}^{v_{k+1}} \mu_{A_i}(x)\mu_{V_k}(x)\mathrm{d}x = \int_{v_k}^{v_{k+1}} \left(\frac{x - x_{i-1}}{x_i - x_{i-1}}\right)\left(\frac{v_{k+1} - x}{v_{k+1} - v_k}\right)\mathrm{d}x = \frac{a^2}{6m^2 b} \tag{5.36}$$

Therefore,

$$\hat{R}_{A_i, V_k} = \hat{R}_{A_i, C_k} \cong \frac{a^2}{6m^2 b^2} \tag{5.37}$$

Otherwise,

$$\hat{R}_{A_i, V_k} = \hat{R}_{A_i, C_k} = 0 \tag{5.38}$$

It can be seen that the precision of the estimated probability functions is determined by the support of the input sets and the gain of the system.

It should be noted that the integrals are most easily evaluated by changing variables to move the origin and simplify the limits, and making use of the properties of even and odd functions to simplify the integrals.

Example 5.6

Consider the case of the fuzzy FRM of static linear system, which has 101 output sets ($a = 0.01$) trained using 10^5 samples. The input is uniformly distributed in the range [0, 1] during training. Figure 5.10 shows the results obtained when the FRM has 6 inputs sets ($b = 0.2$), the gain of the system is 1 and the value of the input is 0.4. The solid line indicates the result obtained from simulation; the dotted line indicates the theoretical result.

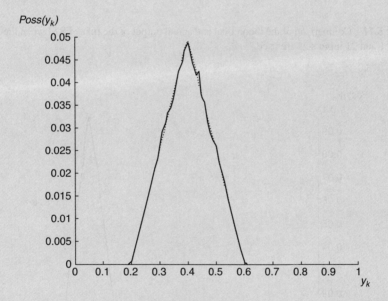

Figure 5.10 Comparison of the theoretical and actual output of the fuzzy FRM when the system gain is 1 and 6 input sets are used.

Figure 5.11 shows the equivalent result obtained when the FRM has 21 inputs sets ($b = 0.05$).

Figure 5.12 shows the equivalent result obtained when the FRM has 21 inputs sets and the gain of the system is 2. As there is no uncertainty associated with these examples, in each case the output should ideally be a fuzzy singleton.

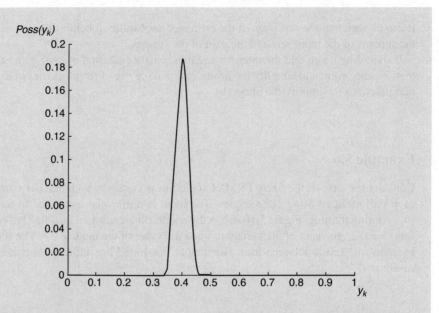

Figure 5.11 Comparison of the theoretical and actual output of the fuzzy FRM when the system gain is 1 and 21 input sets are used.

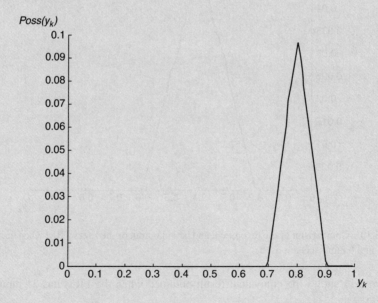

Figure 5.12 Comparison of the theoretical and actual output of the fuzzy FRM when the system gain is 2 and 21 input sets are used.

Figure 5.13 shows the results obtained when the FRM has 21 inputs sets, the gain of the system is 4 and the value of the input is 0.2.

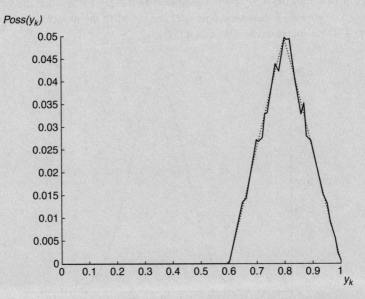

Figure 5.13 Comparison of the theoretical and actual output of the fuzzy FRM when the system gain is 4 and 21 input sets are used.

It can be seen that the support of the fuzzy output resulting from the finite granularity of the fuzzy FRM depends on both the gain of the system and the number of reference sets used to describe its inputs. Note that the support of the fuzzy output is approximately equal to the product of the system gain and the support of the input reference sets.

5.4.3 Estimation Based on Limited Amounts of Training Data

The amount of training data needed to cover the input space of an FRM with a large number of input sets can be significant. In practice, much smaller amounts of training data will be used; it is important to understand the effect of this on the fuzzy output of the resulting FRM and to consider ways in which the resulting problems can be alleviated.

Example 5.7

Consider the case where the training data are generated by adding noise, which is normally distributed with zero mean and standard deviation 0.1, to the output of a static linear system with unity gain. The input is uniformly distributed in the range [0, 1] during training. A fuzzy FRM, which has 21 inputs sets ($b = 0.05$) and 51 output sets

($a = 0.02$), is identified using 10^6 data pairs. The effect on the output of the FRM of changing the number of input sets is first examined.

Figure 5.14 shows the fuzzy output produced by a fuzzy FRM when the value of the input is 0.5. Figure 5.15 shows the equivalent result when the number of output sets of the fuzzy FRM is increased to 201 ($a = 0.005$).

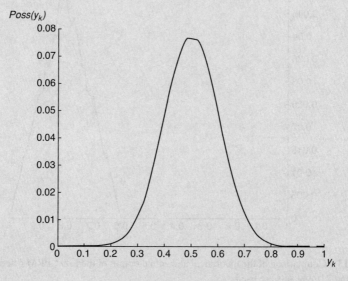

Figure 5.14   Fuzzy output when the fuzzy FRM has 21 input sets and 51 input sets.

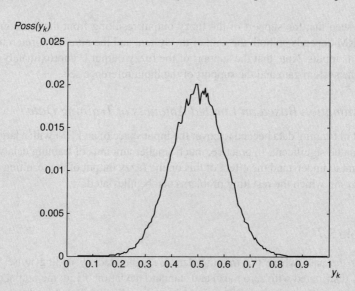

Figure 5.15   Fuzzy output when the fuzzy FRM has 21 input sets and 201 output sets.

It can be seen that the use of too many output sets can result in a fuzzy output, which is a poor estimate of the PDF.

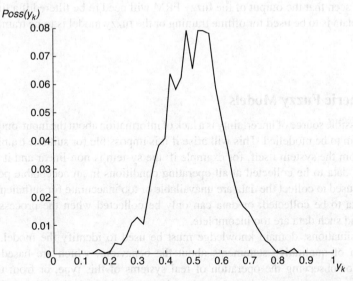

Figure 5.16   Fuzzy output when the original fuzzy FRM is trained using 10^4 data pairs.

The effect of reducing the amount of training data is examined next. Figure 5.16 shows the results obtained when the original fuzzy FRM is trained with only 10^4 data pairs. The results show that the use of too little training data can also produce unacceptable results.

One way of alleviating the problem is to post-process the fuzzy output using a median filter (Arakawa, 1996). Figure 5.17 compares the previous result with that obtained after the fuzzy

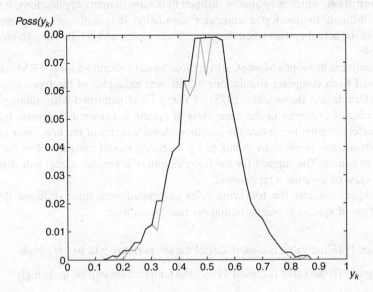

Figure 5.17   Fuzzy output before filtering (grey line) and after filtering (black line).

output has been processed using a simple filter that replaces the value of the possibility at points of inflection with the largest of the adjacent possibility values.

It can be seen that the output of the fuzzy FRM will need to be filtered if a limited amount of training data is to be used for offline training or the fuzzy model is to be trained online (see Chapter 10).

5.5   Generic Fuzzy Models

Another possible source of uncertainty is a lack of information about the input-output behaviour of the system to be modelled. This will arise if it is impossible for suitable training data to be obtained from the system itself, for example if: the system is non-linear and it is impossible for training data to be collected at all operating conditions in an acceptable period of time; the sensors used to collect the data are unavailable or too inaccurate for sufficiently consistent training data to be collected; or data can only be collected when the process is operating normally and such data are too incomplete.

In such situations, domain knowledge must be used to identify the model. This can be in the form of qualitative statements about the behaviour, which are based on previous experience of observing the operation of real systems of this type, or from the behaviour of a computer simulation of the system based on a mathematical model derived from the laws of physics. The problem with the former approach is that it is usually unclear to what extent the qualitative statements about the behaviour actually apply to the real system. The problem with the latter approach is twofold. Firstly, first-principles mathematical models of complex systems are inevitably based on simplifying assumptions. Secondly, the values of some of the parameters of such approximate models cannot be determined from first principles and the simulation model must be 'calibrated' using data from the actual system or design information, which may also be difficult to obtain. In many applications, it is therefore extremely difficult to develop a computer simulation that will accurately represent the behaviour of the actual system when the process is complex and detailed design information is not available.

An alternative is to adopt a Monte Carlo approach and to identify a fuzzy FRM using training data obtained from computer simulations of different examples of the type of system to be modelled (Dexter and Benouarets, 1995). A fuzzy FRM identified with training data taken from a number of examples of the same class of system is known as a generic fuzzy model. Generic models attempt to capture the common characteristics of the behaviour of a class of similar systems; any predictions produced by a generic model must therefore be qualitative (or fuzzy) in nature. The support for the fuzzy output of a generic model will depend on the size of the class of systems it represents.

For example, consider the following rules associated with three different designs of a particular type of system at one particular operating condition:

Design 1: IF the valve is closed THEN the temperature will be very high

Design 2: IF the valve is closed THEN the temperature will be quite high

Design 3: IF the valve is closed THEN the temperature will be reasonably high

The equivalent rule in the generic model might be:

IF the valve is closed THEN the temperature will be high

where the fuzzy sets 'very high', 'quite high' and 'reasonably high' are all subsets of the fuzzy set 'high'.

5.5.1 Identification of Generic Fuzzy Models

A generic fuzzy model is identified by using the same mathematical model to simulate the behaviours of different designs of the type of system to be modelled. If the resulting generic model is to capture the common characteristics of the different designs, the variations in the behaviours of the different designs must be greater than the accuracy of the mathematical model that is used. Suitable mathematical models can be derived from first principles, taken from the scientific literature or, in many applications, found in the library of a commercially available computer simulation package.

A schematic of the methodology used to generate a generic model based on N_m different designs is shown in Figure 5.18. A fuzzy identification scheme is used to generate a FRM of the behaviour of each design from training data obtained from its associated computer simulation. These models are then aggregated to produce a generic fuzzy model that produces

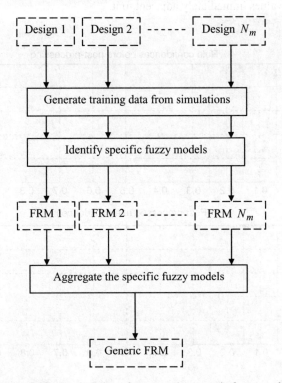

Figure 5.18   Methodology for generating generic fuzzy model.

a possibility distribution whose support depends on the extent to which the behaviours of the different designs differ. If the design parameters for each design are very different, the class of systems which the generic model describes will be large and the output of the model will be very fuzzy.

The generic model describes the behaviour of the first design OR the second design OR the third design, and so on. Since the MAX operator is commonly used to represent fuzzy OR, the generic model is produced by taking the maximum value of the rule confidences of the equivalent rules in the fuzzy reference models of the individual designs. Hence, the rule confidences of the generic FRM R_{generic} are given by

$$R_{\text{generic}} = \underset{m=1}{\overset{N_m}{Max}}(R_m) \tag{5.39}$$

As a result, the FRMs describing the individual designs are fuzzy subsets of the associated generic fuzzy model.

Post-processing of the resultant fuzzy relational array may be necessary because a lack of information about the behaviours of some designs in the class may cause 'holes' to occur in the FRM and the output of the identified generic fuzzy model may not always be convex.

Figure 5.19 shows an example of the rule confidences generated using this method. The set of non-convex rule confidences (top plot) is converted into a set of convex rule confidences (bottom plot) by replacing the possibility value at a local point of inflection with the larger of the two possibility values immediately adjacent to it.

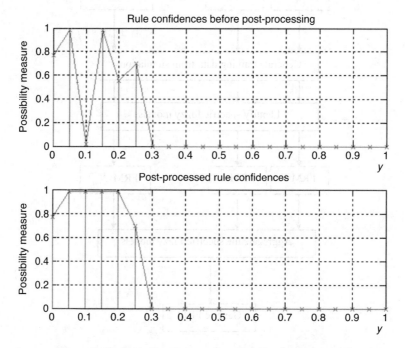

Figure 5.19   Example of post-processing the rule confidences.

The behaviour of the actual system will always lie somewhere within the fuzzy behaviour predicted by the generic model if a max-product fuzzy identification scheme is used to identify the individual FRMs and the values of the parameters of the model of the actual system are a subset of the parameters of the simulated systems.

Example 5.8

Consider the case where the training data are generated from five examples of a static linear system which has a gain and an offset. In this example, all five members of the class of systems to be modelled have unity gain but each has a different offset (-0.1, -0.05, 0.0, 0.05 and 0.1). A fuzzy FRM, which has 11 inputs sets and 41 output sets, is identified using 10^4 data pairs. The input is a uniformly distributed in the range $[0, 1]$ during training. Figure 5.20 shows the fuzzy output produced by the generic fuzzy model before post-processing, when the values of the input are 0.3 and 0.7.

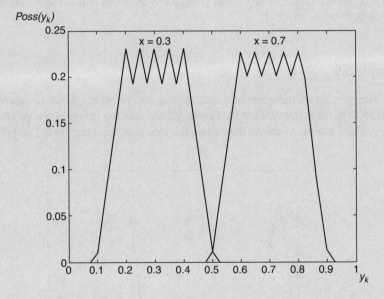

Figure 5.20 Fuzzy output produced by the generic model of a class of linear systems with an uncertain offset, before post-processing.

Figure 5.21 shows the fuzzy output produced by the generic fuzzy model after post-processing, when the values of the input are 0.3 and 0.7.

The results show that the possibility distribution of the output of the fuzzy FRM captures the variations in the behaviours of the different members of the class of system being modelled.

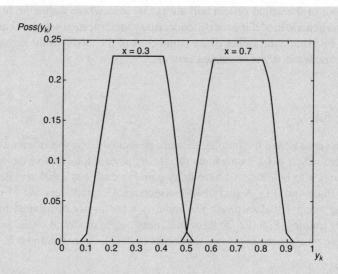

Figure 5.21 Post-processed fuzzy output produced by the generic model of a class of linear systems with an uncertain offset.

Example 5.9

In this example, all six members of the class have zero offset but different values of the gain (0.75, 0.8, 0.85, 0.9, 0.95, 1.0). Figure 5.22 shows the fuzzy output produced by the fuzzy FRM after post-processing, when the values of the input are 0.1 and 0.9.

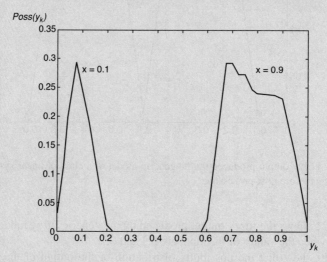

Figure 5.22 Post-processed fuzzy output produced by the generic model of a class of linear systems with an uncertain gain.

The variations in the height of the possibility distribution generated by the fuzzy FRM when the input is 0.9 are a result of differences in the output possibility distributions of the individual models, which will occur when the same type of FRM is used to model systems with different gains (see Section 5.4.2).

5.5.2   Reducing the Time Required to Generate the Training Data

A model which is able to predict the behaviour of a non-linear dynamic process at all operating conditions, without any errors, could be described as an ideal model of the process. However, such a model is usually impossible to create in practice and, consequently, models have been proposed that only match the behaviour of the system at certain operating conditions. For example, it can be shown that a two-input one-output FRM, with triangular fuzzy membership functions, is able to model exactly a piecewise linear system (Wu and Dexter, 2003). Consequently, such a fuzzy relational model might be described as an 'ideal' model if it has no prediction errors when the values of the inputs are at the centres of the fuzzy input sets.

If a fuzzy identification scheme is used to estimate the rule confidences of the FRM, the trained model will converge to the 'ideal' model if it is trained at the apex of each of the input fuzzy sets. Full firing of each input fuzzy reference set is therefore indicative of high quality or 'ideal' training data. In contrast, training data which result in a model that has prediction errors at the centre of the fuzzy input sets or is incomplete are referred to as 'poor' training data (Sing and Postlethwaite, 2000).

The problems associated with training neurofuzzy models of non-linear dynamic systems with 'poor' training data are well known (Tan, 2007) and fuzzy identification schemes have been proposed that can take into account the uneven distribution of the training data (Sing and Postlethwaite, 2000). A systematic technique for generating uniform training data has also been suggested (Laukonen and Passino, 1995), although it is admitted that there are problems with the computational complexity of the proposed algorithm. In practice, random or pseudo-random inputs are frequently used to generate training data, which are pre-processed to select only those datasets with maximal activation of each combination of input sets (Thompson and Dexter, 2005). The main disadvantage of this approach is that it can take a very long time before every rule is close to being fully fired.

Many open-loop stable high-order dynamic systems can be represented by a first-order plus time-delay dynamic model. The following scheme can reduce the time required to generate the training data required to generate a model for such a system (Wu and Dexter, 2009). A sequence of values of the control signal u is applied to the simulated system, while each of the other inputs is held constant at a value which is at the centre of one of its fuzzy reference sets. The control signal takes on two values for each combination of the other inputs. The first value is chosen in order to guarantee that the simulated steady-state value of the output at the $(n-1)$th sample $y(n-1)$ is at the centre of one of its fuzzy references sets (these values are first found by slowly stepping the control signal over its complete range). The second value is chosen in order to move the control signal to the centre of one of its own reference sets at the $(n-d)$th sample time (see Figure 5.23), where d is the time delay expressed in sampling intervals. The procedure is then repeated for every other combination of values of the output

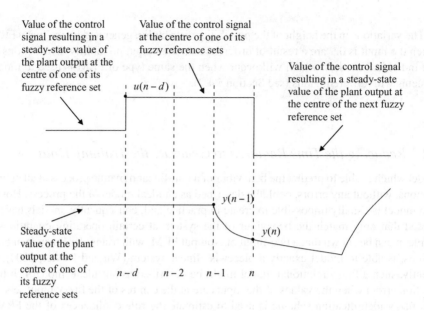

Figure 5.23   Control sequences used to generate the 'ideal' training data.

and control signal, which are at the centres of the associated fuzzy reference sets. Only the 'ideal' values of the training data, $u(n - d)$, $y(n-1)$ and $y(n)$, are used to identify the FRM.

It should be noted that this method of generating high-quality training data will be less efficient when it is used on systems whose dynamic, as well as static, behaviour is highly non-linear because there will be significant variations in the effective time delay of the system.

5.6   Summary

The fuzzy relation approach to fuzzy modelling has been explained and methods of identifying fuzzy relational models with both numeric and fuzzy outputs have been reviewed. The relationship between the fuzzy output of an FRM identified using a computationally efficient fuzzy identification scheme and inconsistencies in the training data resulting from un-modelled disturbances or measurement noise has been derived. The effect of structural errors and the availability of limited amounts of training data have also been explained. Finally, the concept of the generic fuzzy model has been introduced and a scheme for identifying generic FRMs from simulation-based training data has been explained. A simple method of generating good-quality training data has been described.

References

Abonyi, J., Babuska, R., Verbruggen, H.B. and Szeifert, F. (2000) Incorporating prior knowledge in fuzzy model identification. *International Journal of Systems Science*, **31**(5), 657–667.

Arakawa, K. (1996) Median filter based on fuzzy rules and its application to image restoration. *Fuzzy Sets and Systems*, **77**, 3–13.

Bishop, C.M. (1995) *Neural Networks for Pattern Recognition*, Oxford University Press, Oxford, UK.

Bourke, M.M. and Grant Fisher, D. (2000a) Identification algorithms for fuzzy relational matrices, Part 1: non-optimizing algorithms. *Fuzzy Sets and Systems*, **109**, 305–320.

Bourke, M.M. and Grant Fisher, D. (2000b) Identification algorithms for fuzzy relational matrices, Part 2: optimizing algorithms. *Fuzzy Sets and Systems*, **109**, 321–341.

Brown, M. and Harris, C. (1994) *Neurofuzzy Adaptive Modelling and Control*, Prentice-Hall, Hertfordshire, UK.

Campello, R.J.G.B. and Caradori do Amaral, W. (1998) Refinement and Identification of fuzzy relational models. *IEEE Transactions on Fuzzy Systems*, **1**, 651–656.

Campello, R.J.G.B. and Caradori do Amaral, W. (2006) Hierarchical fuzzy relational models: linguistic interpretation and universal approximation. *IEEE Transactions on Fuzzy Systems*, **14**(3), 446–453.

Chen, J.Q., Lu, J.H. and Chen, L.J. (1994) An on-line fuzzy identification algorithm for fuzzy systems. *Fuzzy Sets and Systems*, **64**(11), 63–72.

Chmielewski, L.J. (2006) Fuzzy histograms, weak fuzzification and accumulation of periodic quantities. *Pattern Analysis Applications*, **9**, 189–210.

Dexter, A.L. and Benouarets, M. (1995) Generic modelling of HVAC plant for fault diagnosis. Proceedings of 4th IBPSA International Conference: Building Simulation '95, Madison, Wisconsin, USA, pp. 339–345.

Herzallah, R. and Lowe, D. (2003) Robust control of nonlinear stochastic systems by modelling conditional distributions of control signals. *Neural Computing & Applications*, **12**, 98–108.

Johnston, L.P.M. and Kramer, M.A. (1994) Probability density function estimation using elliptical basis functions. *Process Systems Engineering, AIChE Journal*, **40**(10), 1639–1649.

Kelkar, B. and Postlethwaite, B. (1998) Enhancing the generality of fuzzy relational models for control. *Fuzzy Sets and Systems*, **100**, 117–129.

Klir, G.J. and Yuan, B. (1995) *Fuzzy Sets and Fuzzy Logic: Theory and Applications*, Prentice Hall, New Jersey.

Laukonen, E.G. and Passino, K.M. (1995) Training systems to perform estimation and identification. *Engineering Applications of Artificial Intelligence*, **8**(5), 499–514.

Loquin, K. and Strauss, O. (2008) Histogram density estimators based upon a fuzzy partition. *Statistics & Probability Letters*, **78**, 1863–1868.

Marakami, K. and Seborg, D.E. (2000) Constrained parameter estimation with applications to blending operations. *Journal of Process Control*, **10**, 195–202.

Pedrycz, W. (1984) An identification algorithm in fuzzy relational systems. *Fuzzy Sets and Systems*, **13**(2), 153–167.

Pedrycz, W. and Gomide, F. (2007) *Fuzzy Systems Engineering*, Wiley-Interscience, New Jersey.

Postlethwaite, B. (1991) Empirical comparison of methods of fuzzy relational identification. *IEE Proceedings, Part D*, **138**(3), 199–206.

Postlethwaite, B.E., Brown, M. and Sing, C.H. (1997) A new identification algorithm for fuzzy relational models and its application in model-based control. *Transactions of Institute of Chemical Engineering, Part A*, **75**, 453–458.

Ridley, J.N., Shaw, I.S. and Kruger, J.J. (1988) Probabilistic fuzzy model for dynamic systems. *IEE Electronic Letters*, **24**(14), 890–892.

Runkler, T.A. (2004) Fuzzy histograms and fuzzy chi-squared tests for independence. IEEE International Conference on Fuzzy Systems, 3, 1361–1366.

Shaw, I.S. and Kruger, J.J. (1992) New fuzzy learning model with recursive estimation for dynamic systems. *Fuzzy Sets and Systems*, **48**(1), 217–229.

Sing, C.H. and Postlethwaite, B. (2000) Identification of fuzzy relational models from unevenly distributed data using optimisation methods. *Transactions of Institute of Chemical Engineering, Part A*, **78**(4), 522–527.

Tan, W-W. (2007) An on-line modified least-mean-square algorithm for training neurofuzzy controllers. *Transactions of the ISA*, **46**, 181–188.

Thompson, R. and Dexter (2005) A fuzzy decision-making approach to temperature control in air conditioning systems. *Control Engineering Practice*, **13**, 689–698.

Wong, C.H., Shah, S.L. and Fisher, D.G. (2000) Fuzzy relational predictive identification. *Fuzzy Sets and Systems*, **113**, 417–426.

Wu, Y. and Dexter, A.L. (2003) Modelling capabilities of fuzzy relational models. Proceedings of IEEE International Conference on Fuzzy Systems (FUZZ-IEEE2003).

Wu, Y. and Dexter, A.L. (2009) A Computationally Efficient Method of Identifying Generic Fuzzy Models. *Fuzzy Sets and Systems*, **160**(17), 2567–2578.

Xu, C-W. (1989) Fuzzy systems identification. *Proceedings of IEE, Part D*, **136**(4), 146–150.

Xu, C-W. and Lu, Y-Z. (1987) Fuzzy model identification and self-learning for dynamic systems. *IEEE Transactions on Systems, Man, and Cybernetics*, **17**(4), 683–689.

Part II

Control of information-poor systems

6

Fuzzy Decision-Making

Decisions about most real problems must be made in a complex setting with uncertain, ambiguous and vague information, and imprecise constraints and goals. There are also differences between decision-making at the management level (non-technical decision-makers) and at the engineering level (Opricovic and Tzeng, 2003).

Decision-making in a fuzzy environment has been defined as a decision process in which the goals and/or constraints, but not necessarily the system under consideration, are fuzzy in nature (Bellman and Zadeh, 1970). There are two basic approaches to decision-making in a fuzzy environment: the 'conventional' approach where defuzzification is performed at an early stage, and the fuzzy approach where the fuzziness is eliminated at a later stage (Opricovic and Tzeng, 2003).

6.1 Risk Assessment in Information-Poor Systems

Proper evaluation of uncertainties has become a major concern in environmental (Darba et al., 2008) and health risk assessment studies (Kentel and Aral, 2007), where it is usually necessary to incorporate imprecise and incomplete knowledge about the process under consideration.

In practice, there may be insufficient data available for statistical methods to be used for all variables and a hybrid approach based on a combination of statistical and fuzzy methods may be more appropriate (Guyonnet et al., 2003). Consider the case where the risk is an uncertain function of a number of crisp variables C_i and a number of fuzzy variables F_j

$$R = f(C_1, C_2, \cdots C_n, F_1, F_2, \cdots F_m) \tag{6.1}$$

The fuzzy variables F are described by fuzzy numbers and fuzzy arithmetic is used to relate risk to these crisp and fuzzy variables.

Decision-making is based on the validity of the proposition P that 'the fuzzy risk is less than or equal to the compliance guideline'. The possibility $Poss(R \leq C_L)$ and the necessity $Nec(R \leq C_L)$ of the proposition (see Section 3.2.2), where R is the risk and C_L is the specified compliance level, can be used as measures of the validity of the proposition.

Monitoring and Control of Information-Poor Systems: An Approach based on Fuzzy Relational Models, First Edition. Arthur L. Dexter.
© 2012 John Wiley & Sons, Ltd. Published 2012 by John Wiley & Sons, Ltd.

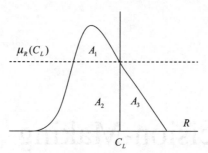

Figure 6.1   Use of weighting parameters to estimate the compliance level.

Upper and lower bounds on the probability of the fuzzy risk exceeding the compliance level are given by

$$Poss(R > C_L) = \sup_{r > C_L}\{\mu_R(r)\} \text{ and } Nec(R > C_L) = 1 - \sup_{r \leq C_L}\{\mu_R(r)\},$$

respectively (Guyonnet *et al.*, 2003).

However, the use of fuzzy arithmetic has been found to overestimate the uncertainty range and an approach based on the theory of evidence has been shown to give less conservative results (Baudrit *et al.*, 2005).

Another measure Γ of whether the risk is less than or equal to the compliance level is given by Kentel and Aral (2007):

$$\Gamma = \begin{cases} 0 & if \quad C_L \leq L_R(0) \\ \dfrac{1}{2}\left(\beta Poss(R \leq C_L) + \gamma Nec(R \leq C_L)\right) & if L_R(0) < C_L < U_R(0) \\ 1 & if \quad C_L \geq U_R(0) \end{cases} \tag{6.2}$$

The weighting parameters β and γ are given by (see Figure 6.1):

$$\beta = \frac{\displaystyle\int_0^{\mu_R(C_L)} \alpha\,[C_L - L_R(\alpha)]\,d\alpha}{\displaystyle\int_0^{\mu_R(C_L)} \alpha\,[U_R(\alpha) - L_R(\alpha)]\,d\alpha} = \frac{A_2}{A_2 + A_3}, \tag{6.3}$$

where β is an indicator of the ratio of the strength of evidence supporting the proposition to the total evidence available within the membership range $[0, \mu_R(C_L)]$, and

$$\gamma = \begin{cases} 0 & if \quad L_R(\mu_R(C_L)) = C_L \\ 1 & if \quad U_R(\mu_R(C_L)) = C_L, \end{cases} \tag{6.4}$$

That is, γ has a value of zero unless $Nec(R \leq C_L)$ is non-zero. $U_R(\alpha)$ and $L_R(\alpha)$ are the upper and lower bounds of the α-cut set of the fuzzy risk.

The value of Γ can take any value in the range $[0, 1]$ and the regulatory authority is left to decide on an appropriate value of Γ such that the resulting risk may be identified as acceptable (Kentel and Aral, 2007).

6.2 Fuzzy Optimization in Information-Poor Systems

Each decision is made from a choice of alternatives (e.g. a list of available cars) defined by the values of one or more attributes (e.g. price, maximum speed, petrol consumption, colour, etc.), some of which may be described in qualitative terms. An objective is defined in terms of one or more of the attributes. The goal of the decision-making (e.g. deciding whether to buy a car) is to maximize the degree of satisfaction of one or more poorly defined objectives (e.g. economy of operation, environmental friendliness, etc.) (Ribeiro, 1996).

6.2.1 Fuzzy Goals and Fuzzy Constraints

Consider a simple example where a decision u is to be taken in order to maximize a fuzzy goal $\mu_G(u)$ and satisfy a fuzzy constraint $\mu_C(u)$. An obvious choice would be the value of $u = u^*$ that maximizes the membership of the goal and the constraint, that is, $\mu_D(u^*) = \sup\{\mu_D(u)\}$ where $\mu_D(u) = \mu_C(u) \wedge \mu_G(u)$ and $\wedge$ is the min operator. Note that fuzzy constraints are handled in the same way as fuzzy goals in fuzzy decision-making (FDM).

6.2.2 Fuzzy Aggregation Operators

Multi-objective decision-making (MODM) is concerned with a set of conflicting goals that cannot be achieved simultaneously. Multi-attribute decision-making (MADM) is concerned with choosing from several alternatives, each of which is characterized by a set of attributes. In this case, both the value and the importance of each attribute must be taken into account.

There are usually two stages in the fuzzy decision-making process for multiple attribute problems: rating the alternatives according to their degrees of satisfaction of the goals and constraints, and ranking of the alternatives (Ribeiro, 1996).

A variety of methods have been suggested for rating alternatives taking the relative importance of the different attributes into account. Weighted aggregation, in which weight factors represent the relative importance of the various criteria, is often used to deal with multiple objectives and multiple attributes in fuzzy decision-making (Mendonca et al., 2003). Because the overall goal of fuzzy optimization is the simultaneous satisfaction of all the objectives and constraints, conjunctive aggregation using t-norms must be used. The weighted aggregation function is then of the general form:

$$\mu_D(\bar{u}, \bar{\omega}) = T\left\{I[\mu_{C_1}(u_1), \omega_{C1}], \ldots, I[\mu_{C_n}(u_n), \omega_{C_n}], \ldots, I[\mu_{G_1}(u_1), \omega_{G_1}], \ldots \right.$$
$$\left. I[\mu_{G_1}(u_n), \omega_{G_m}]\right\} \tag{6.5}$$

where $\bar{u}$ is a vector of the optimization variables, $\bar{\omega}$ is a vector of the weighting factors, T is a t-norm and I is a function of the membership function and the associated weight. For example, the extended Yager t-norm could be used. Then,

$$\mu_D(\bar{u}, \bar{\omega}) = Max\left\{0, 1 - \sqrt[s]{\sum_{l=1}^{r} \omega_l[1 - \mu_l(u_l)]^s}\right\} \tag{6.6}$$

where $s > 0$.

Ranking the alternatives is trivial if the results of rating them are crisp. Otherwise, a fuzzy ranking scheme must be used to determine the best design.

6.2.3   Fuzzy Ranking

The ranking of fuzzy numbers to determine which fuzzy set most satisfies a fuzzy goal is central to fuzzy attribute decision-making. Desirable properties of fuzzy ranking methods are the ability to express their decisions in linguistic rather than numerical terms, to reflect human behaviour in terms of the consistency and coherence of their decisions, to distinguish between meaningful differences in the fuzzy alternatives and to tolerate minor errors in estimating the membership functions of the fuzzy numbers (Yuan, 1991). A number of ranking methods have been suggested. The least computationally demanding method is to use one of the classical defuzzification algorithms (see Section 4.6) to convert each fuzzy set into a real number for which natural order exists so that conventional arithmetic can be used to determine which fuzzy set is closest to the other fuzzy set. The main disadvantage of this approach is that most defuzzification schemes fail to differentiate between fuzzy sets with different membership functions but the same mean value (Opricovic and Tzeng, 2003).

Another option is to generate a measure of the equality of two fuzzy sets A and B by calculating the degree to which fuzzy set A is a proper subset of fuzzy set B (i.e. $A \subset B$) and the degree to which fuzzy set B is a proper subset of fuzzy set A (i.e. $B \subset A$), where $A \subset B$ if $\forall x \in X : \mu_A(x) < \mu_B(x)$ and so on. The grade of equality of A and B is then given by

$$\{A = B\}(x) = \min\{[\mu_A(x) \, \beta \, \mu_B(x)], [\mu_B(x) \, \beta \, \mu_A(x)]\}$$

where the inclusion operator β is defined as follows:

$$\mu_A(x) \, \beta \, \mu_B(x) = 1 \qquad \text{if} \quad \mu_A(x) < \mu_B(x)$$
$$\mu_A(x) \, \beta \, \mu_B(x) = \mu_B(x) \quad \text{if} \quad \mu_A(x) \geq \mu_B(x)$$

It can be shown that an optimistic overall grade of equality of the two fuzzy sets is then given by

$$\|A = B\| = \sup_x [\{A = B\}(x)] = \sup_x [\min\{\mu_A(x), \mu_B(x)\}], \qquad (6.7)$$

the possibility of A given B.

A distance measure can also be used to find how close two fuzzy sets are to each other, for example, one of the Minkowski r-metrics (Szmidt and Kacprzyk, 2006):

$$\text{the normalized Hamming distance } d(A, B) = \frac{1}{n} \sum_{i=1}^{n} |\mu_A(x_i) - \mu_B(x_i)|$$

$$\text{the normalized Euclidean distance } d(A, B) = \sqrt{\frac{1}{n} \sum_{i=1}^{n} [\mu_A(x_i) - \mu_B(x_i)]^2}$$

or the Hausdorff distance (see Section 8.2).

Another approach is to use a fuzzy similarity measure. For example, the similarity of two fuzzy sets could be defined as the ratio of the common area between the two membership functions to the total area underneath the membership function of one of the fuzzy sets. Thus,

$$S(A, B) = \frac{\int\limits_{UoD} \min[\mu_A, \mu_B]dx}{\int\limits_{UoD} \mu_A dx} \tag{6.8}$$

This simple measure does not take account of the fact that, if the non-overlapping area is associated with high possibility values, the degree of similarity should be lower than if it had been associated with low possibility values. This can be taken into account by weighting the elements in the overlap area according to their associated possibility values. For example, one method is to project the areas onto a plane at an angle to the vertical, and use the enclosed volume (instead of the area) in the similarity measure (Smith and Verma, 2004). Another method based on area measures is to use fuzzy subtraction to find the difference between the two fuzzy numbers, and to compare the areas under the part of the membership function associated with positive values and with negative values on the Universe of Discourse (UoD) (Chen and Klein, 1997).

Alternatively, a measure of the degree of dissemblance (or dissimilarity) of two fuzzy numbers A and B can be used (Kacprzyk, 1995) where the degree of dissemblance $diss(A, B)$ is given by

$$diss(A, B) = \int_0^1 \frac{1}{2} \left(|a_L^\alpha - b_L^\alpha| + |a_U^\alpha - b_U^\alpha| \right) d\alpha \tag{6.9}$$

where $[a_L^\alpha, a_U^\alpha]$ and $[b_L^\alpha, b_U^\alpha]$ are the α-cuts of A and B $\forall \alpha \in (0, 1)$.

6.3 Multi-Stage Decision-Making

Many human activities involve the management and control of dynamic systems (e.g. production management, traffic control, online control of industrial processes or the operation of water supply networks). Such activities require the solution of multi-stage decision-making problems (Kacprzyk, 1997). For example, consider an N-stage process governed by the equation $x_{s+1} = f[x_s, u_s]$ where x_s is the state of the system at stage s, u_s is the decision taken at stage s, and $s = 0, 1, 2 \ldots N - 1$.

Let us assume that at each stage s the decision (u_s) is subject to a fuzzy constraint $\mu_{C_s}(u_s)$ and a fuzzy goal $\mu_{G_s}(x_{s+1})$. The degree of satisfaction of the goals and constraints over the N stages is given by:

$$\mu_D(u_0, u_1, \ldots u_{N-1}) = \mu_{C_0}(u_0) \wedge \mu_{G_0}(x_1) \wedge \mu_{C_1}(u_1) \wedge \mu_{G_1}(x_2) \ldots$$
$$\mu_{C_{N-1}}(u_{N-1}) \wedge \mu_{G_{N-1}}(x_N) \tag{6.10}$$

where $\wedge$ is a suitable aggregation operator (the min operator is often used).

The optimal sequence of decisions $[u_0^*, u_1^*, \ldots u_{N-1}^*]$ is the sequence that maximizes the overall degree of satisfaction μ_D. Note that in practice it may be impossible to find a sequence that satisfies the goals and constraints at all stages. One way of avoiding this problem is to 'soften' this requirement by introducing a fuzzy linguistic quantifier so that 'almost all' of the goals and constraints are satisfied (Kacprzyk, 1986).

In practice, finding the optimal sequence of decisions is a non-convex optimization problem that must be solved numerically.

6.3.1 Fuzzy Dynamic Programming

The dynamic programming approach to finding the optimum sequence (fuzzy dynamic programming) was originally proposed by Bellman and Zadeh in 1970. Because only the last two terms of the equation for the overall degree of satisfaction $\mu_D(u_0, u_1, \ldots u_{N-1})$ depend on the value of u_{N-1}, the function to be optimized can be written in recursive form and solved using dynamic programming.

The degree of satisfaction of the goals and constraints over the N stages is given by

$$\mu_D(u_0, \ldots u_{N-1} | x_0) = \mu_{C_0}(u_0) \wedge \mu_{G_0}(x_1) \wedge \mu_{C_1}(u_1) \wedge \mu_{G_1}(x_2)$$

$$\wedge \ldots \wedge \mu_{C_{N-1}}(u_{N-1}) \wedge \mu_{G_{N-1}}(x_N)$$

or

$$\mu_D(u_0, \ldots u_{N-1} | x_0) = \mu_{C_0}(u_0) \wedge \mu_{G_0}(f[x_0, u_0]) \wedge \mu_{C_1}(u_1) \wedge \mu_{G_1}(f[x_1, u_1]) \wedge \ldots$$
$$\ldots \wedge \mu_{C_{N-1}}(u_{N-1}) \wedge \mu_{G_{N-1}}(f[x_{N-1}, u_{N-1}])$$

$$(6.11)$$

The optimization problem is to find the sequence of fuzzy decisions $[u_0^*, u_1^*, \ldots u_{N-1}^*]$ such that

$$\mu_D(u_0^*, \ldots u_{N-1}^* | x_0) = \max_{u_0, \ldots u_{N-1}} \{\mu_D(u_0, \ldots u_{N-1} | x_0)\}$$

$$= [\max_{u_0}\{\mu_{C_0}(u_0) \wedge \mu_{G_0}(f[x_0, u_0])\}] \wedge [\max_{u_1}\{\mu_{C_1}(u_1) \wedge \mu_{G_1}(f[x_1, u_1])\}] \wedge \ldots \quad (6.12)$$

$$\ldots \wedge [\max_{u_{N-1}}\{\mu_{C_{N-1}}(u_{N-1}) \wedge \mu_{G_{N-1}}(f[x_{N-1}, u_{N-1}])\}]$$

The equation can be written in recursive form, as follows,

$$\mu_D(u_0^*, \ldots u_{N-1}^* | x_0) = \max_{u_0}\{\mu_{C_0}(u_0) \wedge \mu_{G_0}(f[x_0, u_0])\} \wedge \mu_{D_1}(x_1) \quad (6.13)$$

where

$$\mu_{D_1}(x_1) = \max_{u_1}\{\mu_{C_1}(u_1) \wedge \mu_{G_1}(f[x_1, u_1])\} \wedge \mu_{D_2}(x_2) \text{ etc.} \quad (6.14)$$

and

$$\mu_{D_{N-1}}(x_{N-1}) = \max_{u_{N-1}}\{\mu_{C_{N-1}}(u_{N-1}) \wedge \mu_{G_{N-1}}(f[x_{N-1}, u_{N-1}])\} \quad (6.15)$$

The optimal sequence of decisions can now be found one stage at a time using backward iteration starting from stage $(N-1)$, using the dynamic programming approach (Kacprzyk, 1986). There is however a major problem: there are, in general, an infinite number of fuzzy

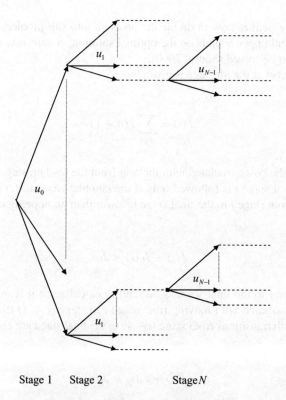

Stage 1   Stage 2   Stage N

Figure 6.2   Branch and bound search tree.

states and fuzzy decisions that must be considered. A way of avoiding this problem is to describe the fuzzy states and fuzzy decisions using a relatively small number of fuzzy reference sets, and to take advantage of fuzzy interpolation (so-called *fuzzy discretization*). However, the use of dynamic programming still requires an exhaustive search through all possible combinations of the fuzzy reference sets and, in practice, this approach still suffers from 'the curse of dimensionality'. The computational demands can therefore be very high unless other measures are taken to ameliorate the problem (Kacprzyk and Esogbue, 1996). As the computational time increases exponentially with the number of stages in the problem, an even more computationally efficient method will be required if the number of stages is large.

6.3.2   Branch and Bound

The branch and bound (B&B) method is a structured search technique that solves the problem by dividing the discretized search space into smaller sub-problems using a tree structure (see Figure 6.2) and eliminating those branches of the tree that do not contain the optimal solution.

A *branching rule* defines how to divide the problem into sub-problems. A *bounding rule* establishes lower and upper bounds on the optimal solution. A *selection rule* determines the next sub-problem to be solved (Sousa, 2000).

The cumulative cost at the ith stage is given by

$$J(i) = \sum_{n=1}^{i} J(n-1) \qquad (6.16)$$

where $J(n-1)$ is the cost associated with moving from the $(n-1)$th stage to the nth stage. A particular branch j at stage i is followed only if the cumulative cost $J(i)$ plus a lower bound on the cost $J_L(i)$ from stage i to the final stage is lower than an upper bound of the total cost J_U, that is,

$$J(i) + J_L(i) < J_U \qquad (6.17)$$

The cost from stage i to the final stage is difficult to calculate so it is often assumed that it is the cost $J_j(i)$ associated with moving from stage i to stage $(i+1)$ (that is the remaining costs associated with transitions from stage $(i+1)$ to the final stage are zero). In this case, the rule becomes

$$J(i) + J_j(i) < J_U \qquad (6.18)$$

The number of new branches to be considered will be smallest if the upper bound is as low as possible and the lower bound is as large as possible. The lower bound is originally chosen conservatively but, during the search, it can be replaced by the lowest overall cost determined so far. The upper bound is often initialized to the overall cumulative cost associated with a path that minimizes the cost $J_j(i)$ at each stage (Mendonca *et al.*, 2004).

Example 6.1

Consider a simple example of a three-stage decision process in which only three values of decision variable are possible. The decision tree is shown in Figure 6.3. The numerical values are the costs associated with particular decisions. The path that minimizes the cost $J_j(i)$ at each stage is indicated by long dashed lines. Branches indicated by short dashed lines have not satisfied the branching condition (it has been assumed that the cost associated with transitions from stage $(i+1)$ to the final stage are zero). Branches indicated by dotted lines are not searched. The optimal sequence of decisions is indicated by solid lines. It can be seen that the use of B&B requires the cost to be evaluated 15 times whereas an exhaustive search would require the cost to be evaluated 39 times.

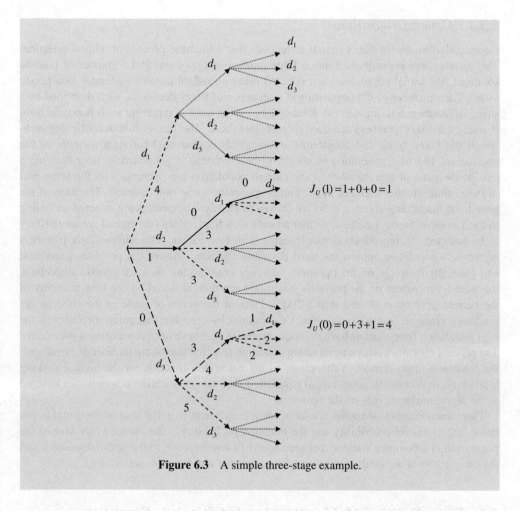

Figure 6.3   A simple three-stage example.

The method is guaranteed to find the global optimum and is not negatively influenced by poor initialization as is the case with iterative methods of optimization. Constraints are dealt with explicitly. In fact, the presence of constraints improves the efficiency of the algorithm by eliminating certain branches (Sousa, 2000).

The B&B method requires that the cumulative cost at the ith stage is less than or equal to the cumulative cost at the $(i + 1)$th stage (Sousa, 2000). To satisfy this condition for fuzzy optimization, the cost must be evaluated using the complement of the degree of satisfaction of the fuzzy goals and constraints, and the aggregation operator must be a t-norm so that the overall degree of satisfaction at the pth stage is never greater than that at the $(p - 1)$th stage. The fitness function is therefore given by

$$\mu_D(u_0, u_1, \ldots u_{N-1}) = \mu_{\tilde{C}_0}(u_0) \wedge \mu_{\tilde{G}_1}(x_1) \wedge \mu_{\tilde{C}_1}(u_1) \wedge \mu_{\tilde{G}_2}(x_2) \ldots \mu_{\tilde{C}_{N-1}}(u_{N-1}) \wedge \mu_{\tilde{G}_N}(x_N)$$

$$(6.19)$$

6.3.3   Genetic Algorithms

Genetic algorithms are direct search techniques that mimic the process of natural selection. The search space is partitioned into a finite number of fuzzy sets and a number of feasible solutions (the initial population) are selected using a random number generator (Kacprzyk, 1998). Each member of the population is a sequence of fuzzy decisions, each described by a string of numbers (chromosomes). Real coding can be used to represent each fuzzy decision if triangular fuzzy numbers are used to represent them. The fitness function (the degree to which the fuzzy goals and constraints are satisfied) is evaluated for each member of the population. The next generation of children (the offspring) is generated by interchanging a part of the gene of one member of the current population (one parent) with the same part of the gene of another member of the current population (the other parent). The parts of the gene to be interchanged (crossover) are chosen randomly. The parents are selected according to the values of fitness function, so that parents with high values (the fittest) are more likely to be selected for reproduction than parents with low values. This ensures that members of the new population inherit the most desirable characteristics of the previous generation and gradually converge on the optimum sequence of decisions. In some genetic algorithms, the weakest members of the previous generation are then replaced by the best members of the current generation (Causa *et al.*, 2008). Parts of the genes of some of the children are randomly changed (mutated) to avoid local optima by ensuring that some members of the next generation (population) have decision strategies that have not been explored previously. The process continues until a terminating condition (e.g. the maximum number of evolutions, the minimum improvement in the values of the evaluation function for the highest ranking individuals in successive generations) is satisfied. The rate of mutation is sometimes lowered as the algorithm converges on the optimum.

There are a number of application-dependent parameters (e.g. the size of the initial population, the crossover probability and the mutation probability). The choice of the size of the population is a complex tradeoff between speed of convergence to the optimal solution and the computational demands associated with generating a new population.

6.4   Fuzzy Decision-Making Based on Intuitionistic Fuzzy Sets

Fuzzy decision-making based on ordinary fuzzy sets assumes that there is no uncertainty associated with the degree to which a goal (or constraint) is satisfied. Fuzzy decision-making based on intuitionistic fuzzy sets allows the degree to which it is believed that the goal (or constraint) may not be satisfied to be taken into account. This allows upper (complement of the non-membership function) and lower bounds on the membership functions to be defined, and an uncertainty interval to be specified (Gurkan *et al.*, 2002).

6.4.1   Definition of an Intuitionistic Fuzzy Set

The idea of intuitionistic fuzzy sets (IFSs) was first suggested by Atanassov as a generalization of ordinary fuzzy sets (Atanassov, 1986). An IFS is defined by two membership functions: the degree of membership of x in fuzzy set A, μ_A, (x), and the degree of non-membership of x in fuzzy set A, $v_A(x)$, where $\mu_A(x) + v_A(x) \leq 1$. The intuitionistic fuzzy set reverts to an ordinary fuzzy set when $v_A(x) = 1 - \mu_A(x)$. A measure of the lack of knowledge about whether x belongs to A, the hesitation margin, is given by $\pi(x) = 1 - \mu_A(x) - v_A(x)$ (Szmidt and Kacprzyk, 2006). As is the case with ordinary fuzzy sets, either expert knowledge or online training can be used to select the membership and non-membership functions defining an IFS (Gurkan *et al.*, 2002).

6.4.2 Multi-Attribute Decision-Making Using Intuitionistic Fuzzy Numbers

Although IFS-based FDM is computationally more demanding than FDM based on conventional fuzzy sets, it takes account of both belief and non-belief and may be a better way of formulating uncertain optimization problems (Angelov, 1997) since it can eliminate negative decisions (Szmidt and Kacprzyk, 2006) and is capable of generating higher degrees of satisfaction of the goal (Angelov, 1997).

Let A be a finite number of n options, G be a finite number of m attributes and W be a set of weights associated with each attribute. Consider the case where each option A_j is described by an intuitionistic fuzzy number (μ_{A_j}, v_{A_j}). A measure of the distance $d(A_j, A^+)$ between the IFNs representing each option and the ideal result A^+ can be used to determine the optimal choice (Xu, 2007):

$$d(A_j, A^+) = \sum_{i=1}^{m} w_i d(r_{i,j}, \alpha^+) \qquad (6.20)$$

where $r_{i,j} = (\mu_{i,j}, v_{i,j})$ is an IFN. $\mu_{i,j}$ indicates the degree of truth that the choice A_j satisfies the attribute G_i, $v_{i,j}$ indicates the degree of truth that A_j does not satisfy the attribute G_i and $\alpha^+ = (1, 0)$ (i.e. complete belief in the attribute being satisfied and no belief in the attribute not being satisfied). The distance measure can be used to rank all of the options as a smaller distance implies a better option.

Ideally the distance measure (see Section 6.2.3) should be based on the degree of membership, the degree of non-membership and the hesitation margin (Hung and Yang, 2004; Szmidt and Kacprzyk, 2006).

Example 6.2

Consider a simple example in which a choice has to be made between six candidates for the post of senior buyer in a major department store selling women's clothing based on three criteria: fashion sense, communication skills and commercial expertise. Suppose the appointment panel has agreed that 50% of the final decision should be based on fashion sense, 30% on communication skills and 20% on commercial expertise. After interviewing each applicant, the scores of individual panel members are used to determine the degrees to which each candidate is thought to satisfy $\mu_{i,j}$ and to not satisfy each of the criteria $v_{i,j}$ (see Table 6.1).

Table 6.1   IFNs for each candidate and attribute.

	A_1	A_2	A_3	A_4	A_5	A_6
G_1	(0.4, 0.3)	(0.5, 0.2)	(0.7, 0.2)	(0.4, 0.3)	(0.6, 0.2)	(0.6, 0.3)
G_2	(0.6, 0.1)	(0.3, 0.4)	(0.3, 0.7)	(0.6, 0.2)	(0.5, 0.1)	(0.7, 0.2)
G_3	(0.5, 0.4)	(0.8, 0.2)	(0.6, 0.2)	(0.7, 0.1)	(0.4, 0.6)	(0.5, 0.4)

Table 6.2 lists the distance measures $d(A_j, A^+)$ obtained if the normalized Hamming distance $d(A_i, A_j) = \frac{1}{2}\left[|\mu_{A_1} \quad \mu_{A_2}| + |\nu_{A_1} - \nu_{A_2}|\right]$ is used to rank the candidates.

Table 6.2   Weighted distance measures for each candidate.

	A_1	A_2	A_3	A_4	A_5	A_6
$d(A_j, A^+)$	0.390	0.380	0.395	0.355	0.360	0.340

Clearly candidate six is the best candidate for the post.

6.5   Summary

This chapter has given a brief introduction to fuzzy decision-making and shown how it can be used to solve risk assessment and optimization problems in information-poor systems. A fuzzy approach to risk assessment in information-poor systems has been described. The ideas behind fuzzy optimization (fuzzy goals and fuzzy constraints, fuzzy aggregation operators) have been discussed and several methods of fuzzy ranking have been considered. Multi-stage fuzzy decision-making based on fuzzy dynamic programming has been explained and two widely used search methods (branch and bound and genetic algorithms) have been described. Fuzzy decision-making based on intuitionistic fuzzy sets has also been discussed.

References

Angelov, P.P. (1997) Optimisation in an intuitionistic fuzzy environment. *Fuzzy Sets and Systems*, **86**(3), 299–306.

Atanassov, K.T. (1986) Intuitionistic fuzzy sets. *Fuzzy Sets and Systems*, **20**, 87–96.

Baudrit, C., Guyonnet, D., and Dubois, D. (2005) Post-processing the hybrid method for addressing uncertainty in risk assessments. *J. Environmental Engineering*, **131**(12), 1750–1754.

Bellman, R.E. and Zadeh, L.A. (1970) Decision-making in a fuzzy environment. *Management Science*, **17**(4), 141–164.

Causa, J., Karer, G., Nunez, A., *et al.* (2008) Hybrid fuzzy predictive control based on genetic algorithms for the temperature control of a batch reactor. *Computers and Chemical Engineering*, **32**(12), 3254–3263.

Chen, C-B. and Klein, C.M. (1997) A simple approach to ranking a group of aggregated fuzzy utilities. *IEEE Transactions on Systems, Man and Machines*, **27**(11), 26–35.

Darba, R.M., Eljarrat, E. and Barcelo, D. (2008) How to measure uncertainties in environmental risk assessment. *Trends in Analytical Chemistry*, **27**(4), 377–385.

Gurkan, E., Erkmen, I. and Erkmen, A.M. (2002) Two-way fuzzy adaptive identification and control a flexible-joint robot arm. *International Journal of Information Sciences*, **145**, 13–43.

Guyonnet, D., Bourgine, B., Dubois, D., *et al.* (2003) Hybrid approach for addressing uncertainty in risk assessments. *Journal of Environmental Engineering*, **129**(1), 68–78.

Hung, W-L. and Yang, M-S. (2004) Similarity measures of intuitionistic fuzzy sets based on Hausdorff distance. *Pattern Recognition Letters*, **25**, 1603–1611.

Kacprzyk, J. (1986) Towards 'human consistent' multistage decision-making and control models using fuzzy sets and fuzzy logic. *Fuzzy Sets & Systems*, **18**, 299–314.

Kacprzyk, J. (1995) Multistage control of a fuzzy system using a genetic algorithm. IEEE International Conference on Fuzzy Systems, 1083–1088.

Kacprzyk, J. (1997) *Multi-Stage Fuzzy Control: A Model-Based Approach to Fuzzy Control and Decision-Making*, Wiley, Chichester, UK.

Kacprzyk, J. (1998) Multistage control of a stochastic system in a fuzzy environment using a genetic algorithm. *International Journal of Intelligent Systems*, **13**, 1011–1023.

Kacprzyk, J. and Esogbue, A.O. (1996) Fuzzy dynamic programming: main developments and applications. *Fuzzy Sets & Systems*, **81**, 31–45.

Kentel, E. and Aral, M.M. (2007) Risk tolerance measure for decision-making in fuzzy analysis: a health risk assessment perspective. *Stochastic Environmental Research Risk Assessment*, **21**, 405–417.

Mendonca, L.F., Sousa, J.M., Kaymak, U. and Sa da Costa, J. (2003) Fuzzy issues in multivariable predictive control. IEEE International Conference on Fuzzy Systems, 506–511.

Mendonca, L.F., Sousa, J.M. and Sa da Costa, J.M.G. (2004) Optimization problems in multivariable fuzzy predictive control. *International Journal of Approximate Reasoning*, **36**, 199–221.

Opricovic, S. and Tzeng, G-H. (2003) Defuzzification within a multi-criteria decision model. *International Journal of Uncertainty, Fuzziness & Knowledge-based Systems*, **11**(5), 635–652.

Ribeiro, R.A. (1996) Fuzzy multiple attribute decision-making: a review and new preference elicitation techniques. *Fuzzy Sets and Systems*, **78**(2), 155–181.

Smith, C. and Verma, D. (2004) Conceptual system design evaluation: rating and ranking versus compliance analysis. *Systems Engineering*, **7**(4), 338–351.

Sousa, J.M. (2000) Optimisation issues in predictive control with fuzzy objective functions. *International Journal of Intelligent Systems*, **15**, 879–899.

Szmidt, E. and Kacprzyk, J. (2006) A model of case-based reasoning using intuitionistic fuzzy sets. IEEE International Conference on Fuzzy Systems, 1769–1776.

Xu, Z.S. (2007) Models for multiple attribute decision-making with intuitionistic fuzzy information. *International Journal of Uncertainty, Fuzziness & Knowledge-based Systems*, **15**(3), 285–297.

Yuan, Y. (1991) Criteria for evaluating fuzzy ranking methods. *Fuzzy Sets and Systems*, **43**(2), 139–157.

7

Predictive Control in Uncertain Systems

7.1 Model-Based Predictive Control

At each sampling instant an explicit mathematical model is used to predict the future behaviour of the plant over a specified prediction horizon in response to a given sequence of changes in the control input over a control horizon (see Figure 7.1). The sequence of changes in the control signal that produces a response that is closest to the desired behaviour is found. Only the first change in the sequence is applied to the plant as the whole process is repeated at the next sampling instant (a so-called *receding horizon control* strategy).

Several different forms of model-based predictive control (MPC) have been proposed (Garcia *et al.*, 1989). In its simplest form the requirement is that the controlled variable is equal to the value of the setpoint at every sampling instant (so-called dead-beat control). A cost function is usually used to define the desired behaviour of the plant and a constrained optimization problem must be solved to find the most appropriate control action.

MPC can handle constraints in a systematic way and is robust to uncertainty in the estimated value of any time delay associated with the response of the plant. The main difficulties associated with the practical application of MPC are identifying a suitable mathematical model and specifying an appropriate control objective (Richalet, 1993).

The optimal value of the control signal must be found at each sampling instant. An analytical (closed-form) solution to the optimization problem exists if the model of the plant is linear, there are no constraints and the cost function is quadratic (Garcia *et al.*, 1989).

Some form of numerical approach to solving the optimization problem is usually necessary if the model is non-linear (Bloemen *et al.*, 2001; Magni and Scattolini, 2004), the optimization is constrained (Demircioglu and Yavuzyilmaz, 2002) and/or one or more of the optimization variables can take only integer values (Borrelli *et al.*, 2005). The key questions arising in the design of non-linear MPC involve the existence and uniqueness of, and the numerical problems associated with finding, the control law.

One way of dealing with non-linearity is to base the control scheme on a set of piece-wise linear models of the plant so that locally linear predictive controllers can be designed

Monitoring and Control of Information-Poor Systems: An Approach based on Fuzzy Relational Models, First Edition.
Arthur L. Dexter.
© 2012 John Wiley & Sons, Ltd. Published 2012 by John Wiley & Sons, Ltd.

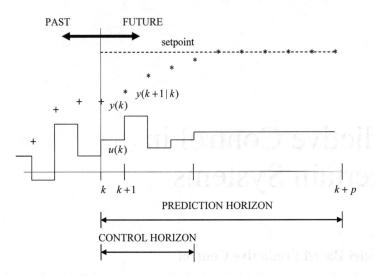

Figure 7.1 Model-based predictive control.

analytically (Mahfouf *et al.*, 2002; He *et al.*, 2005; Feng, 2006). Another approach is to use a non-linear system (e.g. a physical model or a neural network) to model the plant and a direct search algorithm to find the optimal control signal (Onnen *et al.*, 1997).

7.2 Fuzzy Approaches to Model-Based Control of Uncertain Systems

One way of dealing with the uncertainty is to use fuzzy decision-making to determine the appropriate control action (Sousa and Kaymak, 2002). One approach (see Section 7.2.1) to taking account of modelling errors, which can be used if the open-loop system and its inverse are stable, is to assume that the plant can be described by one or more linear difference equations whose parameters are fuzzy intervals (a so-called *fuzzy interval system*). Another approach (see Section 7.2.2) is to use fuzzy decision-making in combination with a fuzzy model that is capable of generating a fuzzy output, whose distribution reflects the uncertainties associated with the prediction of the future behaviour of the plant.

7.2.1 Inverse Control of Fuzzy Interval Systems

First consider the solution of the fuzzy equation $BX + A = C$ where A, B, C and X are fuzzy intervals and B is either entirely positive or entirely negative. Let $a^\alpha = [a^{\alpha-}, a^{\alpha+}]$, $b^\alpha = [b^{\alpha-}, b^{\alpha+}]$ and $c^\alpha = [c^{\alpha-}, c^{\alpha+}]$ be the α-cuts of the fuzzy intervals. It then follows that $b^\alpha x^\alpha + a^\alpha = c^\alpha$. It can be shown that, if a solution exists, this equation can be solved using a two-step approach based on α-cuts and interval arithmetic (Boukezzoula *et al.*, 2006).

Let

$$d^\alpha = b_\alpha . x^\alpha = c^\alpha - a^\alpha \tag{7.1}$$

We then have that

$$d^\alpha = [d^{\alpha-}, d^{\alpha+}] = [(c^{\alpha-} - a^{\alpha-}), (c^{\alpha+} - a^{\alpha+})] \tag{7.2}$$

and

$$x^\alpha = \frac{d^\alpha}{b^\alpha} = [x^{\alpha-}, x^{\alpha+}] \tag{7.3}$$

where

$$x^{\alpha-} = \frac{Mid\{d^\alpha\} - Rad\{d^\alpha\}.sign[Mid\{b_\alpha\}]}{Mid\{b^\alpha\} - Rad\{b^\alpha\}.sign[Mid\{d^\alpha\} - Rad\{d^\alpha\}.sign[Mid\{b_\alpha\}]]} \tag{7.4}$$

and

$$x^{\alpha+} = \frac{Mid\{d^\alpha\} + Rad\{d^\alpha\}.sign[Mid\{b^\alpha\}]}{Mid\{b^\alpha\} - Rad\{b^\alpha\}.sign[Mid\{d^\alpha\} + Rad\{d^\alpha\}.sign[Mid\{b_\alpha\}]]} \tag{7.5}$$

where

$$Mid\{d^\alpha\} = \frac{(d^{\alpha-} + d^{\alpha+})}{2} \quad \text{and} \quad Rad\{d^\alpha\} = \frac{(d^{\alpha+} - d^{\alpha-})}{2} > 0$$

Now consider a stable time-invariant single-input single-output linear uncertain system which can be represented by an autoregressive moving average (ARMA) model with parameters that are fuzzy intervals. Thus,

$$y(k+1) = \sum_{i=1}^{n} A_i y(k-i+1) + \sum_{j=0}^{m-1} B_j u(k-j) \tag{7.6}$$

where $y(k)$ and $u(k)$ correspond to the output and input of the plant and A_i and B_j are fuzzy intervals representing the uncertain parameters of the plant.

Let the desired value of $y(k+1)$ be $Y_d(k+1)$, a fuzzy interval. The control objective can be viewed as determining a fuzzy control interval such that values of u in its α-cut are able to maintain the controlled variable within the α-cut of the desired value. The control problem is solved by finding the solution to the fuzzy equation for all values of α:

$$y_d^\alpha(k+1) = \sum_{i=1}^{n} a_i^\alpha y^\alpha(k-i+1) + \sum_{j=1}^{m-1} b_j^\alpha u^\alpha(k-j) + b_0^\alpha u^\alpha(k) \tag{7.7}$$

using the method described by Equations (7.1)–(7.5).

The inverse controller has been applied to a first-order linear discrete-time system with parameters that are triangular fuzzy intervals and a fuzzy desired value that is also a triangular fuzzy interval (Boukezzoula et al., 2006). The actuator was assumed to saturate when $u = \pm 7.5$, and the time-varying modal value of the desired value was given by $y_d^m = \frac{2}{3}[\sin(2\pi k/50) + \sin(2\pi k/75)]$. Simulation results showed that the fuzzy output tracked the fuzzy desired value perfectly, without saturating the actuator. It should be noted that the control output is a fuzzy interval, which must be defuzzified if this control strategy is to be implemented on a real plant.

7.2.2   Fuzzy Model-Based Predictive Control

There are three basic types of fuzzy MPC schemes: those whose design is based on goals
and constraints that are fuzzy, but use a conventional mathematical model to predict the plant
output (Kaymak *et al.*, 1997; Belarbi and Megri, 2007); those that are based on a fuzzy model,
fuzzy goals and fuzzy constraints, but assume that the output of the fuzzy model is crisp or
defuzzified (Da Costa Sousa and Kaymak, 2001); and those that are designed assuming that
the model predictions, the goals and the constraints are all fuzzy (Thompson and Dexter, 2001;
Vukovic, 2001).

Fuzzy model-based predictive control (FMPC) is an application of multi-stage decision-
making (see Section 6.3) where the set of alternatives constitute the different possible control
actions, the decision criteria are the control performance criteria and the termination time is a
generalization of the prediction horizon used in conventional MPC (Sousa and Kaymak, 2002).

A common control goal is the minimization of the tracking error subject to a fuzzy constraint
on the control signal or rate of change of the control signal.

The control policy is a sequence of control actions (which may include making no change
to the control signal if the control horizon is shorter than the prediction horizon) for the entire
prediction horizon H_p of the control scheme. Each fuzzy goal and fuzzy constraint constitutes
a decision criterion μ_{d_j}, where $j = 1, \ldots N$ and N is the total number of goals and constraints
at each time step. The total number of decision criteria is given by the product of N and H_p.
The membership value for the entire control sequence μ_π is given by:

$$
\begin{aligned}
\mu_\pi = {} & (\mu_{d_{11}} \otimes_g \ldots \otimes_g \mu_{d_{1q}}) \otimes (\mu_{d_{1(q+1)}} \otimes_c \ldots \otimes_c \mu_{d_{1N}}) \otimes \ldots \\
& \otimes (\mu_{d_{H_p1}} \otimes_g \ldots \otimes_g \mu_{d_{H_pq}}) \otimes (\mu_{d_{H_p(q+1)}} \otimes_c \ldots \otimes_c \mu_{d_{H_pN}})
\end{aligned}
\tag{7.8}
$$

where q is the number of goals, $\otimes_g$ is the aggregation operator for combining the goals, $\otimes_c$ is
the aggregation operator for combining the constraints and $\otimes$ is the aggregation operator for
combining the aggregated goals and constraints at each time step over the prediction horizon.

The use of various aggregation operators has been proposed (the min operator, the product
operator, parametric t-norms such as the Yager t-norm and the generalized mean). Averaging
operators are sometimes used for aggregation in some applications. However, a t-norm must
be used for aggregation if all goals and constraints must be satisfied simultaneously to some
extent. The use of a single aggregation operator reduces the complexity, but it can constrain
the designer in terms of his/her ability to make tradeoffs between the different goals and
constraints (Da Costa Sousa and Kaymak, 2001).

The optimal sequence of control actions is found by maximizing μ_π. However, as explained
in Section 6.3, the optimization problem is usually non-convex and a numerical technique
must be used to find the optimum. It should be noted that the computation time associated with
finding the optimal sequence of control actions grows exponentially with the control horizon.

Conventional MPC schemes usually use a quadratic cost function. More complex objective
functions, which can more closely describe the true control objectives and constraints, may
however be more appropriate in non-linear applications with non-convex constraints. For
example, the goal shown in Figure 7.2 might be used if the actuator is rate limited and even
modest changes in its position are to be avoided. Note that an asymmetric goal might be more
appropriate in applications in which the physical limitations imposed on the operation of the
plant vary, depending on whether the control signal is increasing or decreasing.

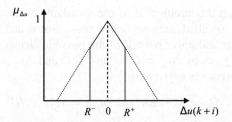

Figure 7.2   Example of a non-quadratic goal.

Different criteria can be used at different time steps. For example, the support for the fuzzy set describing the error goal might become smaller as the end of the prediction horizon is approached.

The use of membership functions to define the control objective reduces the effort required to choose appropriate numerical values for weighting factors for the cost function used in conventional MPC.

It should be noted that, in this type of fuzzy controller, human knowledge is used to specify the control objectives (goals) and constraints rather than the control rules themselves.

7.3   Practical Issues Associated with Multi-Step Fuzzy Decision-Making

In practice, there are a number of important issues that must be taken into account when designing a control scheme of this type.

7.3.1   Limiting the Accumulation of Uncertainty

Short-term predictions must normally be used in predictive control when the available model is not very accurate and cannot predict plant behaviour a large number of steps ahead with any accuracy. Two approaches to avoiding this problem have been suggested: minimize the prediction horizon (Da Costa Sousa and Kaymak, 2001) or base the control on one-step-ahead prediction (Vukovic, 2001).

7.3.2   Avoiding Excessive Computational Demands When Using Enumerative Search Optimization

Enumerative search methods such as branch and bound (B&B) involve searching for the best control policy in a discretized control space. Computational demands will grow with the size of the prediction and control horizons and with the number of goals and constraints. The size of the search space is a tradeoff between the number of control alternatives and computational complexity. Oscillations (limit cycles or chattering) will occur at constant setpoints if too few control alternatives are available, whereas a large number of control alternatives may result in unacceptably long search times.

One way of alleviating this problem is to use an adaptive set rather than a fixed set of control alternatives (the so-called *fuzzy predictive filter*) (Sousa and Setnes, 1999; Mendonca *et al.*, 2004). If the upper and lower bounds of the possible changes in the control signal at the kth sample time are given by $\Delta u_k^+ = U^+ - u(k-1)$ and $\Delta u_k^- = U^- - u(k-1)$, the set of allowable control alternatives is given by

$$\Omega_k = \{0, \lambda_l \Delta u_k^+, \lambda_l \Delta u_k^- | l = 1, 2, \dots N\} \tag{7.9}$$

where l is an application-dependent parameter that specifies the number of control alternatives and λ determines the allowable values of the changes.

Example 7.1

Only the following changes to the control signal are permitted if $l = 3$ and $\lambda_l = \frac{1}{4^{l-1}}$:

$$\Omega_k = \left\{ \Delta u_k^-, \frac{\Delta u_k^-}{4}, \frac{\Delta u_k^-}{16}, 0, \frac{\Delta u_k^+}{16}, \frac{\Delta u_k^+}{4}, \Delta u_k^+ \right\} \tag{7.10}$$

These values are multiplied by a scaling factor or gain to generate the adaptive set

$$\Omega_k^* = \gamma(k).\Omega_k \tag{7.11}$$

where $0 \leq \gamma(k) \leq 1$ depends on the size of the predicted error and the rate of change of the predicted error. One approach (Sousa and Setnes, 1999) is to base $\gamma(k)$ on the degree of membership $\mu_e(k)$ of the predicted value of the error at the end of the current prediction horizon in the fuzzy set 'error is small' and on the degree of membership $\mu_{\Delta e}(k)$ of the current value of the rate of change of the error in the fuzzy set 'change in error is small'.

For example, let $\gamma(k) = 1 - Min\{\mu_e(k), \mu_{\Delta e}(k)\}$ so that the gain becomes small when both the error and the rate of change of error are small (that is when the system is close to steady-state), and approaches unity when either the error or the rate of change of error are large.

7.3.3 Avoiding Excessive Computational Demands When Using Evolutionary Algorithms

The main disadvantage of evolutionary algorithms such as genetic or particle swarm algorithms (Solis *et al.*, 2006) is that they need a certain number of generations to evolve before they converge to the optimum control strategy (Nunez *et al.*, 2009); they therefore cannot be applied to systems with fast dynamics and short sampling times (Sarimveis and Bafas, 2003).

Various techniques have been proposed to speed up the genetic algorithms. For example, past evolutions can be used to help to initialize the next population. One approach is to keep the best individuals of the previous generation and randomly initialize the others. Because the genes correspond to predictions at the different time steps into the future, they should be shifted before they are used to initialize the chromosome of an individual in the next population (Onnen *et al.*, 1997). Another way of speeding up the search is to remember the best control policy determined for previously encountered situations. However, in order to maintain genetic diversity, identical solutions must only be stored once. Another approach is to assess the quality of the results online and to check a terminating condition (e.g. absolute fitness, rate of convergence of the maximum fitness or the fitness of the individual to be used as the current control signal) so that the evolution can be aborted once a predefined level of optimality has been reached or a real-time constraint is about to be exceeded (Onnen *et al.*, 1997).

Alternatively, special genetic operators such as niching methods (Solis *et al.*, 2006) can also be used to increase the efficiency of the search (Onnen *et al.*, 1997), which is always a tradeoff between the computational demands and control performance. Computation times can also be improved by minimizing the number of individuals in the population, and limiting the maximum number of evolutions. It should be noted that genetic algorithms have been shown to outperform the B&B method when the population size is small (Nunez *et al.*, 2009) or the control horizon is long (Onnen *et al.*, 1997; Causa *et al.*, 2008).

7.3.4 Handling Infeasibility

One method of ensuring that the control policy is always feasible is to define the fuzzy goals in such a way that the membership grades can never be zero, indicating that all states of the system are allowable though not necessarily desirable (Da Costa Sousa and Kaymak, 2001). Another approach is to increase the supports of the fuzzy sets describing the goals until a feasible control strategy can be found (Thompson and Dexter, 2005).

Alternatively, unfeasible situations can be handled as a special case. For example, a check can made to detect situations when even the extreme values of the control signal cannot satisfy the goal, in which case the control signal can be set to its maximum or minimum value as appropriate (see Section 7.5).

The control policy may also appear to be unfeasible if the goal is not very fuzzy and a small number of alternatives are used for the value of the control signal, or one or more of the rules in the fuzzy model are missing as a result of incomplete training (see Section 10.2). If either of these situations occurs, the only sensible option is to leave the control signal at its previous value. This can, however, result in limit-cycle behaviour.

7.3.5 Choosing the Weighting in Multi-Criteria Cost Functions

Weights are sometimes used to represent the relative importance the decision-maker must attach to the different goals and constraints (Mendonca *et al.*, 2006). As the weights represent the preferences of the user in terms of the influence of each criterion on the choice of the optimal control strategy, it is usually easier to choose appropriate weights if the specification of the criteria is separated from the specification of the aggregation operation. Weighted aggregation of fuzzy sets can be achieved by modifying the membership functions before they are aggregated. For example, using the weighted extension of the product t-norm:

$$\mu_\pi = \prod_{j=1}^{N} [\mu_{d_j}]^{\omega_j} \tag{7.12}$$

The weights are usually normalized in order to find a common metric for defining the importance of the various criteria. The most common normalization is to require the sum of the weights to be unity. Clearly an increase in the importance of one criterion must then be accompanied by a decrease in the importance of the other criteria. A 'greedy' heuristic is often used to choose appropriate values for the weights. For example, the values of all of the weights are initially set to the same value. Each weight is then decreased in turn to determine which criterion has the greatest effect on improving the control performance. The weight that affected the performance most is then further reduced. If the performance improves, that weight is reduced again until no further improvement is observed. That weight is then fixed at its optimal value. The process is then repeated for the next most important weight, and so on.

7.3.6   Dealing with Hard Constraints

Care must be taken to make sure that the use of particular aggregation operators does not lead to the violation of any hard constraints. One approach is to use t-norms for aggregation and membership functions with bounded support to define the fuzzy constraints that must not be violated (Da Costa Sousa and Kaymak, 2001).

7.4   A Simplified Approach to Fuzzy FRM-Based Predictive Control

A block diagram of the control scheme is shown in Figure 7.3. The fuzzy FRM generates a fuzzy description of the k-step-ahead predicted process output $Y_m(n + k)$ from: candidate

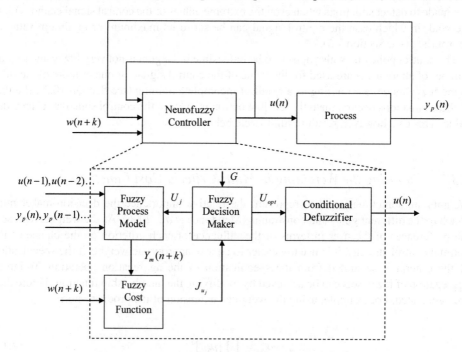

Figure 7.3   Neurofuzzy model-based predictive control.

fuzzy values of the current control signal U_j generated by the fuzzy decision-maker; the previous values of the control signal $u(n-1)$, $u(n-2)$ and so on; and measured values of the current and previous process output $y_p(n)$, $y_p(n-1)$, and so on. (Note that upper case is used to denote a fuzzy variable or quantity.) As is not the case in most applications of neurofuzzy models, the output from the model is not defuzzified but remains in the fuzzy domain to retain information on the uncertainties associated with predicting the behaviour of the process under control. A fuzzy cost J_{u_j} is computed by considering the similarity of the setpoint trajectory $W(n+k)$, which may or may not be fuzzy, and the fuzzy predictions. The fuzzy decision-maker generates the optimum fuzzy control signal U_{opt} by determining the degree to which the fuzzy cost associated with each of the candidate fuzzy values of the control signal satisfies a pre-defined fuzzy goal G. A conditional defuzzification scheme then determines what, if any, changes are to be made to the position of the control actuator. The entire decision-making process takes place in the fuzzy domain.

7.4.1   The Fuzzy Decision-Maker

The main function of the fuzzy decision-maker is to choose the optimal control signal that will drive the process. The extent to which the fuzzy cost function satisfies the fuzzy goal is used to determine the fuzzy set describing the optimal control signal U_{opt}. For example, consider the trapezoidal fuzzy goal function $G(e)$ shown in Figure 7.4.

At each time step, a discrete membership function for the optimal control signal is found by comparing the similarity of $G(e)$ with the fuzzy cost functions $J_{u_j}(e)$ generated by exhaustively stepping through each of the candidate fuzzy values of the control signal U_j. The degree of similarity between the fuzzy set describing the fuzzy cost and the fuzzy set describing the goal is taken as the ratio of the common area between the two membership functions to the total area under the membership function describing the fuzzy cost.

The support of the fuzzy goal establishes the importance of the criterion upon which the goal is defined. A wider membership function for the goal implies that criterion is less important and dissatisfaction of the goal is more tolerable. Specification of the width allows the control

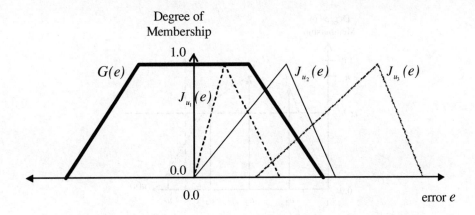

Figure 7.4   Use of fuzzy similarity in the fuzzy decision-making scheme.

engineer to stipulate directly the importance of a particular criterion and can be used to establish its relative importance in multi-criteria control problems (Thompson and Dexter, 2001).

The fuzzy decision-maker automatically adjusts the width of the membership function of the goal so that, for the particular control problem, there always exists at least one candidate value of the control signal which results in a fuzzy cost that satisfies the goal to some extent. In applications involving multiple goals, maintaining a constant ratio of the widths preserves the relative importance of the goals when they are relaxed to make all of them achievable.

The simplified FRM-based control scheme uses one-step-ahead prediction (Vukovic, 2001), a single goal based on the tracking error and exhaustive search based on a small number of candidate values of the control signal.

7.4.2 Conditional Defuzzification

The optimum fuzzy control signal generated by the fuzzy decision-maker must be defuzzified to produce a crisp value of the control signal. One way of avoiding unnecessary changes to the control signal produced by a fuzzy FRM-based controller is to use a conditional defuzzification scheme.

Conditional defuzzification schemes first check whether actuator movement is justified, and defuzzify only if this condition is met. If repositioning of the actuator is not justified, the previous value of the control signal is used as the current value and no actuator movement occurs. This type of scheme has a greater potential for reducing control activity and actuator wear than traditional approaches.

The conditional defuzzification scheme proposed here is illustrated in Figure 7.5. The actuator will not be repositioned if $\mu_{opt}(u(n-1))$, the degree of membership of the fuzzy optimum control signal given the previous value of the control signal, is greater than a user-specified fraction α of the height h of the discrete membership function describing the optimum fuzzy control signal. If $\mu_{opt}(u(n-1)) < \alpha h$, a new value of the control signal is calculated from the optimal fuzzy control signal using a conventional defuzzification scheme. The choice of the conditional defuzzification threshold α is determined by the level of control activity that is acceptable in a particular application.

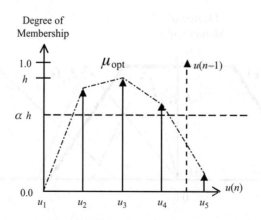

Figure 7.5   Conditional defuzzification based on linear interpolation.

The values of μ_{opt} are available only at discrete points u_j along the universe of discourse, which correspond to the apexes of the fuzzy reference sets used to describe the candidate fuzzy values of the control signal ($j = 1, 2, \ldots, 5$ in this simple example). Since the current value of the control signal $u(n{-}1)$ may lie between the apexes, an estimate of $\mu_{\text{opt}}(u(n-1))$ is obtained by linearly interpolating between the adjacent values of $\mu_{\text{opt}}(u_j)$.

The value of the conditional defuzzification threshold and the uncertainty associated with the optimal control signal will determine the level of the resulting control activity.

It should be noted that most of the conditional defuzzification schemes used in the examples given in this and later chapters use height defuzzification to defuzzify the optimal fuzzy control signal. It may however be better to use the value of the control signal with the highest probability (or possibility), rather than an estimate of the mean value of the distribution, if the probability (or possibility) distribution of the optimal control signal is highly asymmetric and/or multi-modal (Herzallah and Lowe, 2003, 2008).

Example 7.2

Figure 7.6 shows the output generated by the conditional defuzzification scheme based on height defuzzification with a threshold of 0.5 when it is used to defuzzify the fuzzy output generated by singleton fuzzification of a sinusoidally varying crisp signal using 11 equally spaced fuzzy reference sets with triangular membership functions. It can be seen that the time interval between successive decisions to defuzzify increases as the rate of change of the input decreases.

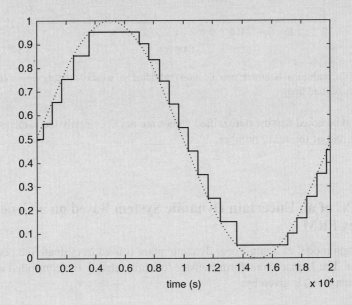

Figure 7.6 Conditional height defuzzification (solid line) of a fuzzified sine wave (dotted line).

Example 7.3

Figure 7.7 shows the output generated by the conditional defuzzification scheme with a threshold of 0.5 when it is used to defuzzify a fuzzy number, which has a triangular membership function with a base of 0.6 units and an apex at a location that varies sinusoidally. It can be seen that, although the output generated during the first half cycle is a mirror image of that generated during the second half cycle, the output generated when the input is increasing is different to that generated when the input is decreasing.

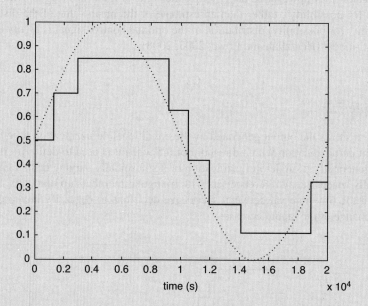

Figure 7.7   Conditional height defuzzification (solid line) of a fuzzy number whose centre varies sinusoidally (dotted line).

It should be noted that the defuzzified values are not necessarily the same as the value at the centroid of the fuzzy number.

7.5   FMPC of an Uncertain Dynamic System Based on a Generic Fuzzy FRM

A Hammerstein model of a non-linear dynamic plant is used to evaluate the performance of the controller. The Laplace transform transfer function relating the controlled output $y(t)$ to the control signal $u(t)$ is given by:

$$Y(s) = L(f(u(t))) \left[\frac{0.899}{0.9d + 0.8} \right] \left[\frac{1}{1 + 200s} \right] \tag{7.13}$$

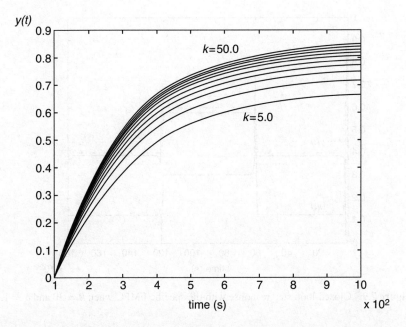

Figure 7.8   Unit step response of each member of the class used to train the generic model.

where the static non-linearity $f(u(t)) = \frac{1}{3.434} \ln(30\,u(t) + 1)$ and d is an input disturbance, which is held at a constant value of 0.275 during testing. The fuzzy goal is defined by a discrete triangular membership function of the form:

$$[0.0/(r - 0.050) \quad 0.5/(r - 0.025) \quad 1.0/r\,0.5/(r + 0.025) \quad 0.0/(r + 0.050)].$$

The position of the apex r of the fuzzy goal is stepped up and down from 0.4 to 0.6 during testing.

The fuzzy FRM used in the MPC is a discrete-time first-order autoregressive model with two inputs, $u(i - 1)$ and $y(i - 1)$, and a sample time of 100 s. The 21 reference sets are used to describe each of the two inputs. The fuzzy output is described by 41 reference sets.

The training data used to identify the specific fuzzy FRM are obtained by simulating a plant with the same dynamics and static non-linearity as the plant used for testing. The training data used to identify a generic fuzzy FRM are obtained by simulating ten plants with the same dynamics but with different static non-linearities, $f(u(t)) = \frac{1}{\ln(k+1)} \ln(ku(t) + 1)$, where the value of k varies from 5 to 50 in steps of 5. Figure 7.8 shows the step responses of each of the simulated plants.

The training data are generated using a control signal in the range 0.0 to 1.0, which is produced by clipping low-pass filtered ($\tau = 100$s) white noise which is uniformly distributed in the range −2.0 to +3.0. The disturbance is held at a constant value of 0.275 during each training period (100 000 samples). The resulting fuzzy models are post-processed five times using the method described in Section 5.5.1.

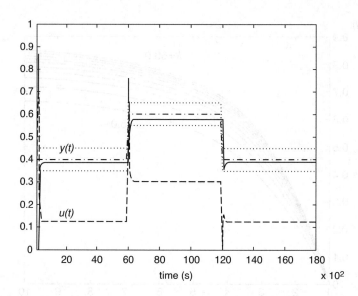

Figure 7.9 Closed-loop step response with the specific FMPC when $k = 30$ and $\alpha = 1.0$.

Sum-product inferencing is used to predict the one-step-ahead fuzzy prediction of the plant output over a Universe of Discourse from 0.0 to 1.0. The fuzzy prediction is normalized so that the sum of the possibilities is unity. If no rules are fired, all values of the next plant output are assumed to be equally possible.

The degree of satisfaction of the goal is the overlap area divided by the area under the normalized fuzzy prediction of the plant output. The fuzzy decision-maker determines the degree of satisfaction of the goal for 11 candidate values of the control signal. If at least one candidate value of the fuzzy control signal satisfies the goal, the fuzzy control signal is conditionally defuzzified using height defuzzification.

If none of the candidate values of the control signal satisfy the goal, the lower and upper bounds of the support of the fuzzy goal are found and u is set to u_{max} or u_{min} (depending on whether the lower bound of the support of the fuzzy prediction for the minimum value of the control signal is closer to the upper bound of the support of the fuzzy goal than the upper bound of the support of the fuzzy prediction for the maximum value of the control signal is to the lower bound of the support of the fuzzy goal). It should be noted that this method assumes that increasing u will increase y.

As can be seen in Figure 7.9, the closed-loop transient behaviour (solid line) is under-damped and there is only a small steady-state error when FMPC is based on the specific model and the fuzzy control signal (dashed line) is defuzzified at every sampling instant.

There is little change in the closed-loop transient behaviour when the FMPC is based on the generic model. However, the size of the steady-state error is larger, and it varies depending on both the operating point and the plant used for testing the controller (see Figures 7.10 and 7.11).

Much fuzzier predictions are generated by the generic model (see Figure 7.12) and, as expected, the transient control activity is reduced when the threshold for conditional defuzzi-fication is lowered to 0.4 (see Figure 7.13).

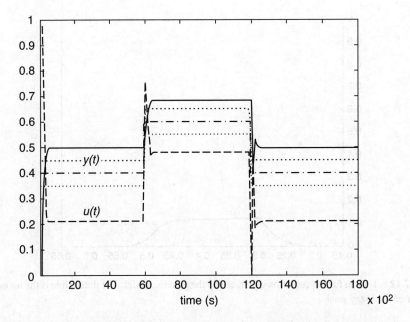

Figure 7.10 Closed-loop step response with the generic FMPC when $k = 30$ and $\alpha = 1.0$.

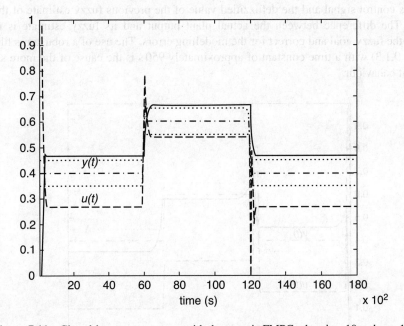

Figure 7.11 Closed-loop step response with the generic FMPC when $k = 10$ and $\alpha = 1.0$.

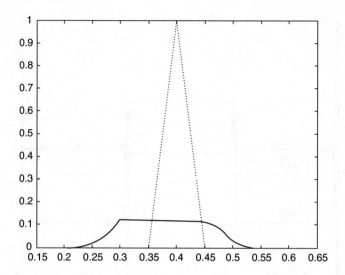

Figure 7.12   Typical fuzzy prediction generated by the generic model. The dotted line is the membership function of the fuzzy goal.

As can be seen in Figures 7.14 and 7.15, the steady-state errors can be reduced by embedding the FMPC in a fuzzy internal model control (FIMC) structure (see Section 9.2). In this design, the internal model generates a fuzzy estimate of the current plant output using the previous control signal and the defuzzified value of the previous fuzzy estimate of the plant output. The difference between the actual plant output and its fuzzy estimate is used to modify the fuzzy goal and correct for the modelling errors. The use of a robustness filter (see Section 9.1.3) with a time constant of approximately 950 s is the cause of the more sluggish transient behaviour.

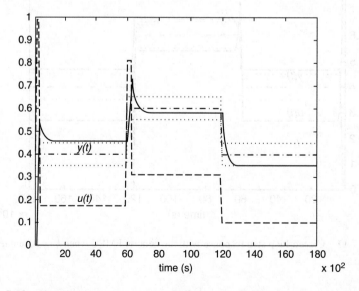

Figure 7.13   Closed-loop step response with the generic FMPC when $k = 30$ and $\alpha = 0.4$.

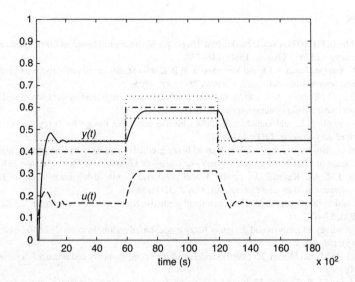

Figure 7.14   Closed-loop step response with the generic FIMC when $k = 30$ and $\alpha = 1.0$.

7.6   Summary

In this chapter, it has been shown how a fuzzy FRM and fuzzy decision-making can be used to design a model-based predictive control scheme that is suitable for controlling information-poor systems. Two fuzzy approaches to the model-based control of uncertain systems have been described and the practical limitations of multi-step fuzzy decision-making (the accumulation of uncertainty, the excessive computational demands and infeasibility issues) have been discussed. A simplified approach to fuzzy FRM-based predictive control has also been described and a method of conditional defuzzification has been suggested.

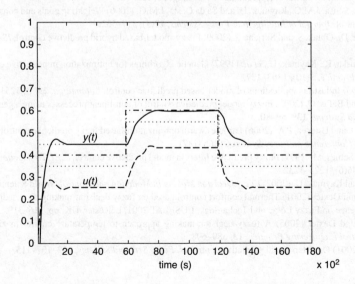

Figure 7.15   Closed-loop step response with the generic FIMC when $k = 10$ and $\alpha = 1.0$.

References

Belarbi, K. and Megri, F. (2007) A stable model-based fuzzy predictive control based on fuzzy dynamic programming. *IEEE Transactions on Fuzzy Systems*, **15**(4), 746–754.

Bloemen, H.H.J., Van Den Boon, T.J.J. and Verbruggen, H.B. (2001) Model-based predictive control of Hammerstein-Wiener systems. *International Journal of Control*, **75**(5), 482–495.

Borrelli, F., Baotic, M., Bemporad, A. and Morari, M. (2005) Dynamic programming for constrained optimal control of discrete-time linear hybrid systems. *Automatica*, **41**(10), 1709–1721.

Boukezzoula, R., Foulloy, L. and Galichet, S. (2006) Inverse controller design for fuzzy interval systems. *IEEE Transactions on Fuzzy Systems*, **14**(1), 111–124.

Causa, J., Karer, G., Nunez, A., *et al.* (2008) Hybrid fuzzy predictive control based on genetic algorithms for the temperature control of a batch reactor. *Computers and Chemical Engineering*, **32**(12), 3254–3263.

Da Costa Sousa, J.M. and Kaymak, U. (2001) Model predictive control using fuzzy decision functions. *IEEE Transactions on Systems, Man, and Cybernetics, Part 2*, **31**(1), 54–65.

Demircioglu, H. and Yavuzyilmaz, C. (2002) Constrained predictive control in continuous time. *IEEE Control Systems Magazine*, **22**(4), 57–67.

Feng, G. (2006) A survey on analysis and design of fuzzy model-based control systems. *IEEE Transactions on Neural Networks*, **14**(5), 676–697.

Garcia, C.E., Prett, D.M. and Morari, M. (1989) Model predictive control: theory and practice – a survey. *Automatica*, **25**(3), 335–348.

He, M., Cai, W-J. and Li, S-Y. (2005) Multiple fuzzy model-based temperature predictive control for HVAC systems. *International Journal of Information Sciences*, **169**, 155–174.

Herzallah, R. and Lowe, D. (2003) Robust control of nonlinear stochastic systems by modelling conditional distributions of control signals. *Neural Computing and Applications*, **122**, 98–108.

Herzallah, R. and Lowe, D. (2008) A Bayesian perspective on stochastic neurocontrol. *IEEE Transactions on Neural Networks*, **19**(5), 914–924.

Kaymak, U., Sousa, J.M. and Verbruggen, H.B. (1997) A comparative study of fuzzy and conventional criteria in model-based predictive control. *IEEE International Conference on Fuzzy Systems*, 907–914.

Magni, L. and Scattolini, R. (2004) Stabilizing model predictive control of nonlinear continuous time systems. *Annual Reviews in Control*, **28**(1), 1–11.

Mahfouf, M., Kandiah, S. and Linkens, D.A. (2002) Fuzzy model-based predictive control using an ARX structure with feedforward. *Fuzzy Sets and Systems*, **125**, 39–59.

Mendonca, L.F., Sousa, J.M. and Sa da Costa, J.M.G. (2004) Optimization problems in multivariable fuzzy predictive control. *International Journal of Approximate Reasoning*, **36**, 199–221.

Mendonca, L.F., Sousa, J.M.C., Kaymak, U. and Sa da Costa, J.M.G. (2006) Weighting goals and constraints in fuzzy predictive control. *Journal of Intelligent & Fuzzy Systems*, **17**(5), 517–532.

Nunez, A., Saez, D., Oblak, S. and Skrjanc, I. (2009) Fuzzy-model-based hybrid predictive control. *ISA Transactions*, **48**(1), 24–31.

Onnen, C., Babuska, R., Kaymak, U., *et al.* (1997) Genetic algorithms for optimization in predictive control. *Control Engineering Practice*, **5**(10), 1363–1372.

Richalet, J. (1993) Industrial applications of model-based predictive control. *Automatica*, **29**(5), 1251–1274.

Sarimveis, H. and Bafas, G. (2003) Fuzzy model predictive control of non-linear processes using a genetic algorithm. *Fuzzy Sets and Systems*, **139**, 59–80.

Solis, J., Saez, D. and Estevez, P.A. (2006) Particle swarm optimization-based fuzzy predictive control strategy. *IEEE International Conference on Fuzzy Systems*, 1866–1871.

Sousa, J.M. and Setnes, M. (1999) Fuzzy predictive filters in model predictive control. *IEEE Transactions on Industrial Electronics*, **46**(6), 1225–1232.

Sousa, J.M.C. and Kaymak, U. (2002) *Fuzzy Decision Making in Modeling and Control*, World Scientific, Singapore.

Thompson, R. and Dexter (2001) Thermal comfort control based on fuzzy decision-making. Proceedings of International Conference on Fuzzy Logic and Technology, EUSFLAT 2001, Leicester, UK, pp. 316–319.

Thompson, R. and Dexter (2005) A fuzzy decision-making approach to temperature control in air conditioning systems. *Control Engineering Practice*, **13**, 689–698.

Vukovic, P.D. (2001) One-step ahead predictive controller. *Fuzzy Sets and Systems*, **122**, 107–115.

8

Incorporating Fuzzy Inputs

The two main causes of the uncertainty associated with the inputs to a controller in an information-poor system are imprecision in the definition of the setpoint and measurement noise. Figure 8.1 shows the location of these inputs (the setpoint, the measured plant output and any measured disturbance) in the control system. This chapter will describe techniques for dealing with uncertainties associated with the setpoint and the measured values of the plant output. Ways of taking measured disturbances into account will be considered in the next chapter.

8.1 Fuzzy Setpoints and Fuzzy Measurements

8.1.1 Fuzzy Setpoints

As can be seen in Figure 8.2, an imprecise setpoint can be considered as a fuzzy control objective.

For example, when controlling thermal comfort in a building, it is more meaningful to specify the control objective as "to maintain the air temperature at about 24°C" rather than give a precise temperature setpoint. The setpoint is then a fuzzy set defined by an appropriate membership function. The grade of membership of this fuzzy set can be used as a measure of the degree of satisfaction of the control objective and the α-cut of the fuzzy setpoint set indicates the range of values of the air temperature that satisfy the control objective by at least α.

8.1.2 Fuzzy Measurements

Direct use of a noisy signal can result in unacceptable levels of control activity, particularly when the controller uses the time derivative of the signal as one of its inputs. One obvious solution is to use a low-pass filter to smooth the noisy signal before it is used by the controller. There are two problems with this approach: (1) it cannot be used to remove measurement noise in the same frequency range as the controlled variable without affecting the control

Monitoring and Control of Information-Poor Systems: An Approach based on Fuzzy Relational Models, First Edition.
Arthur L. Dexter.
© 2012 John Wiley & Sons, Ltd. Published 2012 by John Wiley & Sons, Ltd.

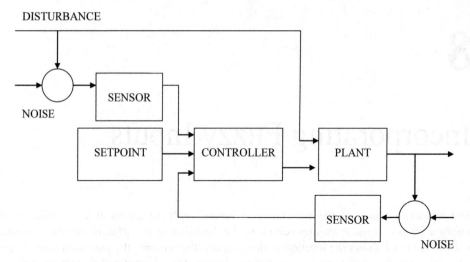

Figure 8.1 Location of the controller's input uncertainties.

performance; and (2) simple low-pass filters do not take the probability distribution of the noise into account.

An alternative might be to estimate the probability distribution from a histogram of the measurement noise and to calculate the modal value of the measurement. However, this approach still fails to take account of the available information about the confidence that should be assigned to particular values of the measured signal. A more attractive option is either to transform the histogram into a fuzzy set and to feedback the fuzzy signal and its derivatives to the input of the controller (Driankov *et al.*, 1994), or to use prior knowledge of the mean and variance of the measurement noise to convert the crisp measured value into a fuzzy set (Foulloy and Galichet, 2003).

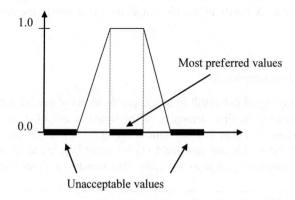

Figure 8.2 Example of a fuzzy setpoint.

In the former case, a histogram is generated from the noisy data and normalized with respect to its maximum likelihood value to convert it into a possibility distribution (Driankov *et al.*, 1994). The resulting membership function indicates the possibility of different values of the measurand, the degree of asymmetry in the distribution resulting from non-linearities in the system and the occurrence of multiple peaks in the distribution.

8.2   Fuzzy Measures of the Tracking Error and its Derivative

The calculation of the error signal used in classical feedback control loops is more complex when either or both the setpoint and the measured value of the controlled variable are fuzzy. Fuzzy matching schemes or distance measures can be used to generate a measure of the extent to which the two fuzzy sets are similar (see Section 6.2.3). However, most controllers also need to differentiate between the case when the measurement is below the setpoint and when it is above it.

Possibility can be used to determine the degree of matching and sign to indicate whether the measurement is greater or less than the setpoint. For example, let $\varepsilon = \text{sign}(e).[1 - \sup_x \min\{\mu_R(x), \mu_M(x)\}]$ where $e = r - m$, the difference between the modal values of $\mu_R(x)$ and $\mu_M(x)$, sign $= +1$ if $e > 0$ and sign $= -1$ if $e \le 0$ and ε is a fuzzy singleton.

Example 8.1

Consider the fuzzy representations of the setpoint and the plant output shown in Figure 8.3. The way in which the resulting error signal ε varies as the difference in the modal values e changes is shown in Figure 8.4. Note that $\varepsilon = 0$ when the fuzzy setpoint is a subset of the fuzzy plant output, and $|\varepsilon| = 1$ if the supports of the fuzzy setpoint and the fuzzy measurement do not overlap.

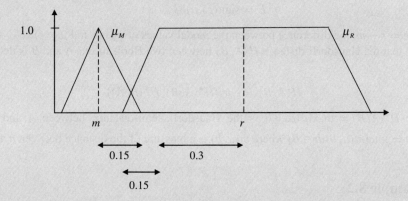

Figure 8.3   Definition of the fuzzy representations of the setpoint and output measurement.

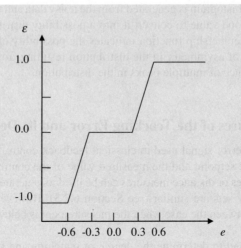

Figure 8.4   Relationship between the resulting crisp values of e and ε.

Alternatively, the extension principle can be used to generate a fuzzy measure of the difference between two fuzzy sets,

$$\mu_E(\varepsilon) = \sup_{\varepsilon = x_1 - x_2} \{\min[\mu_R(x_1), \mu_M(x_2)]\} \tag{8.1}$$

or a fuzzy distance measure can be applied to the α-cuts of the two sets.

For example, an extension of the Hausdorff distance (Chaudhuri and Rosenfeld, 1999) can be used to determine the fuzzy distance between $H(R, M)$ the fuzzy setpoint and the fuzzy measurement and sign to indicate whether the measurement is greater or less than the setpoint, in which case

$$E = \text{sign}(e).\mu_{H(R,M)}(\varepsilon) \tag{8.2}$$

where $e = r - m$, the difference between the modal values of $\mu_R(x)$ and $\mu_M(x)$.

Note that the Hausdorff distance $H(A, B)$ between two Boolean sets A and B is defined as follows:

$$H(A, B) = \max\{H^*(A, B), H^*(B, A)\} \tag{8.3}$$

where $H^*(A, B) = \max_{a \in A}[d(a, B)]$ is the Hausdorff semi-distance between A and B and $d(a, B) = \min[m(a, b)|b \in B]$ where $m(a, b)$ is a measure of the distance between a and b.

Example 8.2

Consider the simple example shown in Figure 8.5 where $A = [a_1, a_2]$ and $B = [b_1, b_2]$ are two crisp intervals and $m(a, b) = |(a - b)|$.

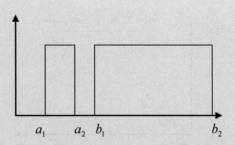

Figure 8.5 Two crisp intervals.

Then $d(a_1, B) = \min[|b_1 - a_1|, |b_2 - a_1|] = |b_1 - a_1|$ and $d(a_2, B) = \min[|b_1 - a_2|, |b_2 - a_2|] = |b_1 - a_2|$;

therefore $H^*(A, B) = \max[|b_1 - a_1|, |b_1 - a_2|] = |b_1 - a_1|$ and similarly $H^*(B, A) = \max[|b_1 - a_2|, |b_2 - a_2|] = |b_2 - a_2|$.

We therefore have $H(A, B) = \max\{|b_1 - a_1|, |b_2 - a_2|\} = |b_2 - a_2|$.

The Hausdorff distance can be extended to fuzzy sets using their α-cut representations (Prade and Testemale, 1987). The membership function of the fuzzy Hausdorff semi-distance is given by

$$\mu_{H^*(A,B)}(d) = \sup_{d=H^*(A,B)}(\alpha) \qquad (8.4)$$

and the membership function of the fuzzy Hausdorff distance is given by

$$\mu_{H(A,B)}(d) = \sup_{d=\max\{d_1,d_2\}} \{\min[\mu_{H^*(A,B)}(d_1), \mu_{H^*(B,A)}(d_2)]\} \qquad (8.5)$$

It should be noted that, in general, a particular value of the Hausdorff semi-distance can be associated with one, two or an infinite number of values of α.

Example 8.3

Consider the fuzzy representations of the measurement and setpoint shown in Figure 8.6.

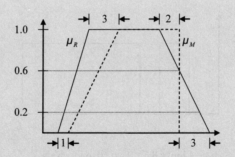

Figure 8.6 Fuzzy representations of the measurement and the setpoint.

The fuzzy semi-Hausdorff distances calculated using the extension principle are shown in Figure 8.7.

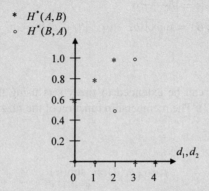

Figure 8.7 Semi-Hausdorff distances between the fuzzy sets.

Details of the calculation of the fuzzy Hausdorff distance (see Figure 8.8) are shown in Table 8.1.

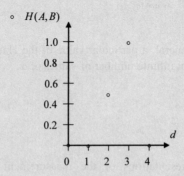

Figure 8.8 Hausdorff distance between the fuzzy sets.

Table 8.1 Calculation of the Hausdorff distance using the extension principle.

d	d_1	d_2	$\min[\mu_{H^*(A,B)}(d_1), \mu_{H^*(B,A)}(d_2)]$	$\mu_{H(A,B)}(d)$
0	0	0	0.0	0.0
1	0	1	0.0	
1	1	1	0.0	0.0
1	1	0	0.0	
2	0	2	0.0	
2	1	2	0.0	
2	2	2	0.5	0.5
2	2	1	0.5	
2	2	0	0.5	
3	0	3	0.0	
3	1	3	0.0	
3	2	3	0.0	
3	3	3	0.0	1.0
3	3	2	1.0	
3	3	1	0.8	
3	3	0	0.6	
4	0	4	0.0	
4	1	4	0.0	
4	2	4	0.0	
4	3	4	0.0	
4	4	4	0.0	0.0
4	4	3	0.0	
4	4	2	0.0	
4	4	1	0.0	
4	4	0	0.0	

It should be noted that the associated computational demands are much reduced if the fuzzy sets have triangular membership functions.

There are several ways of finding the rate of change of the tracking error when the error is represented by a fuzzy set (Palm and Driankov, 1995). One approach is to assume that the fuzzy set $\dot{E}$ (see Figure 8.9) is defined as follows:

$$\mu_{\dot{E}}(\dot{\varepsilon}^i(t)) = \sup_k \{\mu_E(\varepsilon^k(t))\} \quad \forall k \quad \dot{\varepsilon}^k(t) = \dot{\varepsilon}^i(t) \tag{8.6}$$

that is, the maximum degree of membership is chosen for the fuzzy set $\dot{E}$ if several values of the error have the same value of the rate of change of error. For example, consider the case where

$$\mu_E(\varepsilon^i(t)) = \exp\left\{\frac{-[\varepsilon^i(t) - \bar{\varepsilon}(t)]^2}{2\sigma^2(t)}\right\}.$$

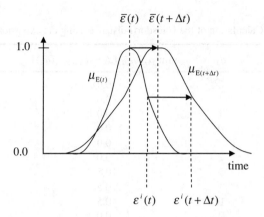

Figure 8.9   Example of a time-varying fuzzy set representing the tracking error.

As can be seen in Figure 8.9, $\mu_E(\varepsilon^i(t)) = \mu_E(\varepsilon^i(t + \Delta t))$ and therefore

$$\frac{\varepsilon^i(t) - \bar{\varepsilon}(t)}{\sigma(t)} = \frac{\varepsilon^i(t + \Delta t) - \bar{\varepsilon}(t + \Delta t)}{\sigma(t + \Delta t)}$$

However, $\varepsilon^i(t + \Delta t) \cong \varepsilon^i(t) + \dot{\varepsilon}^i(t)\Delta t$, $\bar{\varepsilon}(t + \Delta t) \cong \bar{\varepsilon}(t) + \dot{\bar{\varepsilon}}(t)\Delta t$ and $\sigma(t + \Delta t) \cong \sigma(t) + \dot{\sigma}(t)\Delta t$.

We therefore have

$$\dot{\varepsilon}^i(t) \cong \dot{\bar{\varepsilon}}(t) + [\varepsilon^i(t) - \bar{\varepsilon}(t)]\frac{\dot{\sigma}(t)}{\sigma(t)} \tag{8.7}.$$

Note that $\dot{\varepsilon}^i(t) \cong \dot{\bar{\varepsilon}}(t)$ $\forall i$ if $\dot{\sigma}(t) \cong 0$, in which case the fuzzy set describing the derivative of the error can be approximated by a fuzzy singleton at $\dot{\bar{\varepsilon}}(t)$.

An estimate of $\dot{\bar{\varepsilon}}(t)$ can be obtained by low-pass filtering the difference between the modal value of the fuzzy setpoint and the measured value of the plant output, and using a simple backward difference approximation to find the derivative (Palm and Driankov, 1995). Thus

$$\dot{\bar{\varepsilon}}(t) = \frac{\varepsilon_{LPF}(t) - \varepsilon_{LPF}(t - h)}{h} \tag{8.8}$$

where h is the sampling interval.

8.3   Inference with Fuzzy Inputs

Most fuzzy logic controllers deal with crisp inputs and therefore use singleton fuzzifiers. If the input to a fuzzy rule is a fuzzy set, the degree of satisfaction of the antecedent of a fuzzy rule is no longer a crisp value but is itself a fuzzy set. For example, consider a fuzzy rule with an antecedent "X is A", where $\mu_A(x)$ is the membership function associated with the fuzzy set A. Let $\mu_{A'}(x)$ be the membership function associated with a fuzzy assertion A'. The degree of satisfaction of the antecedent of the rule is the intersection of A' and A. If minimum inferencing is used, $\mu_{\text{dos}}(x) = \min\{\mu_{A'}(x), \mu_A(x)\}$ where $\mu_{\text{dos}}(x)$ is the membership function of the fuzzy set describing the degree of satisfaction.

General results for non-singleton fuzzification and inference in fuzzy logic systems with fuzzy inputs can be derived (Mouzouris and Mendel, 1997a, b). Consider a rule-base in which the lth rule is of the general form

$$R^l: \text{IF } X_1 \text{ is } A_1^l \text{ and } X_2 \text{ is } A_2^l \text{ and } \ldots \text{ and } X_p \text{ is } A_p^l \text{ THEN } Y \text{ is } B^l$$

and the discrete case where the membership functions of the ith antecedent A_i^l and the consequent B^l have non-zero values at discrete points $x_{i,1}, \ldots x_{i,n_i}$ and $y_1, \ldots y_m$, respectively. Let the p-dimensional fuzzy assertion be given by

$$A = \sum_{i_1=1}^{n_1} \cdots \sum_{i_p=1}^{n_p} \mu_{X_1}(x_{1,i_1}) * \ldots * \mu_{Xp}(x_{p,i_p})/(\mathbf{x}), \tag{8.9}$$

where $\mathbf{x} \equiv (x_{1,i_1}, x_{2,i_2}, \ldots x_{p,i_p})$, * denotes a t-norm and sigma denotes union. Using a compositional rule of inference (see Section 4.2.3) and the properties of t-norms, the discrete output of the lth rule is given by

$$Y^l = \sum_{j=1}^{m} \mu_{B^l}(y_j) * \tag{8.10}$$

$$\sup_{\mathbf{x}} \left\{ \sum_{1}^{n_1} \cdots \sum_{1}^{n_p} [\mu_{X_1}(x_{1,i_1}) * \ldots * \mu_{Xp}(x_{p,i_p})] * [\mu_{A_p^l}(x_{p,i_p}) * \ldots * \mu_{A_p^l}(x_{p,i_p})]/(\mathbf{x}) \right\} /(y_j)$$

Example 8.4

Consider inference based on a single-input single-output fuzzy logic system with only one rule: IF X is A THEN U is B, where the membership functions describing the fuzzy input X, the antecedent set A and the consequence set B are given by:

$$\mu_X(x) = \exp\left(-\frac{(x - m_X)^2}{2\sigma_X^2}\right),$$

$$\mu_A(x) = \exp\left(-\frac{(x - m_A)^2}{2\sigma_A^2}\right),$$

and

$$\mu_B(x) = \exp\left(-\frac{(x - m_B)^2}{2\sigma_B^2}\right)$$

respectively. Using product inferencing, $U(u) = \mu_B(u) . \sup_x\{\mu_X(x).\mu_A(x)\}$. By differentiating to find the value of x at which the maximum value occurs, it is easily shown that

$$U(u) = \mu_B(u).\mu_X(x_{\max}).\mu_A(x_{\max}), \quad \text{where } x_{\max} = \frac{\sigma_X^2 m_A + \sigma_A^2 m_X}{\sigma_X^2 + \sigma_A^2}.$$

Note that if there is no input uncertainty $\sigma_X^2 = 0$, $x_{\max} = m_x$ and

$$U(u) = \mu_B(u).\mu_X(m_x).\mu_A(m_x) = \mu_B(u).\mu_A(m_x),$$

which is the same as the result that would have been obtained using singleton fuzzifica-tion. However, if there is uncertainty (unless the centre of the fuzzy input is at the centre of the antecedent set), $x_{\max} \neq m_x$ and the result is not the same as that which would have been obtained using singleton fuzzification.

Alternatively, if max-min inferencing is used, $U(u) = \mu_B(u).\sup_{x}\{\min[\mu_X(x),$ $\mu_A(x)]\} = \mu_B(u).Poss(A|X)$ and the result is the same as that obtained by using the possibility of the antecedent set given the fuzzy input signal (see Section 4.2.3). Therefore, $U(u) = \mu_B(u).\mu_A(x_{\max})$, where $x_{\max} = \frac{\sigma_X m_A + \sigma_A m_X}{\sigma_X + \sigma_A}$ is found by equating the expressions for the two membership functions.

When the input uncertainty is large, the defuzzified output of a non-singleton fuzzy system may be significantly different from that which would be obtained from the corresponding output of a fuzzy singleton system (Mouzouris and Mendel, 1997b). Greater uncertainty in the input not only fires rules at a higher level than is the case in a fuzzy singleton system; it also usually fires more rules.

The results of simulating a second-order linear process controlled by a fuzzy controller based on a noisy measurement of the controlled variable have shown that the fuzzy output of the controller reflects, to a certain degree, the shape and location of the fuzzy input (Palm and Driankov, 1995). However, the use of max-min inferencing will not preserve the noise characteristics of the input signal at the output of the controller (Driankov et al., 1994).

The output of the controller should reflect the input uncertainties so that large apparent changes in the input to the controller should result in smaller changes in the control output than is the case for small changes. However, simulations of a first-order linear process con-trolled by a T-S fuzzy logic controller, which is equivalent to a proportional-plus-integral (PI) controller for precise inputs, have shown that accounting for uncertainty in both the setpoint and the measurement of the process output has little effect on the control performance when the output from the fuzzy controller is defuzzified (Foulloy and Galichet, 2003). Whether there is a paradox in explicitly taking into account fuzzy representations of the control objec-tive and/or of the measurements and then losing the information in the defuzzification process is questionable (Foulloy and Galichet, 2003).

8.4 Fuzzy Output Neural Networks

Some fuzzy neural networks (see Section 4.4) can handle fuzzy inputs and generate fuzzy outputs. Some have weights and biases that are real numbers (Ishibuchi et al., 1993); others have weights and biases that are fuzzy numbers (Ishibuchi and Tanaka, 1992; Buckley and Hayashi, 1994; Dunyak and Wunsch, 1999).

Two kinds of information are often available to train this type of fuzzy neural network (FNN): numerical information from measuring instruments and linguistic information from experts. Fuzzy supervised learning (Lin and Lu, 1996) or fuzzy reinforcement learning (Lin and Lu, 1995) schemes can be used to train the network.

One approach (Keller *et al.*, 1992; Keller and Tahani, 1992a, b) is to represent the fuzzy inputs by their possibility values over discrete universes of discourse and to use real number arithmetic to generate a fuzzy output in the form of possibility values over a discrete universe of discourse. The main advantage of this approach is that the computational demands are relatively low and the network can be trained using back-propagation to minimize

$$e = \frac{1}{2} \sum_j (t_j - u_j)^2$$

where t_j is the target value of the possibility of the jth output and u_j is the possibility value of the jth output predicted by the network.

Another approach (Ishibuchi *et al.*, 1993; Lin and Lu, 1995; Lin and Lu, 1996; Juang and Lee, 2007) is to map a fuzzy input vector, whose elements are fuzzy numbers, into a fuzzy output. The consequents of the rules are fuzzy numbers (Ishibuchi *et al.*, 1993), or combinations of fuzzy numbers and crisp linear functions of the fuzzy inputs (Juang and Lee, 2007). Computational demands can be reduced if the fuzzy inputs are fuzzy numbers with triangular membership functions (Ishibuchi *et al.*, 1993).

The fuzzy inputs are decomposed into α-cut sets so that interval arithmetic can be used to generate a fuzzy output in the form of α-cut sets. The network can be trained using either crisp data (represented as fuzzy singletons), fuzzy data (in the form of fuzzy numbers) or a mixture of crisp and fuzzy data. The learning scheme selects values for the weights and biases that minimize a cost function of the form:

$$J = \frac{1}{N} \sum_\alpha \alpha \left[\left(\hat{y}_{L_\alpha} - y_{L_\alpha} \right) + \left(\hat{y}_{U_\alpha} - y_{U_\alpha} \right)^2 \right] \tag{8.11}$$

where $[y_{L_\alpha}, y_{H_\alpha}]$ is the α-cut set of the fuzzy output training data Y and N is the number of values of α used in the summation. It should be noted that the cost function reduces to the conventional sum of errors squared expression in the case of a non-fuzzy input vector and a non-fuzzy target output (that is when $y_{L_\alpha} = y_{H_\alpha}$ and $\hat{y}_{L_\alpha} = \hat{y}_{H_\alpha}$) if the values of α are equally spaced in the range [0, 1]. Once trained, this type of FNN requires a much smaller number of numerical values to define each of the fuzzy inputs and is computationally undemanding if the interval arithmetic is based on only a small number of α-cut sets.

The uses of both numerical and linguistic information to train the network can improve its generality. For example, consider the problem of using a FNN controller to backup a truck (Juang and Lee, 2007) so that it finally moves in the direction ($x = 0.0$, $\phi = 90°$) where x is the current lateral distance from the final direction in which the truck should reverse and ϕ is the current angle of the truck with respect to the lateral direction. Results have shown that the controller fails if the initial position of the truck is ($x = -8.0$, $\phi = -80°$) or ($x = 5.0$, $\phi = 30°$) when the network is trained using numerical data obtained by recording x, ϕ and θ (the angle of the front wheels with respect to the current direction of travel), as an expert reverses the truck from only two initial states: $(x = -3.0, \phi = 90°)$ or $(x = 3.0, \phi = 90°)$. The controller however succeeds for all initial conditions when both the numerical data and the linguistic information provided by the following generic expert rules are used to train the network:

> IF x is very negative AND ϕ is positive small THEN θ is negative;

and

> IF x is very positive AND ϕ is negative small THEN θ is positive.

Training a network from fuzzy rules can also be used to reduce the number of rules in the rule-base (Lin and Lu, 1996).

FNNs that generate fuzzy outputs are nearly always trained using linguistic information or a mixture of fuzzy and numerical data. As was shown in Section 5.2, fuzzy FRMs are capable of generating fuzzy outputs even if they are trained with purely numerical data.

8.5   Modelling Input Uncertainty Using a Fuzzy FRM

The estimated rule confidences $\hat{R}_{X_iY_q}$ of a fuzzy relational model estimated using the RSK fuzzy identification scheme are proportional to the probability that the output is in the range $[y_q, y_{q+1}]$ when the input is in the range $[x_{i-1}, x_{i+1}]$ (see Section 5.4.1). Hence,

$$\hat{R}_{X_iY_q} \cong p(y_q|x_i)\,[a] = \Pr(y_q < y \leq y_{q+1}|x_i) \tag{8.12}$$

Consider the case when, during training, the output variable $y = x$ where x is the true value of the input, and the input variable $x_m = x + n$ is the measured value of the input where n is the measurement noise. Because $n = x_m - x$ and $p(n) = p(x_m - x) = p_n(x_m - y)$, where $p_n(n)$ is the probability density function of the measurement noise, we have

$$\hat{R}_{X_i,Y_q} \cong p(y_q|x_i)\,[a] = p_n(x_i - y_q)\,[a] \tag{8.13}$$

If the model is used to predict the output y when the input $x_i = x_m$, the estimated rule confidences will be given by

$$\hat{R}_{X_m,Y_q} \cong p(y_q|x_m)\,[a] = p_n(x_m - y_q)\,[a]\,\forall q \tag{8.14}$$

Example 8.5

Consider the identification of a fuzzy FRM of a simple first-order linear dynamic system when the measured input is corrupted by measurement noise. The measurement noise is generated by passing normally distributed random noise (zero mean, with a standard deviation of 2.0) through a first-order low-pass filter with time constant of 20 s. The time constant of the system is 10 s and the sampling interval of the model is 1 s. In this example, it is assumed that an accurate measurement of the output is available. The input signal used to generate the training data is a sinusoid consisting of several periods of different frequencies ($T = 10, 50, 100, 500, 1000, 2000$ s). The FRM has two inputs: the measured values of the input $x(i)$ and the previous value of the output $y(i-1)$. The input and the current and previous values of the output are each described by 51 (equally spaced) fuzzy

sets with triangular membership functions. Figure 8.10 shows the output produced by
the fuzzy FRM in response to a step change (from 0.0 to 0.8) in the input signal. The
possibility distributions have been normalized to simplify interpretation of the results.

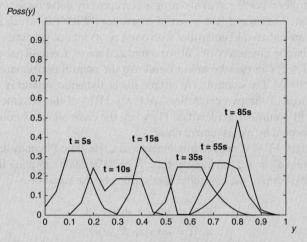

Figure 8.10 Fuzzy output of the FRM generated in response to a step change in the input from
0.0 to 0.8.

It can be seen that the output of the fuzzy FRM gives a reasonably good indication
of the uncertainty resulting from the measurement noise on the input signal that was
used to identify the model. A longer training period would probably result in a better
representation of the probability distribution of the noise.

It can also be seen (see Figure 8.11) that height defuzzification of the fuzzy output
gives a reasonably accurate prediction of the system's transient behaviour.

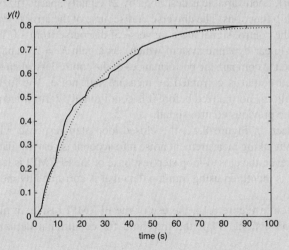

Figure 8.11 Comparison of the defuzzified step response of the fuzzy FRM and that of the
dynamic system.

Example 8.6

The performance of a fuzzy FRM-based PI controller is compared to that of a conventional PI controller when the tracking error is corrupted by noise on the measurement of the plant output. The fuzzy FRM-based PI controller (FFRMPI), which is identified by simulating a conventional PI controller with noise on its inputs, generates a fuzzy output that depends on the characteristics of the simulated noise. Conditional defuzzification (see Section 7.4.2) can then be used to improve the control performance by reducing the control activity. The control of a simple linear dynamic system is used to assess the performances of the two controllers. A fuzzy FRM of the incremental form of a discrete-time PI controller is identified (1) when the error signal is noise-free and (2) when it is corrupted by measurement noise.

Here, the fuzzy FRM is used to implement a discrete-time PI controller, which has a proportional gain $K_p = 0.2$, an integral time $T_i = 20.0$s and a sampling interval $h = 1$s. The incremental form of the controller is described by the equations:

$$\Delta u(n) = u(n) - u(n-1),$$

$$u(n) = u_i(n) + K_p e(n),$$

where $u_i(n) = \alpha_i u_i(n-1) + (1 - \alpha_i) u(n-1)$ and $\alpha_i = e^{-h/T_i}$.

The test signal used to generate the training data is a uniformly distributed random signal in the range ± 1.0, superimposed on a sinusoid with unit amplitude and a period of 1000 s. The measurement noise is a low-pass filtered, normally distributed, random signal, which has zero mean and a standard deviation of 5.0. The time constant of the low-pass filter is $\tau_N = 100.0$s. The fuzzy FRM has two inputs, the current value of the error $e(n)$ and previous value of the error $e(n - 1)$ and one output, the change in the control signal $\Delta u(n)$. Each variable is described by 21 equally spaced fuzzy sets with triangular membership functions. The universe of discourse of the error is from -1.0 to $+1.0$. The change in the control signal has a universe of discourse from -0.125 to $+0.125$.

A first-order linear dynamic system which has a gain $K = 2.0$ and a time constant $\tau = 50.0$ s is used to compare the performances of the controllers when the measurement of the controlled output is corrupted by measurement noise. The fuzzy output of the FFRMPI controller is normalized before it is conditionally defuzzified and added to the crisp value of the previous control signal.

As can be seen in Figure 8.12, the closed-loop step response with the FFRMPI identified without taking measurement noise into account is very similar to that of the PI controller. However, the closed-loop step response of the FFRMPI is less damped when the fuzzy FRM is identified using training data that is corrupted by measurement noise (see Figure 8.13).

As can be seen in Figure 8.14, the use of the FFRMPI results in much less control activity than conventional PI if the threshold for conditional defuzzification is less than 1.0.

The results shown in Table 8.2 compare the performance of the two controllers when the setpoint is held constant at 0.5, the standard deviation of the measurement

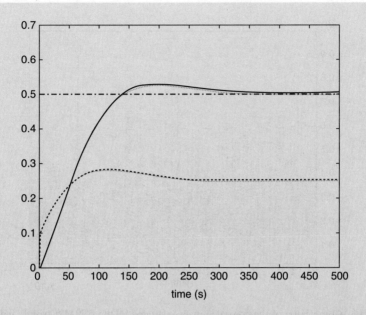

Figure 8.12 Comparison of the closed-loop step response with FFRMPI (solid and dashed lines) and PI control (dotted lines), when there is no measurement noise present during both training and testing.

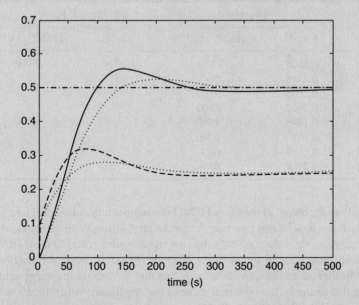

Figure 8.13 Comparison of the closed-loop step response with FFRMPI (solid and dashed lines) and PI control (dotted lines) when the fuzzy FRM is identified using noisy data but there is no measurement noise during testing.

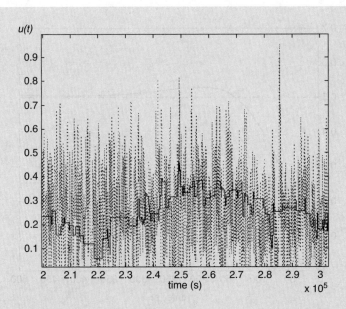

Figure 8.14 Comparison of the control signals generated by the FFRMPI (solid) and conventional PI (dotted) ($\alpha = 0.5$).

Table 8.2 Comparison of the performance of the FFRMPI and conventional PI.

	FFRMPI			PI	
α	RMSE	MADU ($\times10^{-4}$)		RMSE	MADU ($\times10^{-4}$)
1.0	0.2865	74.00		0.2776	75.00
0.9	0.2958	50.0			
0.8	0.2908	26.00			
0.7	0.2602	10.00			
0.6	0.1893	1.80			
0.5	0.1780	0.56			
0.4	0.1819	0.21			
0.3	0.2418	0.06			

noise is 5.0 and the output of the fuzzy FRMPI is conditionally defuzzified using different thresholds. It can be seen that lowering the conditional defuzzification threshold (α) from 1.0 to 0.5 reduces the values of both the root-mean-square tracking error (RMSE) and the mean absolute change in the control signal (MADU). Further reductions in the threshold further reduce the control activity, but they also result in greater values of the RMSE. In this example, the optimum value of the conditional defuzzification threshold is approximately 0.5. Note that the RSME is greater than that resulting from conventional PI control if the threshold for conditional defuzzification is greater than 0.7.

Example 8.7

Consider the control of a non-linear dynamic system using FMPC with feedforward of a noisy measurement of an input disturbance. The Hammerstein model of the same non-linear dynamic system introduced in Chapter 7 is used to evaluate the performance of the controller. The Laplace transform transfer function relating the controlled output $y(t)$ to the control signal $u(t)$ is given by:

$$Y(s) = L(f(u(t))) \left[\frac{0.899}{0.9d + 0.8} \right] \left[\frac{1}{1 + 200s} \right] \qquad (8.15)$$

where the static non-linearity $f(u(t)) = \frac{1}{3.434} \ln(30\,u(t) + 1)$ and d is an input disturbance.

The disturbance is a sinusoid with a mean $M = 0.275$, an amplitude $A = 0.165$ and a period $T = 100\,000$s (see Figure 8.15). The measured value of the disturbance (z) is generated by adding low-pass filtered $(\tau = 950\,\text{s})$ normally distributed white noise $(m = 0.0, \sigma = 0.5)$ to the value of the disturbance (see Figure 8.16).

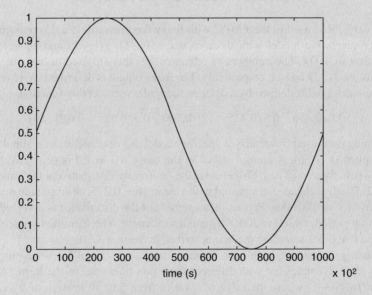

Figure 8.15 Actual value of the disturbance.

It is assumed that the measured disturbance is a fuzzy set which has a triangular membership function with its apex at the measured value of the disturbance and a base six times the assumed value of the standard deviation of the disturbance (0.5). The possibilities of each of the fuzzy reference sets that the fuzzy FRM uses to describe the disturbance, given the triangular fuzzy set describing the measured disturbance, are calculated by finding the appropriate intersection points.

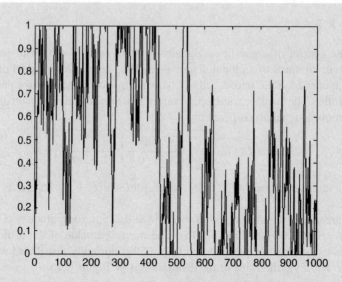

Figure 8.16 Measured value of the disturbance.

The fuzzy FRM used in the FMPC with fuzzy feedforward is a discrete-time first-order autoregressive model with three inputs: $u(i-1)$, $y(i-1)$ and $z(i-1)$ and a sample time of 100 s. The numbers of reference sets that are used to describe each of the inputs are 21, 21 and 11, respectively. The fuzzy output is described by 41 reference sets. The fuzzy goal is defined by a discrete triangular membership function:

$$[0.0/0.450 \quad 0.5/0.475 \quad 1.0/0.500 \quad 0.5/0.525 \quad 0.0/0.550].$$

The training data used to identify a specific model are obtained from a simulation of the test plant ($k = 30$). A control signal in the range 0.0 to 1.0 is generated by clipping low-pass filtered ($\tau = 100$s) white noise, uniformly distributed in the range –24.8 to +25.2. The disturbance is a sinusoid with a mean $M = 0.275$, an amplitude $A = 0.165$ and a period $T = 100\,000$s. Perfect measurement of the disturbance is assumed during the training period, which is 200 000 samples in length. The identified model is post-processed five times using the method described in Section 5.5.1.

To identify a generic model, the training data are obtained by simulating plants with the same dynamics but with different static non-linearities of the form $f(u(t)) = \frac{1}{\ln(k+1)} \ln(ku(t) + 1)$ where the value of k varies from 5 to 50 in steps of 5. A control signal in the range 0.0 to 1.0 is generated by clipping low-pass filtered ($\tau = 100$ s) white noise uniformly distributed in the range -24.8 to $+25.2$. The disturbance is a sinusoid with a mean $M = 0.275$, an amplitude $A = 0.165$ and a period $T = 100\,000$s. As before, perfect measurement of the disturbance is assumed during each training period, which is 10^5 samples in length. The identified model is post-processed five times using the method described in Section 5.5.1.

Sum-product inferencing is used to predict the one-step-ahead fuzzy prediction of the plant output over a Universe of Discourse from 0.0 to 1.0. The fuzzy prediction is normalized so that the sum of the possibilities is unity. If no rules are fired, all values of the next plant output are assumed to be equally possible.

The degree of satisfaction of the goal is the overlap area divided by the area under the fuzzy prediction of the output. The fuzzy decision-maker determines the degree of satisfaction of the goal for 11 candidate values of the control signal. If at least one candidate value of the fuzzy control signal satisfies the goal, the fuzzy control signal is conditionally defuzzified using height defuzzification.

If none of the candidate values of the control signal satisfy the goal, the lower and upper bounds of the support of the fuzzy goal are found and u is set to u_{max} or u_{min} (depending on whether the lower bound of the support of the fuzzy prediction for the minimum value of the control signal is closer to the upper bound of the support of the fuzzy goal than the upper bound of the support of the fuzzy prediction for the maximum value of the control signal is to the lower bound of the support of the fuzzy goal). It should be noted that this method assumes that increasing u will increase y.

The crisp values of the control signal are filtered using a first-order discrete-time low-pass filter with a time constant of 950 s before they are applied to the plant.

Not surprisingly, comparison of the results presented in Figures 8.17 and 8.18 shows that fuzzifying the disturbance measurement results in worse disturbance rejection when there is no measurement noise and no modelling errors.

As might be expected, a comparison of the results presented in Figures 8.18 and 8.19 shows that there is a larger steady-state error and worse disturbance rejection when the FMPC uses the generic model.

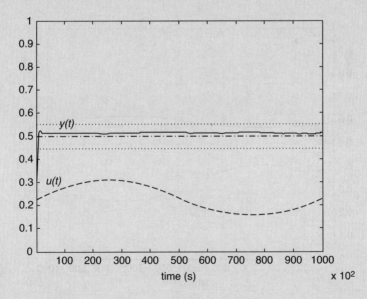

Figure 8.17 Response under FMPC with feedforward using singleton fuzzification and no measurement noise (using the specific model and $\alpha = 1.0$).

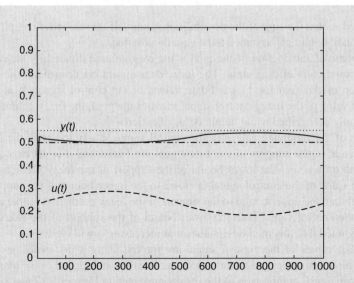

Figure 8.18 Response under FMPC with fuzzy feedforward and no measurement noise (using the specific model and $\alpha = 1.0$).

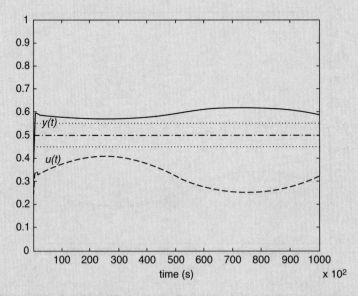

Figure 8.19 Response under FMPC with fuzzy feedforward and no measurement noise (using the generic model and $\alpha = 1.0$).

However, a comparison of the results presented in Figures 8.20 and 8.21 shows that fuzzy number fuzzification does reduce the impact of the measurement noise on the control activity.

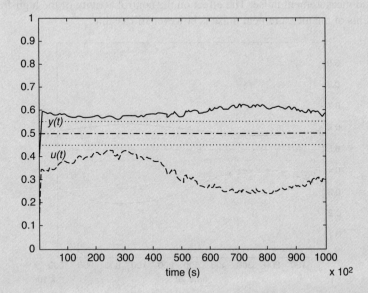

Figure 8.20 Response under FMPC with feedforward using singleton fuzzification and measurement noise and (using the generic model and $\alpha = 1.0$).

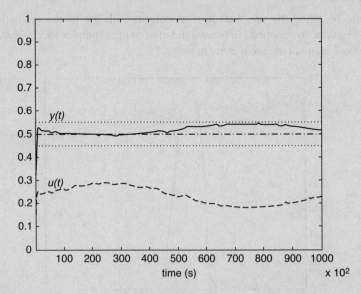

Figure 8.21 Response under FMPC with fuzzy feedforward and measurement noise (using the specific model and $\alpha = 1.0$).

The results presented in Figures 8.21 and 8.22 show that the controllers are able to partially reject the disturbance when only a noisy measurement of the disturbance is available, even though the fuzzy FRMs used in the controllers are trained assuming there is no measurement noise. The effect on the control activity of the high-frequency components of the measurement noise is however still visible.

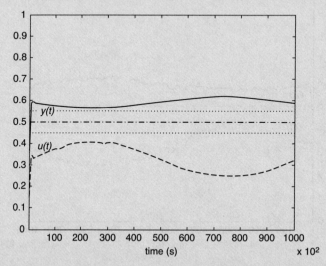

Figure 8.22   Response under FMPC with fuzzy feedforward and measurement noise (using the generic model and $\alpha = 1.0$).

It should be noted that the fuzziness associated with the model predictions (see Figure 8.23) is due to variations in behaviour between the examples used to generate the generic model, and not the measurement noise.

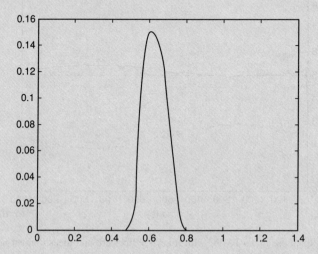

Figure 8.23   Typical output prediction generated by the generic fuzzy FRM.

Nevertheless, as can be seen in Figure 8.24, the use of conditional defuzzification is still able to eliminate the effect of the measurement noise when the disturbance is varying slowly.

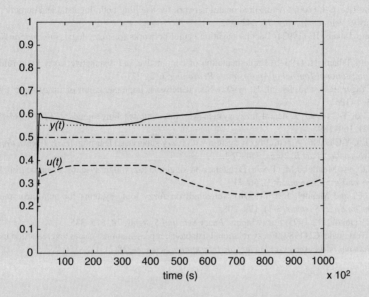

Figure 8.24 Response under FMPC with fuzzy feedforward and measurement noise (using the generic model and $\alpha = 0.8$).

8.6 Summary

This chapter has shown how inaccurate measurements can be incorporated into fuzzy control schemes. The use of fuzzy references and fuzzy measurements has been discussed and different ways of finding fuzzy measures of the tracking error and its derivative have been described. A fuzzy inference scheme that can deal with fuzzy inputs has been explained. Fuzzy neural networks that are capable of generating fuzzy outputs when presented with fuzzy inputs have also been described. A method of modelling input uncertainty using a fuzzy FRM has been proposed and examples of its use have been presented.

References

Buckley, J.J. and Hayashi, Y. (1994) Fuzzy neural networks: a survey. *Fuzzy Sets and Systems*, **66**(1), 1–13.

Chaudhuri, B.B. and Rosenfeld, A. (1999) A modified Hausdorff distance between fuzzy sets. *Information Sciences*, **118**, 159–171.

Driankov, D., Palm, R. and Hellendorn, H. (1994) Fuzzy control with fuzzy inputs: the need for new rule semantics. *Proceedings of IEEE International Conference on Fuzzy Systems*, pp. 13–21.

Dunyak, J. and Wunsch, D. (1999) Fuzzy number neural networks. *Fuzzy Sets and Systems*, **108**(1), 49–58.

Foulloy, L. and Galichet, S. (2003) Fuzzy control with fuzzy inputs. *IEEE Transactions on Fuzzy Systems*, **11**(4), 437–449.

Ishibuchi, H. and Tanaka, H. (1992) Fuzzy regression analysis using neural networks. *Fuzzy Sets and Systems*, **50**(3), 257–265.

Ishibuchi, H., Fujioka, R. and Tanaka, H. (1993) Neural networks that learn from fuzzy If-Then rules. *IEEE Transactions on Fuzzy Systems*, **1**(1), 85–97.

Juang, C-F. and Lee, C-I. (2007) A fuzzified neural network for handling both linguistic and numerical information simultaneously. *Neurocomputing*, **71**, 342–352.

Keller, J.M. and Tahani, H. (1992a) Backpropagation neural networks for fuzzy logic. *Information Science*, **62**(3), 205–221.

Keller, J.M. and Tahani, H. (1992b) Implementation of conjunctive and disjunctive fuzzy logic rules with neural networks. *International Journal of Approximate Reasoning*, **6**(2), 221–240.

Keller, J.M., Yager, R.R. and Tahani, H. (1992) Neural network implementation of fuzzy logic. *Fuzzy Sets and Systems*, **45**, 1–12.

Lin, C-T and Lu, Y-C (1995) A neural fuzzy system with linguistic teaching signals. *IEEE Transactions on Fuzzy Systems*, **3**(2), 169–189.

Lin, C-T and Lu, Y-C (1996) A neural fuzzy system with fuzzy supervised learning. *IEEE Transactions on Systems, Man and Cybernetics, Part B*, **26**(5), 744–763.

Mouzouris, G.C. and Mendel, J.M. (1997a) Dynamic non-singleton fuzzy logic systems: theory and application. *IEEE Transactions on Fuzzy Systems*, **5**(2), 56–71.

Mouzouris, G.C. and Mendel, J.M. (1997b) Non-singleton fuzzy logic systems for non-linear modeling. *IEEE Transactions on Fuzzy Systems*, **5**(1), 199–208.

Palm, R. and Driankov, D. (1995) Fuzzy inputs. *Fuzzy Sets and Systems*, **70**, 315–335.

Prade, H. and Testemale, C. (1987) Fuzzy relational databases: representation issues and reduction using similarity measures. *Journal of the American Society for Information Sciences*, **38**(2), 118–126.

9

Disturbance Rejection in Information-Poor Systems

External disturbances acting on non-linear systems may be classified into three types: input (or actuator) disturbances, plant disturbances and output disturbances.

In general, an external disturbance d can affect both the states $\mathbf{x}$ and the measured outputs $\mathbf{y}$, and the relationship between the states and the control signal u:

$$\dot{\mathbf{x}} = f(\mathbf{x}, u, d) \quad \text{and} \quad \mathbf{y} = g(\mathbf{x}, d)$$

Various approaches to disturbance rejection in non-linear dynamic systems have been suggested (Mukhopadhyay and Narendra, 1993), although simplifying assumptions concerning the plant have to be made in most cases. For example,

1. the disturbance has no direct effect on the measured outputs:

$$\mathbf{y} = g(\mathbf{x});$$

2. the disturbance does not affect the relationship between the states and the control signal:

$$\dot{\mathbf{x}} = f_1(\mathbf{x}, d) + f_2(\mathbf{x}, u);$$

3. the state is fully observable:

$$\mathbf{y} = \mathbf{x};$$

4. the system is affine in u:

$$\dot{\mathbf{x}} = f_1(\mathbf{x}, d) + f_2(\mathbf{x})u;$$

5. the external disturbance is an output disturbance:

$$\dot{\mathbf{x}} = f(\mathbf{x}, u) \quad \text{and} \quad \mathbf{y} = g(\mathbf{x}, d);$$

Monitoring and Control of Information-Poor Systems: An Approach based on Fuzzy Relational Models, First Edition. Arthur L. Dexter.
© 2012 John Wiley & Sons, Ltd. Published 2012 by John Wiley & Sons, Ltd.

6. the matching condition is satisfied:

$$\dot{\mathbf{x}} = f(\mathbf{x}, u + d).$$

The presence of uncertainty further complicates the design problem.

9.1 Rejecting Unmeasured Disturbances in Uncertain Systems

Complete disturbance rejection is not achievable in an uncertain system but there are three main ways of reducing the effect of unknown disturbances: robust fuzzy control, feedback linearization using a fuzzy model-based disturbance estimator and fuzzy model-based internal model control.

9.1.1 Robust Fuzzy Control

Most of the robust fuzzy control schemes use (1) a T-S fuzzy model, whose rule consequences are local linear dynamic systems, to describe the global behaviour of the system to be controlled; and (2) a non-fuzzy norm-bounded representation of the parametric uncertainties and external disturbances (e.g. Lo and Lin, 2006). The use of T-S fuzzy models allows conventional robust control design methods (e.g. Lyapunov stability analysis, H-infinity, LMI optimization) to be applied to non-linear dynamic systems.

For example, consider the single-input single-output case where the ith plant local linear model of the T-S fuzzy model is given by:

$$\text{IF } v_1(t) \text{ is } M_{i1} \text{AND } v_2(t) \text{ is } M_{i2} \text{ AND } \ldots \text{ AND } v_p(t) \text{ is } M_{ip} \text{ THEN}$$
$$\dot{\mathbf{x}}(t) = [A_i + \Delta A_i]\mathbf{x}(t) + B_{1_i}w(t) + [B_{2_i} + \Delta B_{2_i}]u(t)$$
$$z(t) = [C_{1_i} + \Delta C_{1_i}]\mathbf{x}(t) + [D_i + \Delta D_i]u(t)$$
$$y(t) = C_2\mathbf{x}(t)$$

where $v_k(t)$ are the premise variables, $\mathbf{x}(t)$ is the state vector, $u(t)$ is the control input, $w(t)$ is a norm-bounded external disturbance, $z(t)$ is the controlled output, $y(t)$ is the measured output and ΔA_i, ΔB_{2_i}, ΔC_{1_i} and ΔD_i represent the uncertainties in the system.

The fuzzy static output feedback controller is given by:

$$u(t) = \sum \mu_i(v(t))k_i y(t) \tag{9.1}$$

where $\mathbf{v} = [v_1 v_2 \ldots v_n]^T$, $\mu_i(v(t))$ is the degree of satisfaction of the antecedent of the ith rule, and k_i is the gain of the controller associated with the ith local linear model.

The control objective is to design the fuzzy controller so that

$$\int z^2(t)dt \leq \gamma^2 \int w^2(t)dt \tag{9.2}$$

where $\gamma < 1$ is the desired level of disturbance attenuation.

To satisfy the control objective and to guarantee asymptotic stability of the closed loop system, each of the controller gains k_i must satisfy a bilinear matrix inequality. Suitable values can be found by iteratively solving a set of linear matrix inequalities (Huang and Nguang, 2006).

The main disadvantage is that the approach is conservative and may often fail to generate a feasible solution.

9.1.2   Feedback Linearization Using a Fuzzy Disturbance Observer

Most control schemes that are based on this approach are only applicable to a particular class of non-linear systems, and regard the external disturbance and modelling errors as being equivalent to an additive input disturbance (Kim, 2002, 2003; Kim and Lee, 2005) or additive output disturbance (Gao et al., 2008). For example, consider the case where the non-linear system can be represented by the following model:

$$\dot{y} = f_1(\mathbf{x}) + f_2(\mathbf{x})u + d \tag{9.3}$$

where $\mathbf{x} = [x_1 x_2 \ldots x_n]^T$, $y = x_n$, $x_n = \dot{x}_{n-1}$ and d is an unmeasured external disturbance. The modelling uncertainties, $\Delta f_1(\mathbf{x}) = f_1(\mathbf{x}) - f_{1n}(\mathbf{x})$ and $\Delta f_2(\mathbf{x})u = f_2(\mathbf{x})u - f_{2n}(\mathbf{x})u$, where $f_{1n}(\mathbf{x})$ and $f_{2n}(\mathbf{x})$ define the nominal model of the system, are combined with the effect of the unmeasured external disturbance and included in an output observer. We therefore have

$$\dot{\hat{y}} = \sigma(y - \hat{y}) + f_{1n}(\mathbf{x}) + f_{2n}(\mathbf{x})u + \hat{\Omega}(\mathbf{x}|u, \hat{\vartheta}) \tag{9.4}$$

where $\hat{\Omega}(\mathbf{x}|u, \hat{\vartheta})$ is an estimate of the combined uncertainty $\Omega(\mathbf{x}, u, d) = \Delta f_1(\mathbf{x}) + \Delta f_2(\mathbf{x})u + d$ and $\hat{\vartheta}$ is the parameter vector of the disturbance estimator.

A T-S fuzzy model that is linear in its consequent parameters (see Section 4.3) can be used to estimate the combined uncertainty $\Omega(\mathbf{x}, u, d)$. If the model is described by $\hat{\Omega}(\mathbf{x}) = \hat{\vartheta}^T \xi(\mathbf{x})$, the vector of consequent parameters ϑ can be estimated using the adaptation scheme $\dot{\hat{\vartheta}} = \gamma(y - \hat{y})\xi(\mathbf{x})$ where γ is a user-defined constant. It can be shown (Kim, 2002) that

$$\hat{\Omega}(\mathbf{x}|u, \hat{\vartheta}) \rightarrow \Omega(\mathbf{x}|u, \vartheta) \quad \text{as} \quad \hat{y} \rightarrow y.$$

Feedback linearization (Boukezzoula et al., 2007) can then be used to remove the effects of the combined uncertainty and linearize the plant so that a linear feedback controller can be designed. Thus the control signal is given by,

$$u = \frac{u^* - f_{1n}(\mathbf{x}) - \hat{\Omega}(\mathbf{x}|u, \hat{\vartheta})}{f_{2n}(\mathbf{x})}$$

where u^* is the output of the linear controller.

It should be noted that the approach assumes that a nominal model is available, the external disturbance d varies slowly or is constant, and full state feedback is possible.

9.1.3   Fuzzy Model-Based Internal Model Control

The difference between the measured output of the plant and the output of a disturbance free model of the plant is subtracted from the reference signal r to compensate for the external disturbance d in an internal model control (IMC) scheme. The approach can be used to reject unmeasured disturbances in non-linear dynamic systems by using a T-S fuzzy system or FRM as the internal model (Sousa et al., 1997; Edgar and Postlethwaite, 2000; Boukezzoula et al., 2003).

Consider the control of the non-linear plant shown in Figure 9.1, in which the disturbance acts additively at the output of the plant. It can be seen that $y = f_1(u) + f_2(d)$, where the double line indicates that f_1 and f_2 are the non-linear dynamic functions relating the plant output to the control input u and the disturbance d, respectively. Note that $f_2(d) = d$ if d is an output disturbance and $f_2(d) = f_1(d)$ if d is an additive input disturbance.

The error signal $e = y - f_M(u) = f_2(d) + f_1(u) - f_M(u)$, where f_M is the non-linear dynamic function describing the internal model.

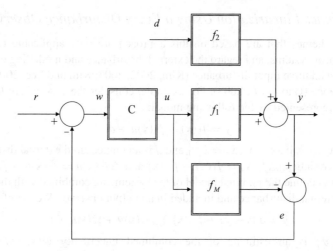

Figure 9.1   Internal model control of a non-linear plant.

Note that if $f_M = f_1$ then $e = f_2(d)$, $w = r - e = r - f_2(d)$ and $u = C(w) = C(r - f_2(d))$ where C is the non-linear dynamic function describing the controller (i.e. the control scheme no longer involves feedback and the stability of the control system is guaranteed if the plant and controller are open-loop stable).

If $C = f_1^{-1}$ then $u = f_1^{-1}(w)$ and $y - f_2(d) = f_1(u) = f_1(f_1^{-1}(w)) = w$. Therefore, $w = y - f_2(d) = r - e = r - f_2(d)$, $y = r$ and the disturbance is rejected totally.

However, it should be noted that a perfect inverse controller $C = f_1^{-1}$ may be impossible to implement in practice, as f_1^{-1} could be ante-causal.

Alternatively, if $C = f_M^{-1}$ then

$$u = f_M^{-1}(w) \quad \text{and} \quad y - e = f_M(u) = f_M(f_M^{-1}(w)) = w.$$

Therefore, $w = r - e = y - e$, $y = r$, and, as before, the disturbance is rejected totally.

Once again however, it may be impossible to implement such a controller as f_M^{-1} could be ante-causal. Note that the steady-state value of the tracking error will be zero if the steady-state gain of the controller is the inverse of the steady-state gain of the internal model and the closed-loop control system is stable (Rivals and Personnaz, 2000).

In practice, it can be seen that the control performance depends on the availability of a good disturbance-free model of the plant and a controller, which ideally should be a perfect inverse of the plant but in practice is a good estimate of the inverse of the internal model over a suitable frequency range. As a result, a low-pass filter is usually inserted at the input to the controller to provide robustness against model mismatch at high frequencies. The design of the filter is a compromise between the control performance and robustness of the controller to modelling errors (Gormandy and Postlethwaite, 2001; Boukezzoula *et al.*, 2003).

It should be noted that there are two basic types of discrete-time autoregressive models that can be used to predict the behaviour of dynamic systems (Balle, 1999): series-parallel models, which use the measured value of the plant output at the previous sampling instant as one of their inputs, and parallel models, which only use the previously estimated values of the plant output. Series-parallel models will only predict the disturbance-free behaviour accurately if the disturbance is constant and its value is known, or if the controller has rejected

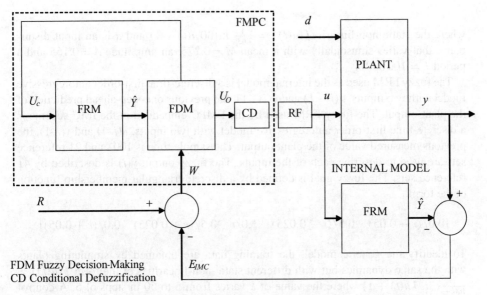

Figure 9.2   Fuzzy FRM-based IMC scheme.

the disturbance totally. Parallel models will only generate accurate predictions if there are no modelling errors. In practice, neither type of model will give accurate results, and the optimal choice will be a tradeoff between the effect of the disturbances and the size of the modelling errors in a particular application.

9.2   Fuzzy IMC Based on a Fuzzy Output FRM

A block diagram of a fuzzy FRM-based IMC (FIMC) scheme is shown in Figure 9.2.

The design of the fuzzy model-based predictive control (FMPC) is identical to that of the controller described in Section 7.5. The only differences are that the fuzzy setpoint R is modified to account for the effect of the disturbance and a robustness filter (RF) is included to reduce the effects of modelling errors at high frequencies. Fuzzy FRMs are used both to estimate the effect of the unmeasured disturbance and to predict the future values of the plant output in the FMPC.

Example 9.1

In this example, a generic FRM is used both as the internal model and in the FMPC. A Hammerstein model of a non-linear dynamic system is used to evaluate the performance of the controller. The Laplace transform transfer function relating the controlled output $y(t)$ to the control signal $u(t)$ is given by:

$$Y(s) = L(f(u(t))) \left[\frac{0.899}{0.9d + 0.8} \right] \left[\frac{1}{1 + 200s} \right] \qquad (9.5)$$

where the static non-linearity $f(u(t)) = \frac{1}{3.434}\ln(30\,u(t) + 1)$ and d is an input disturbance that varies sinusoidally with a mean $M = 0.275$, an amplitude $A = 0.165$ and a
period $T = 100\,000$ s.

The fuzzy FRM used as the internal model is a discrete-time first-order autoregressive
model with two inputs, $u(i - 1)$ and $\hat{y}(i - 1)$, the previous one-step-ahead prediction of
the plant output. The fuzzy FRM used by the FMPC embedded in the IMC scheme is
a discrete-time first-order autoregressive model with two inputs, $u(i-1)$ and $y(i-1)$, the
previous measured value of the plant output. The sample time is 100 s and 21 reference
sets are used to describe each of the inputs. The fuzzy output $\hat{y}(i)$ is described by 41
reference sets. The fuzzy goal is defined by a discrete triangular membership function
of the form:

$$[0.0/(r - 0.05) \quad 0.5/(r - 0.025) \quad 1.0/r \quad 0.5/(r + 0.025) \quad 0.0/(r + 0.05)].$$

To identify the generic model, the training data are obtained by simulating plants
with the same dynamics but with different static non-linearities of the form $f(u(t)) = \frac{1}{\ln(k+1)}\ln(ku(t) + 1)$ where the value of k varies from 5 to 50 in steps of 5. A control
signal in the range 0.0 to 1.0 is generated by clipping low-pass filtered ($\tau = 100$ s) white
noise, uniformly distributed in the range –2 to +3. The disturbance is held constant at
0.275 over the training period (10 000 samples). After identification, the fuzzy model is
post-processed five times using the method described in Section 5.5.1.

Sum-product inferencing is used to predict the one-step-ahead fuzzy prediction of the
plant output over a discrete Universe of Discourse (UoD) from 0.0 to 1.0. The fuzzy
prediction is normalized so that the sum of the possibilities is unity. If no rules are fired,
all values of the next plant output are assumed to be equally possible.

The error signal obtained from the internal model $E_{imc} = Y(i) - \hat{Y}(i)$ is calculated
over a discrete UoD from -1.0 to $+1.0$ using the extension principle. The modified
fuzzy goal $G_{mod} = G - E_{imc}$ is calculated over a discrete UoD from -1.0 to $+2.0$
again using the extension principle.

The design of the fuzzy MPC is the same as that used in Chapter 7. The degree of
satisfaction of the goal is the overlap area divided by the area under the fuzzy output
prediction. The fuzzy decision-maker determines the degree of satisfaction of the goal
for 11 (equally spaced) candidate values of the control signal. If at least one candidate
value of the fuzzy control signal satisfies the goal, then the fuzzy control signal is
conditionally defuzzified using height defuzzification. If none of the candidate values
of the control signal satisfy the goal, the lower and upper bounds of the support of the
fuzzy goal are found and u is set to u_{max} or u_{min} depending on whether the lower bound
of the support of the fuzzy prediction for the minimum value of the control signal is
closer to the upper bound of the support of the fuzzy goal than the upper bound of the
support of the fuzzy prediction for the maximum value of the control signal is to the
lower bound of the support of the fuzzy goal. It should be noted that this method assumes
that increasing u will increase y.

The crisp values of the control signal are filtered using a first-order discrete-time
low-pass filter with a time constant of 950 s before they are applied to the plant.

As can be seen in Figures 9.3–9.5, the disturbance rejection is good at a range of the operating points.

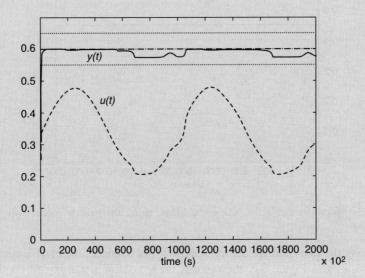

Figure 9.3 Disturbance rejection using the FIMC based on the generic model with $\alpha = 1.0$ and $r = 0.6$.

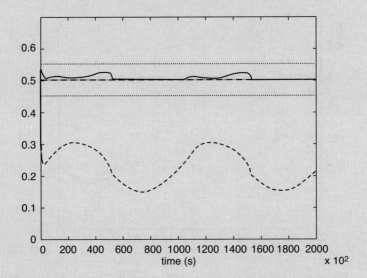

Figure 9.4 Disturbance rejection using the FIMC based on the generic model with $\alpha = 1.0$ and $r = 0.5$.

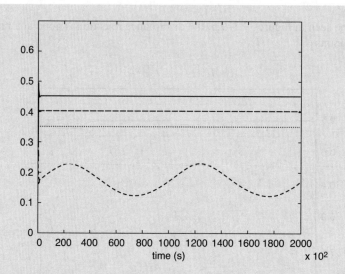

Figure 9.5 Disturbance rejection using the FIMC using the generic model with $\alpha = 1.0$ and $r = 0.4$.

Lowering the threshold for defuzzification to 0.45 reduces the control activity but, as can be seen in Figure 9.6, the optimal fuzzy control signal is still continuously defuzzified for much of time. This occurs because the uncertainty introduced by using generic models of the plant is so great that, for large periods of time, the defuzzified value of the current optimal control signal, and therefore the previous value of the control signal, have a relatively low membership of the optimal fuzzy control signal.

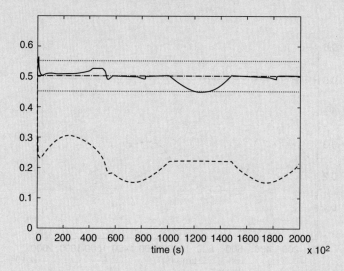

Figure 9.6 Disturbance rejection using the FIMC using the generic model with $\alpha = 0.45$ and $r = 0.5$.

9.3 Rejecting Measured Disturbances in Non-Linear Uncertain Systems

IMC has been used to reject both input and output disturbances (Martin and Haber Guerra, 2009) but most applications assume that only unmeasured output disturbances are present (Rivals and Personnaz, 2000; Haber and Alique, 2004), that any input disturbances can be measured accurately (Nahas *et al.*, 1992; Sousa *et al.*, 1997) or that the input disturbances can be treated as if they were a combination of unmeasured output disturbances and modelling errors (Economou *et al.*, 1986; Brown *et al.*, 1997).

External disturbances cannot always be assumed to be equivalent to a disturbance acting additively on the output of a non-linear system. For example, a particular problem arises if an external disturbance changes the relationship between the plant output and the control signal. Complete disturbance rejection then requires that the internal model is not only perfect but also includes the effect of all of the external disturbances other than output disturbances (Economou *et al.*, 1986). However, a perfect internal model can only be created if these disturbances can be measured accurately, in which case a better approach might be to let the inverse controller take them into account using feedforward (Nahas *et al.*, 1992), as shown in Figure 9.7.

The use of feedforward for disturbance rejection is usually based on the assumption that a perfect measurement of the disturbance and an accurate model of the effect of the disturbance on the behaviour of the plant are available (Petersson *et al.*, 2003). Some disturbances are however difficult to measure accurately (Tan and Dexter, 2006) and, in many applications, the effects of disturbances are difficult to model from first principles (Ren *et al.*, 2009). One option is to low-pass filter the disturbance measurements. However, measurement noise within the bandwidth of the closed-loop control system can lead to unnecessary control activity as it cannot be filtered out without affecting the control performance (Boukezzoula *et al.*, 2003).

IMC schemes that incorporate disturbance measurements have been proposed (Xu, 1990; Nahas *et al.*, 1992; Sousa *et al.*, 1997; Sarimveis and Bafas, 2003; Molov *et al.*, 2004; Flores *et al.*, 2005) but it is assumed that there are no uncertainties associated with the measurements.

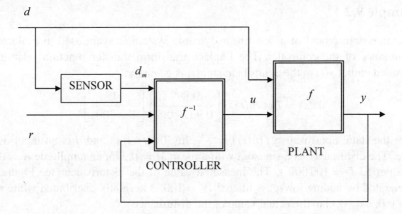

Figure 9.7 Block diagram showing the use of feedforward for disturbance rejection.

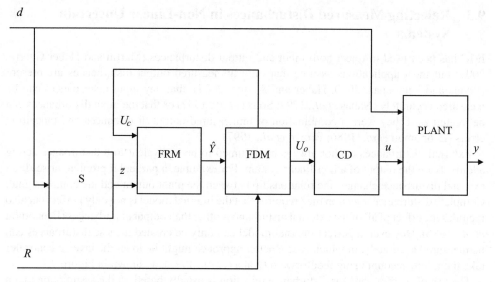

Figure 9.8 FMPC scheme using feedforward of a measured disturbance.

9.4 Fuzzy MPC with Feedforward

A block diagram of the fuzzy FRM-based MPC scheme incorporating feedforward of the measured disturbance is shown in Figure 9.8.

The design of the fuzzy MPC incorporating feedforward is similar to that used in Example 8.7. The only differences are that the fuzzy FRM used by the controller is identified from training data that includes noisy measurements of the disturbance, and feedforward is based on singleton defuzzification of the crisp measured value of the disturbance z generated by the sensor S.

Example 9.2

A Hammerstein model of a non-linear dynamic system is again used to evaluate the performance of the controller. The Laplace transform transfer function relating the controlled output $y(t)$ to the control signal $u(t)$ is given by:

$$Y(s) = L(f(u(t))) \left[\frac{0.899}{0.9d + 0.8} \right] \left[\frac{1}{1 + 200s} \right] \qquad (9.6)$$

where the static non-linearity $f(u(t)) = \frac{1}{3.434} \ln(30\,u(t) + 1)$ and d is an input disturbance. The disturbance is a sinusoid with a mean $M = 0.275$, an amplitude $A = 0.165$ and a period $T = 100\,000$ s. The measured value of the disturbance (see Figure 9.9) is generated by adding low-pass filtered ($\tau = 100$ s) normally distributed white noise ($m = 0.0$, $\sigma = 0.1$) to the actual value of the disturbance.

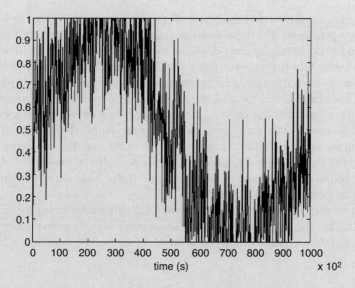

Figure 9.9 The measured value of the disturbance.

The FRM used in the FMPC is a discrete-time first-order autoregressive model with three inputs, $u(i-1)$, $y(i-1)$ and $z(i-1)$, and a sample time of 100 s. The numbers of reference sets that are used to describe the inputs are 21, 21 and 11, respectively. The fuzzy output $y(i)$ is described by 41 reference sets. The fuzzy goal, which is held constant during testing, is defined by a discrete triangular membership function:

$$[0.0/0.450 \quad 0.5/0.475 \quad 1.0/0.500 \quad 0.5/0.525 \quad 0.0/0.550].$$

In this example, the FMPC is based on a specific model so that the uncertainty resulting from modelling errors is small compared to the uncertainty arising from inaccurate measurement of the disturbance. A simulation of the test plant ($k = 30$) is used to generate the training data for identifying the fuzzy model. A control signal in the range 0.0 to 1.0 is generated by clipping low-pass filtered ($\tau = 100$ s) white noise, uniformly distributed in the range -2.0 to 3.0. The disturbance is a sinusoid with a mean $M = 0.275$, an amplitude $A = 0.165$ and a period $T = 100\,000$ s over the training period (100 000 samples). The measured value of the disturbance is generated by adding low-pass filtered ($\tau = 100$ s) normally distributed white noise ($m = 0.0$, $\sigma = 0.2$) to the value of the disturbance. After identification, the fuzzy model is post-processed five times using the method described in Section 5.5.1.

Sum-product inferencing is used to predict the one-step-ahead fuzzy prediction of the plant output over a discrete UoD from 0.0 to 1.0. The measured value of the disturbance is fuzzified using singleton fuzzification. The fuzzy prediction is normalized so that the sum of the possibilities is unity. If no rules are fired, all values of the next plant output are assumed to be equally possible.

The degree of satisfaction of the goal is the overlap area divided by the area under the fuzzy output prediction. The fuzzy decision-maker determines the degree of satisfaction of the goal for 11 (equally spaced) candidate values of the control signal. If at least one candidate value of the fuzzy control signal satisfies the goal then the fuzzy control signal is conditionally defuzzified using height defuzzification. If none of the candidate values of the control signal satisfy the goal, the lower and upper bounds of the support of the fuzzy goal are found and u is set to u_{max} or u_{min} depending on whether or not the lower bound of the support of the fuzzy prediction for the minimum value of the control signal is closer to the upper bound of the support of the fuzzy goal than the upper bound of the support of the fuzzy prediction for the maximum value of the control signal is to the lower bound of the support of the fuzzy goal. It should be noted that this method assumes that increasing u will increase y.

As can be seen in Figure 9.10, the controller does partially reject the disturbance (upper plot) but the measurement noise causes small high-frequency changes in the control signal (lower plot) when the conditional defuzzification threshold is 1.0.

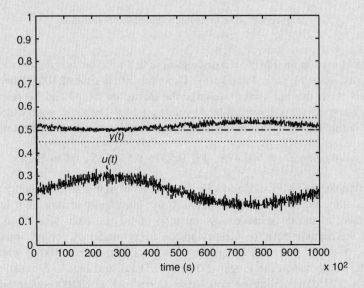

Figure 9.10 Disturbance rejection using the FMPC based on the specific model using feedforward ($\alpha = 1.0$).

Lowering the conditional defuzzification threshold significantly reduces the control activity, but it has relatively little effect on the controller's ability to reject the disturbance (see Figures 9.11 and 9.12).

The fuzziness of the predictions (see Figure 9.13) is almost entirely due to the measurement noise introduced while generating the training data used to identify the FRM of the plant.

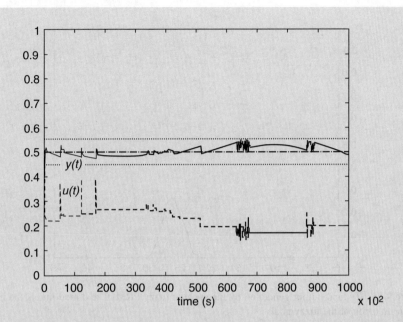

Figure 9.11 Disturbance rejection using the FMPC based on the specific model using feedforward ($\alpha = 0.8$).

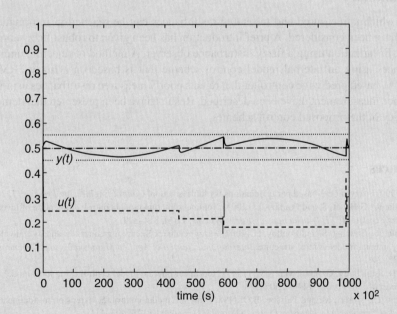

Figure 9.12 Disturbance rejection using the FMPC based on the specific model using feedforward ($\alpha = 0.7$).

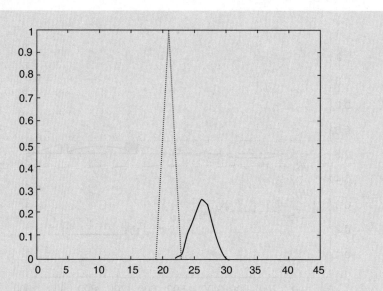

Figure 9.13 Typical output generated by the specific fuzzy FRM. The dotted line is the membership function of the fuzzy goal.

9.5 Summary

Ways in which unmeasured and measured disturbances can be rejected in information-poor systems have been considered. A brief introduction has been given to robust fuzzy control and feedback linearization using a fuzzy disturbance observer. A method of rejecting unmeasured disturbances using an internal model control scheme that is based on a fuzzy FRM, and a fuzzy FRM-based predictive controller that rejects poorly measured disturbances using a fuzzy disturbance measurement, have been described. Results have been presented that demonstrate the efficacy of the proposed control schemes.

References

Balle, P. (1999) Fuzzy-model-based parity equations for fault isolation. *Control Engineering Practice*, **7**, 261–270.

Boukezzoula, R., Galichet, S. and Foulloy, L. (2003) Nonlinear internal model control: application of inverse model based fuzzy control. *IEEE Transactions on Fuzzy Systems*, **11**(6), 814–829.

Boukezzoula, R., Galichet, S. and Foulloy, L. (2007) Fuzzy feedback linearizing controller and its equivalence with the fuzzy internal model control structure. *International Journal of Applied Mathematics and Computer Science*, **17**(2), 233–248.

Brown, M.D., Lightbody, G. and Irwin, G.W. (1997) Non-linear internal model control using local model networks. *Proceedings of IEE, Part D*, **144**(6), 505–514.

Economou, C.G., Morari, M. and Palsson, B.O. (1986) Internal model control. 5. Extension to nonlinear systems. *Industrial & Engineering Chemistry Process Design & Development*, **25**, 403–411.

Edgar, C.R. and Postlethwaite, B.E. (2000) MIMO fuzzy internal model control. *Automatica*, **34**, 867–877.

Flores, A., Saez, D., Araya, J., *et al.* (2005) Fuzzy predictive control of a solar power plant. *IEEE Transactions on Fuzzy Systems*, **13**(1), 58–68.

Gao, Z., Shi, X. and Ding, S.X. (2008) Fuzzy state/disturbance observer design for T-S fuzzy systems with application to sensor fault estimation. *IEEE Transactions on Systems, Man and Cybernetics, Part B*, **38**(3), 875–880.

Gormandy, B.A. and Postlethwaite, B.E. (2001) Model-based control using fuzzy relational models. IEEE International Conference on Fuzzy Systems, 1581–1584.

Haber, R.E. and Alique, J.R. (2004) Nonlinear internal model control using neural networks: an application for machining process. *Neural Computing & Applications*, **13**, 47–55.

Huang, D. and Nguang, S.K. (2006) Robust H-infinite static output feedback control of fuzzy systems: an ILMI approach. *IEEE Transactions on Systems, Man and Cybernetics, Part B*, **36**(1), 216–222.

Kim, E. (2002) A fuzzy disturbance observer and its application to control. *IEEE Transactions on Fuzzy Systems*, **10**(1), 77–84.

Kim, E. (2003) A discrete-time fuzzy disturbance observer and its application to control. *IEEE Transactions on Fuzzy Systems*, **11**(3), 399–410.

Kim, E. and Lee, S. (2005) Output feedback tracking control of MIMO systems using a fuzzy disturbance observer and its application to the speed control of a PM synchronous motor. *IEEE Transactions on Fuzzy Systems*, **13**(6), 725–741.

Lo, J-C. and Lin, M-L. (2006) Robust H-infinity control for fuzzy systems with Frobenius norm-bounded uncertainties. *IEEE Transactions on Fuzzy Systems*, **14**(1), 1–15.

Martin, A.G. and Haber Guerra, R.E. (2009) Internal model control based on a neurofuzzy system for network applications. A case study on the high performance drilling process. *IEEE Transactions on Automation Science and Engineering*, **6**(2), 367–372.

Molov, S., Babuska, R., Abonyi, J. and Verbruggen, H.B. (2004) Effective optimisation for fuzzy model predictive control. *IEEE Transactions on Fuzzy Systems*, **12**(5), 661–675.

Mukhopadhyay, S. and Narendra, K.S. (1993) Disturbance rejection in nonlinear systems using neural networks. *IEEE Transactions on Neural Networks*, **4**(1), 63–72.

Nahas, E.P., Henson, M.A. and Seborg, D.E. (1992) Nonlinear internal model control strategy for neural networks. *Computers & Chemical Engineering*, **16**(12), 1039–1057.

Petersson, M., Arzen, K. E. and Hagglund, T. (2003) A comparison of two feedforward control structure assessments methods. *International Journal of Adaptive Control and Signal Processing*, **17**, 609–624.

Ren, X., Lai, C.Y., Venkataramanan, V., *et al.* (2009) Feedforward control based on neural networks for disturbance rejection in hard disk drives. *IET Control Theory and Applications*, **3**(4), 411–418.

Rivals, I. and Personnaz, L. (2000) Nonlinear internal model control using neural networks: application to processes with delay and design issues. *IEEE Transactions on Neural Networks*, **11**(1), 80–90.

Sarimveis, H. and Bafas, G. (2003) Fuzzy model predictive control of non-linear processes using a genetic algorithm. *Fuzzy Sets and Systems*, **139**, 59–80.

Sousa, J.M., Babuska, R. and Verbruggen, H.B. (1997) Fuzzy predictive control applied to an air-conditioning system. *Control Engineering Practice*, **5**(10), 1395–1406.

Tan, H. and Dexter, A.L. (2006) Estimating airflow rates in air-handling units from actuator control signals. *Building and Environment*, **41**(10), 1291–1298.

Xu, C-W. (1990) Analysis and feedback/feedforward control of fuzzy relational systems. *Fuzzy Sets and Systems*, **35**, 105–113.

Part III

Online learning in information-poor systems

10

Online Model Identification in Information-Poor Environments

10.1 Online Fuzzy Identification Schemes

A number of online fuzzy identification schemes have been proposed but few of them can be used to identify FRMs that are capable of generating a fuzzy output (Xu and Lu, 1987; Shaw and Kruger, 1992; Chen *et al.*, 1994; Bourke and Fisher, 2000). The two main approaches to identifying fuzzy FRMs online are recursive fuzzy least-squares and variants of the recursive RSK identification scheme (see also Section 5.3).

10.1.1 Recursive Fuzzy Least-Squares

The rule confidences of a fuzzy FRM that minimize a fuzzy cost function of the form (see Section 5.3)

$$J = \sum_{n=1}^{L} \sum_{k=1}^{K} \left[Poss_{Y_k}(n) - P\hat{o}ss_{C_k}(n) \right]^2$$

can be estimated using a recursive least-squares parameter estimator (Wong *et al.*, 2000; Campello and Amaral, 2001, 2003).

Example 10.1

Consider the online identification of the fuzzy FRM of a first-order dynamic system:

$$Y(n) = Y(n-1) \circ U(n-1) \circ R \qquad (10.1)$$

Let the difference between the kth element of the output data at the $(n+1)$th sample time and the kth element of the fuzzy prediction generated by the FRM be $\varepsilon_k(n)$, defined

$$\varepsilon_k(n) = Poss_{Y_k}(n+1) - Poss_{C_k}(n+1) \qquad (10.2)$$

Monitoring and Control of Information-Poor Systems: An Approach based on Fuzzy Relational Models, First Edition.
Arthur L. Dexter.
© 2012 John Wiley & Sons, Ltd. Published 2012 by John Wiley & Sons, Ltd.

Note that $\varepsilon_k(n)$ is only a function of the rule confidences associated with rules that have the conclusion 'Y is C_k'. If it is assumed that max-product inferencing is used for one-step-ahead prediction,

$$\hat{Poss}_{C_k}(n+1) = \underset{i=1}{\overset{N}{Max}}\,\underset{j=1}{\overset{M}{Max}}\{\mu_{A_i}(u[n])\mu_{B_j}(y[n])R_{i,j,k}\}$$

and the partial derivative of $\varepsilon_k(n)$ with respect to the rule confidence associated with the rule IF $u[n]$ is A_i AND $y[n]$ is B_j THEN $y[n+1]$ is C_k is given by:

$$\frac{\partial \varepsilon_k(n)}{\partial R_{i,j,k}} = \mu_{A_i}(u[n])\mu_{B_j}(y[n]) \tag{10.3}$$

if

$$\mu_{A_i}(u[n])\mu_{B_j}(y[n])R_{i,j,k} > Max_{p \neq i}Max_{q \neq j}\mu_{A_p}(u[n])\mu_{B_q}(y[n])R_{p,q,k}$$

or

$$\frac{\partial \varepsilon_k(n)}{\partial R_{i,j,k}} = 0 \tag{10.4}$$

if $\mu_{A_i}(u[n])\mu_{B_j}(y[n])R_{i,j,k} \leq Max_{p \neq i}Max_{q \neq j}\mu_{A_p}(u[n])\mu_{B_q}(y[n])R_{p,q,k}$.
Defining the parameter vector

$$\theta_k(n) = \left[R_{1,1,k}(n), R_{2,1,k}(n), \ldots R_{N,M,k}(n)\right]^T \tag{10.5}$$

and the data vector

$$\psi_k(n) = \left[\frac{\partial \varepsilon_k(n)}{\partial R_{1,1,k}}, \frac{\partial \varepsilon_k(n)}{\partial R_{2,1,k}}, \ldots \frac{\partial \varepsilon_k(n)}{\partial R_{N,M,k}}\right]^T \tag{10.6}$$

it can be shown (Wong et al., 2000) that the well-known recursive least-squares algorithm:

$$K_k(n) = \frac{P_k(n-1)\psi_k(n)}{[1 + \psi_k^T(n)P_k(n-1)\psi_k(n)]} \tag{10.7}$$

$$P_k(n) = P_k(n-1) - K_k(n)\psi_k^T(n)P_k(n-1) \tag{10.8}$$

$$\theta_k(n) = \theta_k(n-1) + K_k(n)\varepsilon_k(n) \tag{10.9}$$

can then be used to update the values of the rule confidences.

Note that it may be necessary to constrain the estimated rule confidences if they are to have meaningful values in the expected range.

10.1.2 Recursive Forms of the RSK Algorithm

Consider a rule-base consisting of fuzzy rules of the form:

IF X_1 is A_{1,k_1} **AND** X_2 is A_{2,k_2} **AND** $\ldots$ X_n is A_{n,k_n} **THEN** Y is B_q

where the fuzzy reference sets for $X_1, X_2, \ldots X_n$ and Y are $A_{1,1}$ to A_{1,r_1}, $A_{2,1}$ to $A_{2,r_2} \ldots$, and B_1 to B_r, respectively. Let the rule confidence associated with this rule be $\hat{R}_{A_{1k_1},\ldots,A_{nk_n},B_q}(t)$.

10.1.2.1 Recursive RSK

The original RSK algorithm (see Section 5.3) can be written in recursive form (Postlethwaite, 1994):

$$\hat{R}_{A_{1k_1},\ldots,A_{nk_n},B_q}(t) = \frac{f_{A_{1k_1},\ldots,A_{nk_n}}(x(t))\mu_{B_q}(y(t)) + \hat{R}_{A_{1k_1},\ldots,A_{nk_n},B_q}(t-1)\cdot F_{A_{1k_1},\ldots,A_{nk_n}}(t-1)}{F_{A_{1k_1},\ldots,A_{nk_n}}(t)}$$

(10.10)

where $f_{A_{1k_1},\ldots,A_{nk_n}}(x(t)) = T\{\mu_{A_{1k_1}}(x_1(t)), \mu_{A_{2k_2}}(x_2(t)), \ldots, \mu_{A_{nk_n}}(x_n(t))\}$, $T\{\}$ is a t-norm and

$$F_{A_{1k_1},\ldots,A_{nk_n}}(t) = f_{A_{1k_1},\ldots,A_{nk_n}}(x(t)) + F_{A_{1k_1},\ldots,A_{nk_n}}(t-1)$$

(10.11)

Note that the elements of the F-matrix are indicators of how strongly and how often each combination of the inputs has occurred in the training data.

10.1.2.2 Recursive RSK with Forgetting

The RSK algorithm can be easily modified to include a forgetting factor (Tan and Dexter, 2000):

$$\hat{R}_{A_{1k_1},\ldots,A_{nk_n},B_q}(t) = \frac{f_{A_{1k_1},\ldots,A_{nk_n}}(x(t))\mu_{B_q}(y(t)) + \lambda\cdot\hat{R}_{A_{1k_1},\ldots,A_{nk_n},B_q}(t-1)\cdot F_{A_{1k_1},\ldots,A_{nk_n}}(t-1)}{F_{A_{1k_1},\ldots,A_{nk_n}}(t)}$$

(10.12)

$$F_{A_{1k_1},\ldots,A_{nk_n}}(t) = f_{A_{1k_1},\ldots,A_{nk_n}}(x(t)) + \lambda F_{A_{1k_1},\ldots,A_{nk_n}}(t-1)$$

(10.13)

where λ $(0 \leq \lambda \leq 1)$ is the forgetting factor.

The scheme allows old data to be discounted as new information is gathered, without discarding information about the behaviour in infrequently encountered parts of the input space.

10.1.2.3 Recursive Modified RSK

A modified RSK scheme has been suggested to filter the effect of non-uniform training data (Tan and Dexter, 2000). When the recursive fuzzy identification algorithm is fed with input-output data that coincide with the apexes of the input sets, the resulting rule confidences will converge to the 'ideal' values. The learning scheme should therefore only update the rule confidences if a rule has been fired fully, so preventing the biasing of a correctly estimated rule confidence. This is achieved by introducing a matrix, size $\Pi_{i=1}^{n}r_i$, whose elements are given by:

$$\bar{f}_{A_{1k_1},\ldots,A_{nk_n}}(t) = \max(\bar{f}_{A_{1k_1},\ldots,A_{nk_n}}(t-1), f_{A_{1k_1},\ldots,A_{nk_n}}(t))$$

(10.14)

The estimated rule confidence is only updated if the associated activation level $f_{A_{1k_1},\ldots,A_{nkn}}(t)$ is higher than or equal to the maximum activation achieved in the past. The rule confidences are therefore updated recursively as follows:

$$\hat{R}_{A_{1k_1},\ldots,A_{nkn},B_q}(t) = \begin{cases} \dfrac{f_{A_{1k_1},\ldots,A_{nkn}}(x(t))\mu_{B_q}(y(t)) + \lambda \cdot \hat{R}_{A_{1k_1},\ldots,A_{nkn},B_q}(t-1) \cdot F_{A_{1k_1},\ldots,A_{nkn}}(t-1)}{F_{A_{1k_1},\ldots,A_{nkn}}(t)} \\ \qquad\qquad \text{if } \; f_{A_{1k_1},\ldots,A_{nkn}}(x(t)) \geq \bar{f}_{A_{1k_1},\ldots,A_{nkn}}(t), \\ \hat{R}_{A_{1k_1},\ldots,A_{nkn},B_q}(t-1) \qquad\qquad \text{otherwise} \end{cases}$$

$$(10.15)$$

where λ $(0 \leq \lambda \leq 1)$ is the forgetting factor.

The F matrix is updated recursively as follows:

$$F_{A_{1k_1},\ldots,A_{nkn}}(t) = \begin{cases} f_{A_{1k_1},\ldots,A_{nkn}}(x(t)) + \lambda F_{A_{1k_1},\ldots,A_{nkn}}(t-1) \\ \qquad \text{if } \; f_{A_{1k_1},\ldots,A_{nkn}}(x(t)) \geq \bar{f}_{A_{1k_1},\ldots,A_{nkn}}(t) \\ F_{A_{1k_1},\ldots,A_{nkn}}(t-1) \qquad\qquad \text{otherwise} \end{cases} \qquad (10.16)$$

A comparison of online fuzzy identification schemes (Wu and Dexter, 2003) has shown that the modified RSK fuzzy identification scheme makes better use of training data in noise-free systems. However, the modified RSK scheme can generate an inaccurate fuzzy relational model if the training data are noisy.

10.1.2.4 Recursive RSK with Rounding

The rounded RSK or RRSK scheme (Wu and Dexter, 2008) rounds the input data to the closest centres of the fuzzy input sets before it identifies the fuzzy relational model. Therefore, none of the training data will result in the identification errors that arise when the training data are not at the centre of the input sets. However, rounding errors will be introduced and these may cause prediction errors to be generated at the centre of the fuzzy input sets.

Consider Figure 10.1 where i represents the position of the centre of the ith fuzzy input set and x is the value of the input. When x is at the centres of the fuzzy inputs sets no rounding

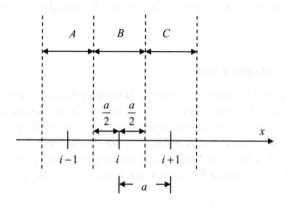

Figure 10.1 Rounding to the closest centres.

occurs, x remains the same value and no identification errors are generated by the identification scheme. If x is not at the centres of the fuzzy input sets and is in the range B, the scheme will round the x to the value at the centre of the ith fuzzy input set. Note that the maximum rounding error is $a/2$ and it depends on the number of fuzzy input sets that are used.

The rounded data will fully fire only rules with the same antecedent if the input training data are rounded to the closest centres of the fuzzy input sets. Consider the case where the rule confidences of the fuzzy relational model are updated recursively using the conventional RSK scheme (Ridley *et al.*, 1988). There are three possible situations as follows.

1. If the training data that fire the rule are also a member of the output fuzzy set, then $f_{A_{1k_1},\dots,A_{nk_n}}(x(t)) = 1$ and $\mu_{B_q}(y(t))$ is the degree of membership of the output set. The updating scheme then becomes:

$$\hat{R}_{A_{1k_1},\dots,A_{nk_n},B_q}(t) = \frac{\mu_{B_q}(y(t)) + \lambda \cdot \hat{R}_{A_{1k_1},\dots,A_{nk_n},B_q}(t-1) \cdot F_{A_{1k_1},\dots,A_{nk_n}}(t-1)}{F_{A_{1k_1},\dots,A_{nk_n}}(t)} \tag{10.17}$$

where

$$F_{A_{1k_1},\dots,A_{nk_n}}(t) = 1 + \lambda F_{A_{1k_1},\dots,A_{nk_n}}(t-1) \tag{10.18}$$

It should be noted that the value of the F matrix will converge to $\frac{1}{1-\lambda}$ after repeated training with these data.

2. If the training data fire the rule but are not members of the output fuzzy set, then $f_{A_{1k_1},\dots,A_{nk_n}}(x(t)) = 1$ and $\mu_{B_q}(y(t)) = 0$. The updating scheme then becomes:

$$\hat{R}_{A_{1k_1},\dots,A_{nk_n},B_q}(t) = \frac{\lambda \cdot \hat{R}_{A_{1k_1},\dots,A_{nk_n},B_q}(t-1) \cdot F_{A_{1k_1},\dots,A_{nk_n}}(t-1)}{F_{A_{1k_1},\dots,A_{nk_n}}(t)}$$

$$= \hat{R}_{A_{1k_1},\dots,A_{nk_n},B_q}(t-1)\left(\frac{\lambda \cdot F_{A_{1k_1},\dots,A_{nk_n}}(t-1)}{F_{A_{1k_1},\dots,A_{nk_n}}(t)}\right) \tag{10.19}$$

where

$$F_{A_{1k_1},\dots,A_{nk_n}}(t) = 1 + \lambda F_{A_{1k_1},\dots,A_{nk_n}}(t-1) \tag{10.20}$$

Since

$$\frac{\lambda \cdot F_{A_{1k_1},\dots,A_{nk_n}}(t-1)}{1 + \lambda F_{A_{1k_1},\dots,A_{nk_n}}(t-1)} < 1,$$

$\hat{R}_{A_{1k_1},\dots,A_{nk_n},B_q}(t) < \hat{R}_{A_{1k_1},\dots,A_{nk_n},B_q}(t-1)$ and repetition of this training data will cause the rule confidence to converge to zero.

3. If the training data do not fire the rule, then the updating scheme becomes:.

$$\hat{R}_{A_{1k_1},...,A_{nk_n},B_q}(t) = \hat{R}_{A_{1k_1},...,A_{nk_n},B_q}(t-1)$$

and

$$F_{A_{1k_1},...,A_{nk_n}}(t) = \lambda F_{A_{1k_1},...,A_{nk_n}}(t-1).$$

10.2   Effect of Poor-Quality and Incomplete Training Data

The main problem associated with online learning is that the training data are unlikely to be uniformly distributed over the input space of the model.

Prior knowledge is often used to pre-select the underlying structure of a model whose parameters are to be estimated online (Abonyi et al., 2000; Milanic et al., 2004) but it can also be used to improve the extrapolation properties of fuzzy models identified from experimental data. One approach is to use the prior knowledge to derive a simple first-principles model, which can be used to constrain the learning scheme. Several approaches to constrained offline identification have been suggested. The difference between time derivatives obtained from a first-principles model and the neurofuzzy model can be included in the objective function used for training with a back-propagation algorithm (Hoo et al., 2002). Data obtained from a simulation based on the first-principles model can be used to augment the training data obtained from the process itself (Linker and Seginer, 2004; Milanic et al., 2004). Alternatively, prior knowledge can be used to define bounds on the steady-state gain of the process, the settling time or the estimated parameters, and quadratic programming rather than least-squares optimization can be used to allow these constraints to be taken into account when the parameters are estimated (Abonyi et al., 2000). Unfortunately, the computational complexity of these techniques prohibits their use in most online applications (Johansen, 1996).

A global model of the behaviour of a non-linear dynamic system can only be identified if training data are obtained for a range of excitation frequencies over the entire input space (Laukonen and Passino, 1995; Sing and Postlethwaite, 2000), and several algorithms for generating system inputs that will produce uniform training data have been proposed. Persistent excitation is a particular problem in adaptive control system where the training data used for online identification must be obtained in closed-loop. One way of ensuring persistent excitation is to introduce dither into the closed-loop identification scheme (Valente de Oliveira and Lemos, 1995). The dither signal is usually a low-power white noise sequence, which is added to the output of the controller. There is a clear trade-off between the requirements of model identification and resulting control performance; the better the control performance the lower the information content of the training data. An alternative approach based on the use of a simple fixed controller during the training period is described in Section 11.4.2.

The fuzzy relational array can be post-processed to 'fill in' any 'holes' resulting from a lack of training data in some regions of the input space. For example, this can be achieved by interpolating rule confidences using weighting that is a function of the amount of training and a measure of similarity of the associated rules (Kelkar and Postlethwaite, 1998).

Data distribution can be taken into account during model identification if the parameter estimation is based on minimizing an objective function that is the sum of local objective functions, which are defined over different crisp or fuzzy regions in the input space. This allows model prediction errors in regions where there are little training data to be weighted so that they are as important as errors in regions where there are a great deal of training data (Sing and Postlethwaite, 2000).

10.3 Ways of Reducing the Computational Demands

There are two basic approaches to reducing the computational demands associated with maintaining a rule-base:

1. reduce the number of input variables used in the antecedent of the rules; and
2. reduce the number of fuzzy sets used to describe each input variable.

It is difficult to identify the optimal structure (the number of inputs, the number of rules, the number of fuzzy reference sets and the shapes of the membership functions) for the model from data unless a complete (fully representative) dataset is available. In practice, this is often not the case as the behaviour of the system to be modelled may change with time and a limited amount of time will be available within which to collect the data. As a result, many fuzzy identification schemes require the model's structure to be known *a priori* (i.e. the rule-base consists of a predefined set of rules) and only the values of the model parameters are changed online.

If there is little prior knowledge, a generic structure must be chosen (e.g. a large number of uniformly distributed reference sets must be defined) and the number of rules in the model will be greater than is necessary for a particular application. Such models often suffer from the 'curse of dimensionality'. One way of avoiding this problem is to generate rules online as new training data are acquired.

10.3.1 Evolving Fuzzy Models

Evolving fuzzy models (Angelov, 2002) gradually modify their structure and change their parameters online as new data become available. New rules are created, existing specific rules are replaced by more general rules and rules that are no longer valid are removed.

A measure of the distance between a new dataset and previous datasets can be used to determine the 'potential' of the new data to be used as the basis of a new rule. Conversely, existing rules with a low potential are candidates for deletion. Two ways of calculating the potential of data or rules have been suggested. Both are based on first-order Cauchy functions. The original definition of potential (Angelov and Buswell, 2002) will always generate low values if there are other rules or datasets that are remote from the rule or data whose potential is to be determined. A measure of the potential of the ith rule $P_i^*[t]$, which is more sensitive to data close to the centre of the rule than data remote from the rule, is given by (De Barros and Dexter, 2007b):

$$P_i^* [t] = 1 - \cfrac{1}{1 + \cfrac{1}{t} \cdot \sum_{v=1}^{t} \left(\cfrac{1}{\sum_{j=1}^{n+1} \left[z_j [v] - z_{ij}^* \right]^2} \right)} \tag{10.21}$$

where z_j is the jth component of the data set $z = [x^T | y] = [x_1 \ x_2 \ldots x_n \ y]^T$ and z_{ij}^* is the jth component of the centre of the ith rule. The potential can be updated recursively.

Let

$$P_i^* [t] = 1 - \cfrac{t}{t + \sum_{v=1}^{t} [D [v]]} \tag{10.22}$$

where

$$D [v] = \cfrac{1}{\sum_{j=1}^{n+1} \left[z_j [v] - z_{ij}^* \right]^2}.$$

By expanding the summation, $P_i^* [t]$ can be expressed as:

$$P_i^* [t] = 1 - \cfrac{t}{t + \sum_{v=1}^{t-1} [D [v]] + D [t]} \tag{10.23}$$

but

$$P_i^* [t - 1] = 1 - \cfrac{(t - 1)}{(t - 1) + \sum_{v=1}^{t-1} [D [v]]} \tag{10.24}$$

and therefore:

$$\sum_{v=1}^{t-1} [D [v]] = \cfrac{P_i^* [t - 1] \cdot (t - 1)}{1 - P_i^* [t - 1]}$$

We therefore have

$$P_i^* [t] = 1 - \cfrac{t}{t + \frac{P_i^* [t-1] \cdot (t-1)}{1 - P_i^* [t-1]} + D [t]} \tag{10.25}$$

or

$$P_i^*[t] = \frac{(t-1) \cdot P_i^*[t-1] + \dfrac{(1-P_i^*[t-1])}{n+1} \sum\limits_{j=1}^{n+1} \left[z_j[t] - z_{ij}^* \right]}{t - P_i^*[t-1] + \dfrac{(1-P_i^*[t-1])}{n+1} \sum\limits_{j=1}^{n+1} \left[z_j[t] - z_{ij}^* \right]}. \tag{10.26}$$

The rule-base can be initialized by using the first dataset as a rule centre for the first rule, or it can be populated with rules based on prior knowledge. In the former case, the potential of the first rule is initially set to zero because the first dataset might be an outlier that is not representative of the true behaviour of the system.

An online clustering scheme, based on a Euclidian measure of the distance between the current data and the closest of the existing rules, can be used to determine whether the new data should be used to generate a new rule (De Barros and Dexter, 2007a). The new data are added as a new point rule if the distance to the closest rule, which is measured using the modal values on the universes of discourse of the antecedent and consequent fuzzy sets, is such that the current data are sufficiently outside the *region of influence* of the nearest rule in the rule-base; otherwise, the rule-base remains unchanged. As the current data might be an outlier, the potential of a new rule is set to zero as soon as it is created.

An existing rule must be deleted before a new rule can be added if the total number of rules would then exceed a user-specified maximum. If this situation occurs, the rule in the current rule-base with the lowest potential is the one chosen for deletion. Note that, because the rule creation scheme is based only on distance in input-output space, it is possible for a data outlier to be added as a rule. However, the potential of an 'outlier rule' will remain low because there is a low probability of encountering similar data. Such a rule will therefore be a strong candidate for deletion when a rule must be deleted. An alternative approach is to base the condition for creation of a new rule on both the closest distance in input-output space and the potential of the new dataset (Angelov and Filev, 2004).

The current data are used to create a new rule in a list of rules to be merged if the nearest input distance is such that the new input data are sufficiently outside the region of influence of the closest rule, and the output distance is such that the new output data are sufficiently similar to the consequent of the closest rule (De Barros and Dexter, 2007a). The potential of the newly created rule is initialized to zero on the assumption that the new data could be an outlier. A new rule is merged with the nearest rule only when its potential becomes high enough for it not to be an outlier. Rules in the merging list are deleted if their potentials have not reached a specified level after a predefined time.

The process of merging (Chen and Linkens, 2003) increases the region of influence of an existing rule, as is shown in Figure 10.2. The membership function μ_B is the triangular membership function of the jth input of the new rule waiting to be merged, μ_A is the jth input membership function of the closest rule and μ_C is the final merged membership function. No change is necessary when μ_B is completely inside μ_A. The region of influence of the rule increases but the output remains similar to the original rule consequent.

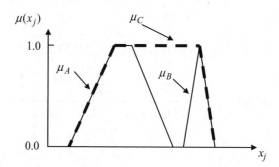

Figure 10.2 Merging of a new rule with an existing rule.

An alternative approach is to check new rules for consistency before they are included in the rule-base (Carmona *et al.*, 2002). The degree of contradiction depends on measures of the similarity of the antecedents A and dissimilarity of the consequents B (see also Section 6.1). Both measures can be calculated using a simplified (assumes trapezoidal sets) geometric measure of dissimilarity based on the degree of dissemblance $\Delta(A, B)$ (see Section 6.2.3):

$$\Delta(A, B) = \int\limits_0^1 \Delta_\alpha(A, B)\, d\alpha \tag{10.27}$$

where $\Delta_\alpha(A, B) = \dfrac{\left|a_1^\alpha - b_1^\alpha\right| + \left|a_2^\alpha - b_2^\alpha\right|}{2\left[\max(\left|a_2^\alpha, b_2^\alpha\right|) - \min(a_1^\alpha, b_1^\alpha)\right]}$.

Example 10.2

Consider two fuzzy rules of the form:

$$R_i: \text{IF } X \text{ is } A_i \text{ THEN } Y \text{ is } B_i$$
$$R_j: \text{IF } X \text{ is } A_j \text{ THEN } Y \text{ is } B_j$$

The degree of contradiction of the two rules $C(R_i, R_j)$ is given by:

$$C(R_i, R_j) = [1 - \Delta(A_i, A_j)]\Delta(B_i, B_j) \tag{10.28}$$

To reduce the effects of noisy data, and because the new data could be an outlier, the rule confidence associated with a newly created rule $R_{M+1, B_q}[t]$ is initialized as follows:

$$R_{M+1, B_q}[t] = \frac{\left(\sum\limits_{i=1}^M \left[\mu_{\mathcal{R}_i}\left(\boldsymbol{x}[t]\right) \cdot R_{i, B_q}[t]\right]\right) + \left(\mu_{\mathcal{R}_{M+1}}\left(\boldsymbol{x}[t]\right) \cdot \mu_{B_q}(y[t])\right)}{\left(\sum\limits_{i=1}^M \left[\mu_{\mathcal{R}_i}\left(\boldsymbol{x}[t]\right)\right]\right) + \mu_{\mathcal{R}_{M+1}}\left(\boldsymbol{x}[t]\right)} \tag{10.29}$$

where $\mu_{\mathcal{R}_i}$ is a multi-dimensional membership function describing the antecedent of the ith rule.

Because the current input $x[t]$ necessarily fully fires the newly created $(M+1)$th rule,

$$\mu_{\mathcal{R}_{M+1}}(x[t]) = 1$$

and

$$R_{M+1,B_q}[t] = \frac{\left(\sum_{i=1}^{M}\left[\mu_{\mathcal{R}_i}(x[t]) \cdot R_{i,B_q}[k]\right]\right) + \mu_{B_q}(y[t])}{\left(\sum_{i=1}^{M}\left[\mu_{\mathcal{R}_i}(x[t])\right]\right) + 1} \qquad (10.30)$$

The consequents of the existing rules must also be updated to take account of the new data. The rule confidence $R_{i,B_q}[t]$ of the ith rule in the existing rule-base is updated as follows:

$$R_{i,B_q}[t] = \frac{\sum_{v=1}^{t}\left[\lambda^{(t-v)} \cdot \mu_{\mathcal{R}_i}(x[v]) \cdot \mu_{i,B_q}y[v]\right]}{\sum_{v=1}^{k}\left[\lambda^{(t-v)} \cdot \mu_{\mathcal{R}_i}(x[v])\right]} \qquad (10.31)$$

where $\mu_{\mathcal{R}_i}[t]$ is the degree to which the current data $x[t]$ fires the ith rule and λ $(0 \leq \lambda \leq 1)$ is a forgetting factor (see Section 10.1.2). A small value of λ causes the estimator to greatly discount the past data, whereas a value of unity causes it to take equal account of all of the past data.

The estimator can be written in a recursive form; we therefore have

$$R_{i,B_q}[t] = \frac{\lambda \cdot R_{i,B_q}[t-1] \cdot F_i[t-1] + \mu_{B_q}(y[t]) \cdot \mu_{\mathcal{R}_i}(x[t])}{F_i[t]} \qquad (10.32)$$

where $F_i[t]$, is given by:

$$F_i[t] = \lambda \cdot F_i[t-1] + \mu_{\mathcal{R}_i}(x[t]) \qquad (10.33)$$

As the new data may be an outlier, the value of F associated with a new rule is initialized to unity when it is created, that is, $F_{R+1}[t] = 1$.

10.3.2   Hierarchical Fuzzy Models

Another way of avoiding the exponential increase in the number of rules as the number of input variables increases is to introduce hierarchy into the rule-base. The number of rules in a hierarchical fuzzy system is a linear rather than exponential function of the number of inputs (Raju *et al.*, 1991). It can be shown that hierarchical fuzzy systems are also universal approximators (Wang, 1999; Wei and Wang, 2000).

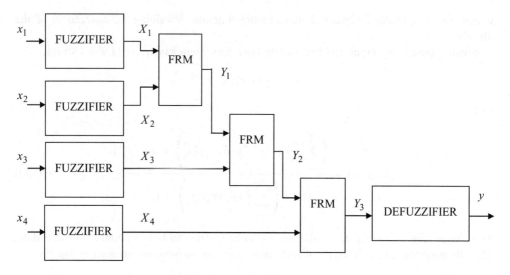

Figure 10.3   Hierarchical FRM with 4 inputs.

The most influential inputs should be used in the first level of the hierarchy and the next most important in the second level and so on. For example, the rules in the first level would have the following form:

$$\text{IF } X_1 \text{ is } A_1 \text{ THEN } Y_1 \text{ is } C_1$$

whereas the rules of the second level would be of the following form:

$$\text{IF } Y \text{ is } C_1 \text{ AND } X_2 \text{ is } B_2 \text{ THEN } Y_2 \text{ is } D_2$$

The output from the first-level rule may be regarded as an approximation, which is subsequently refined by the second-level rule. This process is repeated for each further level of rules in the hierarchy. It can be shown that the total number of rules is minimized when only one new input variable is added at each level (Raju *et al.*, 1991; Raju and Zhou, 1993).

For example, the curse of dimensionality in FRMs can be alleviated by using a hierarchical FRM consisting of fuzzy FRM submodels each with only two inputs (Meleiro *et al.*, 2006) (see Figure 10.3). A hierarchical FRM of this type has been shown to be equivalent to a single-level FRM (Campello and Amaral, 2006).

The introduction of hierarchy can also prioritize the rules (see Figure 10.4) if a *prioritized aggregation operator* is used to aggregate their outputs (Yager, 1993).

For example, the outputs of the rules at the jth and $(j-1)$th levels could be combined using the aggregation operator: $G_j(z) = G_{j-1}(z) + (1 - \alpha_{j-1})F_j(z)$, where z is the variable described by the fuzzy sets in the consequents of the rules and $\alpha_{j-1} = Max_z G_{j-1}(z)$, the largest degree of membership in G_{j-1}. G_j is the output produced when the output of the rules at the jth level are combined with those at the previous level, G_{j-1} is the output produced by the previous

INPUTS

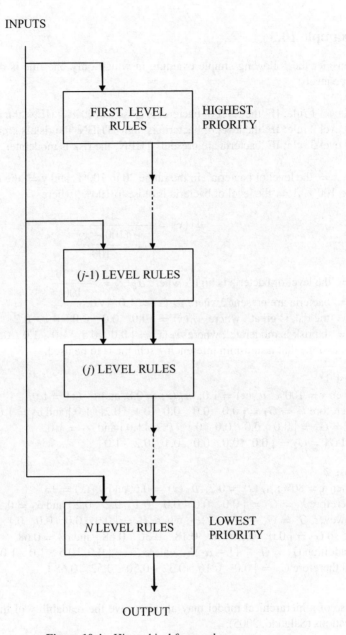

Figure 10.4 Hierarchical fuzzy rules.

$(j-1)$th level and F_j is the output inferred from the rules at the jth level. In this case the more specific the rules are, the higher is their level in the hierarchy. Generic rules are therefore at the lowest level in the hierarchy. The prioritized aggregation operator ensures that the more specific rules are tried first and that, if they produce a highly possible solution, the more generic rules are not used.

Example 10.3

Consider the following simple example in which only one rule is defined at each of three levels:

- Level 1 rule: IF 'the level of bacteria is close to 100%' THEN 'the risk is great';
- Level 2 rule: IF 'the level of bacteria is high' THEN 'the risk is great';
- Level 3 rule: IF 'bacteria are present' THEN 'the risk is moderate.

Let $x =$ 'the level of bacteria', in the range 20 to 100%, and y = 'the risk', in the range 0 to 100%, $A =$ 'the level of bacteria is close to 100%' where

$$\mu_A(x) = \frac{1}{1 + \dfrac{(100 - x)^2}{100}},$$

$B =$ 'the level of bacteria is high', where $\mu_B(x) = \dfrac{(x - 20)}{100}$,

$C =$ 'bacteria are present', where $\mu_C(x) = 1.0 \quad \forall x,$

$D =$ 'the risk is great', where $\mu_D(x) = \begin{bmatrix} 0.0 & 0.0 & 0.0 & 0.1 & 0.2 & 1.0 \end{bmatrix}$,

$E =$ 'the risk is moderate', where $\mu_E(x) = \begin{bmatrix} 0.0 & 0.5 & 1.0 & 1.0 & 0.5 & 0.0 \end{bmatrix}$,

and assume that a sum-min inferencing scheme is to be used.

Case 1

When $x = 100\%$: $\mu_A(x) = 1.0$, $\mu_B(x) = 0.8$, and $\mu_C(x) = 1.0$.

Therefore $F_1 = G_1 = \begin{bmatrix} 0.0 & 0.0 & 0.0 & 0.1 & 0.2 & 1.0 \end{bmatrix}$ and $\alpha_1 = 1.0$,

$G_2 = G_1 = \begin{bmatrix} 0.0 & 0.0 & 0.0 & 0.1 & 0.2 & 1.0 \end{bmatrix}$ and $\alpha_2 = 1.0$,

and $G_3 = G_2 = \begin{bmatrix} 0.0 & 0.0 & 0.0 & 0.1 & 0.2 & 1.0 \end{bmatrix}$

Case 2

When $x = 80\%$: $\mu_A(x) = 0.2$, $\mu_B(x) = 0.6$, and $\mu_C(x) = 1.0$

Therefore $F_1 = G_1 = \begin{bmatrix} 0.0 & 0.0 & 0.0 & 0.1 & 0.2 & 0.2 \end{bmatrix}$ and $\alpha_1 = 0.2$.

However, $G_2 = G_1 + (1 - \alpha_1)F_2$, where $F_2 = \begin{bmatrix} 0.0 & 0.0 & 0.0 & 0.1 & 0.2 & 0.6 \end{bmatrix}$

and so $G_2 = \begin{bmatrix} 0.0 & 0.0 & 0.0 & 0.18 & 0.36 & 0.68 \end{bmatrix}$ and $\alpha_2 = 0.68$.

In addition $G_3 = G_2 + (1 - \alpha_2)F_3$, where $F_3 = \begin{bmatrix} 0.0 & 0.5 & 1.0 & 1.0 & 0.5 & 0.0 \end{bmatrix}$

and therefore $G_3 = \begin{bmatrix} 0.0 & 0.16 & 0.32 & 0.50 & 0.52 & 0.68 \end{bmatrix}$

The use of a hierarchical model may also improve the readability of the rule-base in some applications (Salgado, 2005).

10.3.2.1   Automatic Generation of Hierarchical Fuzzy Models

In most cases, expert knowledge is used to determine the number of levels, the inputs at each level and the number and membership functions for the fuzzy sets describing each input. An iterative scheme based on evolutionary computing that can automatically generate the optimal hierarchical structure for a T-S fuzzy model has however been proposed (Chen *et al.*, 2007).

An evolving approach has also been suggested for constructing hierarchical fuzzy models online (Yager, 1998). The structure of the rule-base is hierarchical based on rule prioritization. The method of generating the rule-base uses a measure of the closeness of the observed and predicted crisp values of the system output to find exceptions to the current rule base.

The strength of the ith exception M_i depends on its closeness to the new exception:

$$M_i(n) = M_i(n-1) + Pe^{-Dist\{(x,y)-(x_i,y_i)\}} \tag{10.34}$$

where $P = 1 - \text{Close}(y, \hat{y})$ and $\text{Close}(y, \hat{y})$ is a measure of the closeness of the observed and predicted values. Each exception is a point rule in the top level of the hierarchy. A new rule is created in the middle level of the hierarchy whenever the strength of an exception exceeds a specified threshold. Point rules that are now accounted for by the new rule are then deleted. The default rules are at the lowest level of the hierarchy.

10.4 Summary

Different ways in which fuzzy FRMs can be identified online have been described and the practical issues that must be taken into account have been explained. Several online fuzzy identification schemes (recursive fuzzy least-squares and recursive forms of RSK) have been described and ways of reducing the effect of poor-quality and incomplete training data (incorporating prior knowledge, improving the information content of the training data or taking data distribution into account) have been explained. Two approaches to reducing the associated computational demands (evolving FRMs and hierarchical FRMs) have also been discussed.

References

Abonyi, J., Babuska, R., Verbruggen, H.B. and Szeifert, F. (2000) Incorporating prior knowledge in fuzzy model identification. *International Journal of Systems Science*, **31**(15), 657–667.

Angelov, P.P. (2002) *Evolving Rule-Based Models*, Physica-Verlag, Heidelberg and New York.

Angelov, P. and Buswell, R. (2002) Identification of evolving fuzzy rule-based models. *IEEE Transactions on Fuzzy Systems*, **10**(5), 667–677.

Angelov, P.P. and Filev, D.P. (2004) An approach to online identification of Takagi-Sugeno fuzzy models. *IEEE Transactions on Systems, Man and Cybernetics*, **34**(1), 484–498.

Bourke, M.M. and Grant Fisher, D. (2000) Identification algorithms for fuzzy relational matrices, Part 2: optimizing algorithms. *Fuzzy Sets and Systems*, **109**, 321–341.

Campello, R.J.G.B. and Amaral, W.C. (2001) Modeling and linguistic knowledge extraction from systems using fuzzy relational models. *Fuzzy Sets and Systems*, **121**(1), 113–126.

Campello, R.J.G.B. and Amaral, W.C. (2003) Towards true linguistic modelling through optimal numerical solutions. *International Journal of Systems Science*, **34**(2), 139–157.

Campello, R.J.G.B. and Amaral, W.C. (2006) Hierarchical fuzzy relational models: linguistic interpretation and universal approximation. *IEEE Transactions on Fuzzy Systems*, **14**(3), 446–453.

Carmona, P., Castro, J.L. and Zurita, J.M. (2002) Contradiction sensitive fuzzy model-based control. *International Journal of Approximate Reasoning*, **30**, 107–129.

Chen, J.Q., Lu, J.H. and Chen, L.J. (1994) An on-line fuzzy identification algorithm for fuzzy systems. *Fuzzy Sets and Systems*, **64**(11), 63–72.

Chen, M.-Y. and Linkens, D.A. (2003) Rule-base self-generation and simplification for data-driven fuzzy models. *Fuzzy Sets and Systems*, **142**(2), 243–265.

Chen, Y., Yang, B., Abraham, A. and Peng, L. (2007) Automatic design of hierarchical Takagi-Sugeno type fuzzy systems using evolutionary algorithms. *IEEE Transactions on Fuzzy Systems*, **15**(3), 385–397.

De Barros, J. C. and Dexter, A.L. (2007a) Evolving Fuzzy Model-based Adaptive Control. Proceedings of IEEE International Conference on Fuzzy Systems, London, UK, article 4295552.

De Barros, J.-C. and Dexter, A.L. (2007b) On-line identification of computationally undemanding evolving fuzzy models. *Fuzzy Sets & Systems*, **158**(18), 1997–2012.

Hoo, K.A., Sinzinger, E.D. and Piovoso, M.J. (2002) Improvements in the predictive capability of neural networks. *Journal of Process Control*, **12**(1), 193–202.

Johansen, T.A. (1996) Identification of non-linear systems using empirical data and prior knowledge – an optimization approach. *Automatica*, **32**(3), 337–356.

Kelkar, B. and Postlethwaite, B. (1998) Enhancing the generality of fuzzy relational models for control. *Fuzzy Sets & Systems*, **100**, 117–129.

Laukonen, E.G. and Passino, K.M. (1995) Training systems to perform estimation and identification. *Engineering Applications of Artificial Intelligence*, **8**(5), 499–514.

Linker, R. and Seginer, I. (2004) Greenhouse temperature modelling: a comparison between sigmoid neural networks and hybrid models. *Mathematics and Computers in Simulation*, **65**(1), 19–29.

Meleiro, L.A.C., Maciel Filho, R., Campello, R.J.G.B. and Amaral, W.C. (2006) Application of hierarchical neural fuzzy models to modelling and control of a bioprocess. *Applied Artificial Intelligence*, **20**(9), 797–816.

Milanic, S., Strmcnik, S., Sel, D., *et al.* (2004) Incorporating prior knowledge into artificial neural networks – an industrial case study. *Neurocomputing*, **62**(1), 131–151.

Postlethwaite, B.E. (1994) A model-based fuzzy controller. *Transactions of Institute of Chemical Engineers*, **72**, 38–46.

Raju, G.V.S. and Zhou, J. (1993) Adaptive hierarchical fuzzy controller. *IEEE Transactions on Systems, Man and Cybernetics*, **23**(4), 973–980.

Raju, G.V.S., Zhou, J. and Kisner, R.A. (1991) Hierarchical fuzzy control. *International Journal of Control*, **54**(5), 1201–1216.

Ridley, J.N., Shaw, I.S. and Kruger, J.J. (1988) Probabilistic fuzzy model for dynamic systems. *IEE Electronic Letters*, **24**(14), 890–892.

Salgado, P. (2005) Clustering and hierarchization of fuzzy systems. *Soft Computing*, **9**(10), 715–731.

Shaw, I.S. and Kruger, J.J. (1992) New fuzzy learning model with recursive estimation for dynamic systems. *Fuzzy Sets and Systems*, **48**(1), 217–229.

Sing, C.H. and Postlethwaite, B. (2000) Identification of fuzzy relational models from unevenly distributed data using optimisation methods. *Transactions of Institute of Chemical Engineers, Part A*, **78**(4), 522–527.

Tan, W.W. and Dexter, A.L. (2000) A self-learning fuzzy controller for embedded applications. *Automatica*, **36**(8), 1189–1198.

Valente de Oliveira, J. and Lemos, J.M. (1995) Long-range predictive adaptive fuzzy relational control. *Fuzzy Sets and Systems*, **70**, 337–357.

Wang, L-X. (1999) Analysis and design of hierarchical fuzzy systems. *IEEE Transactions on Fuzzy Systems*, **7**(5), 617–624.

Wei, C. and Wang, L-X. (2000) A note on universal approximation by hierarchical fuzzy systems. *Information Sciences*, **123**(3), 241–248.

Wong, C.H., Shah, S.L. and Fisher, D.G. (2000) Fuzzy relational predictive identification. *Fuzzy Sets and Systems*, **113**, 417–426.

Wu, Y. and Dexter, A.L. (2003) Modelling capabilities of fuzzy relational models. Proceedings of IEEE International Conference on Fuzzy Systems, **1**, 430–435.

Wu, Y. and Dexter, A.L. (2008) Adaptive fuzzy model-based predictive control using fuzzy decision-making. UK Automatic Control Conference, Paper 47.

Xu, C W. and Lu, Y-Z. (1987) Fuzzy model identification and self-learning for dynamic systems. *IEEE Transactions on Systems, Man and Cybernetics*, **17**(4), 683–689.

Yager, R.R. (1993) On a hierarchical structure for fuzzy modelling and control. *IEEE Transactions on Systems, Man and Cybernetics*, **23**(4), 1189–1197.

Yager, R.R. (1998) On the construction of hierarchical fuzzy systems models. *IEEE Transactions on Systems, Man and Cybernetics, Part C*, **28**(1), 55–66.

11

Adaptive Model-Based Control of Information-Poor Systems

11.1 Robust Adaptive Fuzzy Control

Nearly all robust adaptive fuzzy control schemes use a fuzzy model (a T-S or defuzzified Mamdani FLS) for function approximation and the control design assumes a non-fuzzy norm-bounded representation of the uncertainties. Some schemes can deal with unstructured uncertainties (Yang and Ren, 2003) but others are only able to deal with structured uncertainties (Zheng *et al.*, 2004). Early schemes require the matching conditions to be satisfied or make simplifying assumptions (e.g. norm-bounded external disturbances, no uncertainties in the controller input matrix, full state feedback) or the T-S fuzzy model is used to approximate the non-linear uncertainties and robust stability of only the associated fuzzy model (not the real system) is guaranteed. More recently however, methods based on backstepping (Yang and Feng, 2004; Zhou *et al.*, 2005; Liu *et al.*, 2009; Tong and Li, 2009; Tong *et al.*, 2009a, b; Tong *et al.*, 2010) have been developed that have no such limitations although their application is restricted to a particular class of non-linear systems (strict feedback systems). A major concern is that all of the robust adaptive fuzzy control schemes are too computationally demanding or require too much prior knowledge to be of much use in practice.

The use of T-S fuzzy models allows robust, and robust adaptive, techniques (feedback linearization, Lyapunov stability analysis, LMI optimization, small gain theory and H-infinity) to be applied to non-linear dynamic systems that can be represented by a functional fuzzy model of this type. Although the stability of the control schemes is usually guaranteed, the stability proofs are of little practical significance as they are too conservative and/or are based on unrealistic simplifying assumptions.

An attractive alternative is an adaptive version of the fuzzy FRM-based predictive controller described in Section 7.4.

Monitoring and Control of Information-Poor Systems: An Approach based on Fuzzy Relational Models, First Edition.
Arthur L. Dexter.
© 2012 John Wiley & Sons, Ltd. Published 2012 by John Wiley & Sons, Ltd.

11.2 Adaptive Fuzzy FRM-Based Predictive Control

The FMPC scheme described in Chapter 7 is based on a generic FRM, which was trained offline using the test data generated by computer simulations of several different designs of the plant to be controlled (see Chapter 5). The behaviour of the generic FRM can be very different to that of the real system under control. Consequently, the control performance can be relatively poor due to the mismatch between the generic model and the process. These problems can be overcome by using an online fuzzy identification scheme (see Chapter 10) to adapt the model using operating data from the actual plant (see Figure 11.1). Online adaptation also has the potential to cope with the control of a time-varying process.

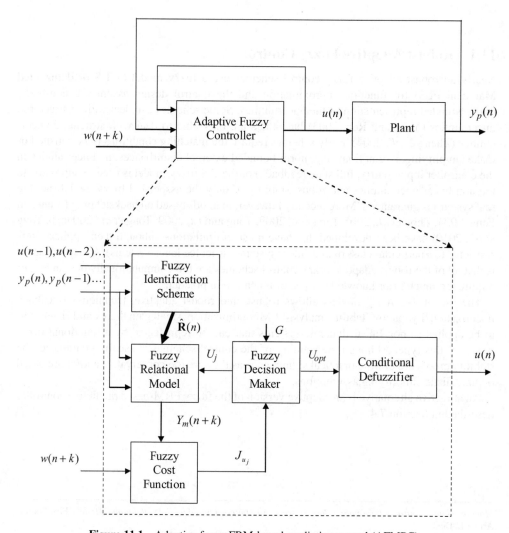

Figure 11.1 Adaptive fuzzy FRM-based predictive control (AFMPC).

11.3 Commissioning the Controller

11.3.1 Methods of Incorporating Prior Knowledge

In most designs, the choice of the sampling rate, the inputs to be used by the controller, the scaling factors to be used to normalize the universes of discourse and the order of the dynamic model are all based on domain knowledge. However, it is also recognized that the inclusion of prior knowledge into model-free (or direct) adaptive (see Chapter 12) and model-based (or indirect) adaptive fuzzy control schemes can help to improve the performance of the controllers (Spooner and Passino, 1996; Ordonez and Passino, 1999; Ordonez *et al.*, 2006). A nominal model based on prior knowledge is often used to initialize the model in an indirect adaptive control scheme or to initialize the controller in a direct adaptive control scheme.

The initialization of model-based adaptive controllers is a widely recognized problem (Ibarrola *et al.*, 2005). It is sometimes suggested that the learnt model is initialized with a model trained offline (Mahouf *et al.*, 2001; Eliasi *et al.*, 2007). However, this may be difficult in practice as collecting a representative training dataset may take an unacceptably long period of time in many applications. More often, a model identified using prior knowledge must be used to initialize the learnt model (Carmona *et al.*, 2002; Wai and Yang, 2008; Wu and Dexter, 2008).

11.3.2 Initialization Using a Generic Fuzzy FRM

In an information-poor system, a generic model is often the most reliable information the controller has before online learning begins, and it is often used to initialize the model used in the controller. When fully trained online, a fuzzy relational model that has a large number of fuzzy inputs sets will have smaller structural errors than a model with fewer inputs sets. However, it will take a long time to generate a generic model with a large number of fuzzy input sets. A scheme for increasing the number of fuzzy input sets used by the FRM is therefore needed to allow a generic fuzzy model with low granularity to be transformed into a generic fuzzy model with high granularity, without the need for additional training.

Consider the use of the following scheme for changing the granularity of a simple single-input example (see Figure 11.2), where the width of the original fuzzy input sets of the fuzzy

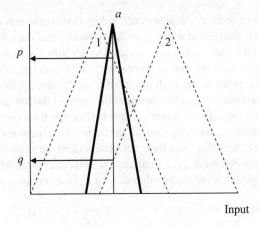

Input

Figure 11.2 Fuzzy input sets.

model is greater than the width of the new fuzzy input sets and the number of output sets remains the same. Let the original rule confidences be R_{1A}, R_{1B}, R_{2A} and R_{2B}, where 1 and 2 are the names of the original fuzzy input sets and A and B are the names of two of the fuzzy output sets. Thus:

IF input is 1 THEN output is A (R_{1A})

IF input is 2 THEN output is A (R_{2A})

IF input is 1 THEN output is B (R_{1B})

IF input is 2 THEN output is B (R_{2B}).

Let the rule confidences associated with the new rules be R_{aA} and R_{aB}, where a is the name of the new fuzzy input set. As can be seen in Figure 11.2, input data that have 100% membership of the new input set a have only partial membership of the original input sets 1 and 2. Let p be the membership of the data in the original fuzzy set 1 and q be the membership of the training data in the original fuzzy set 2. The new rule confidences R_{aA} and R_{aB} can be calculated by weighting the original rule confidences by the memberships of the data in the original fuzzy input sets, that is,

$$R_{aA} = \frac{R_{1A} \times p + R_{2A} \times q}{p + q}$$
$$R_{aB} = \frac{R_{1B} \times p + R_{2B} \times q}{p + q} \tag{11.1}$$

If an increase in the number of sets that are used to describe more than one of the inputs of a multi-input FRM is required, the procedure is simply applied to each of the inputs in turn.

It can be shown that when sum-product inferencing is used the high granularity model will generate the same fuzzy output as the original model if the original fuzzy input sets have evenly spaced, triangular membership functions with a partition of unity, and the new input sets have a partition of unity and are evenly spaced between the centres of the original fuzzy input sets.

For example, consider the fuzzy outputs generated by two single-input single-output FRMs using sum-product inferencing, one a low granularity model and the other a high granularity model. Assume that (1) the fuzzy reference sets used to describe the input to both models have equally spaced triangular membership functions with 50% overlap; (2) the fuzzy reference sets used to describe the input in the high granularity model are equally spaced between the centres of the fuzzy reference sets used to describe the input in the low granularity model; and (3) both models used the same fuzzy reference sets to describe their outputs.

If, without loss of generality, the crisp value of the input x fires the rules in the low granularity model with the input sets A_i and A_{i+1} in their antecedents and the rules in the high granularity model with the input sets B_k and B_{k+1} in their antecedents, the possibility that the conclusion of the low granularity model is the fuzzy reference set C_j is given by:

$$Poss_{low}(C_j/x) = \mu_{A_i}(x) R_{A_i,C_j} + \mu_{A_{i+1}}(x) R_{A_{i+1},C_j}$$
$$= \mu_{A_i}(x) R_{A_i,C_j} + [1 - \mu_{A_i}(x)] R_{A_{i+1},C_j} \tag{11.2}$$

and the possibility that the conclusion of the high granularity model is the fuzzy reference set C_j is given by:

$$Poss_{high}(C_j/x) = \mu_{B_k}(x)R_{B_k,C_j} + \mu_{B_{k+1}}(x)R_{B_{k+1},C_j}$$

$$= \mu_{B_k}(x)R_{B_k,C_j} + [1 - \mu_{B_k}(x)]R_{B_{k+1},C_j} \quad (11.3)$$

where $\mu_{A_i}(x)$, $\mu_{A_{i+1}}(x)$, $\mu_{B_k}(x)$ and $\mu_{B_{k+1}}(x)$ are the degrees of membership of x in the fuzzy sets A_i, A_{i+1}, B_k and B_{k+1}, respectively, and R_{A_i,C_j}, R_{A_{i+1},C_j}, R_{B_k,C_j} and R_{B_{k+1},C_j} are the rule confidences associated with the rules that are fired.

If the apex of the set A_i is at x_i, the apex of the set B_k is at x_k and so on, then

$$\mu_{A_i}(x) = \frac{x_{i+1} - x}{x_{i+1} - x_i} = \frac{x_{i+1} - x}{\delta} \quad (11.4)$$

and

$$\mu_{B_k}(x) = \frac{x_{k+1} - x}{x_{k+1} - x_k} = \frac{x_{k+1} - x}{\lambda} \quad (11.5)$$

It is required that $Poss_{low}(C_j/x) = Poss_{high}(C_j/x)$ for all values of x in the range $x_k \leq x \leq x_{k+1}$. Hence:

$$\mu_{A_i}(x)R_{A_i,C_j} + [1 - \mu_{A_i}(x)]R_{A_{i+1},C_j} = \mu_{B_k}(x)R_{B_k,C_j} + [1 - \mu_{B_k}(x)]R_{B_{k+1},C_j} \quad (11.6)$$

Substituting in the expressions for $\mu_{A_i}(x)$ and $\mu_{B_i}(x)$ and equating the constant coefficients and the coefficients of terms in x, we have

$$\left(\frac{x_{i+1}}{\delta}\right)R_{A_i,C_j} + \left[1 - \left(\frac{x_{i+1}}{\delta}\right)\right]R_{A_{i+1},C_j} = \left(\frac{x_{k+1}}{\lambda}\right)R_{B_k,C_j} + \left[1 - \left(\frac{x_{k+1}}{\lambda}\right)\right]R_{B_{k+1},C_j} \quad (11.7)$$

and

$$-\left(\frac{R_{A_i,C_j}}{\delta}\right) + \left(\frac{R_{A_{i+1},C_j}}{\delta}\right) = -\left(\frac{R_{B_k,C_j}}{\lambda}\right) + \left(\frac{R_{B_{k+1},C_j}}{\lambda}\right) \quad (11.8)$$

Therefore:

$$R_{B_{k+1},C_j} = R_{B_k,C_j} + \frac{\lambda}{\delta}\left(R_{A_{i+1},C_j} - R_{A_i,C_j}\right) \quad (11.9)$$

and

$$\left(\frac{x_{i+1}}{\delta}\right)R_{A_i,C_j} + \left[1 - \left(\frac{x_{i+1}}{\delta}\right)\right]R_{A_{i+1},C_j} = \left(\frac{x_{k+1}}{\lambda}\right)R_{B_k,C_j} + \left[1 - \left(\frac{x_{k+1}}{\lambda}\right)\right]$$

$$\times \left[R_{B_k,C_j} + \frac{\lambda}{\delta}\left(R_{A_{i+1},C_j} - R_{A_i,C_j}\right)\right] \quad (11.10)$$

Hence,

$$R_{B_k,C_j} = \left\{\frac{x_{i+1}}{\delta} + \frac{\lambda}{\delta}\left[1 - \left(\frac{x_{k+1}}{\lambda}\right)\right]\right\}R_{A_i,C_j} + \left\{1 - \frac{x_{i+1}}{\delta} - \frac{\lambda}{\delta}\left[1 - \left(\frac{x_{k+1}}{\lambda}\right)\right]\right\}R_{A_{i+1},C_j}$$

$$= \left\{\frac{x_{i+1} + \lambda - x_{k+1}}{\delta}\right\}R_{A_i,C_j} + \left\{1 - \left[\frac{x_{i+1} + \lambda - x_{k+1}}{\delta}\right]\right\}R_{A_{i+1},C_j}$$

$$= \left\{ \frac{x_{i+1} - x_k}{\delta} \right\} R_{A_i, C_j} + \left\{ 1 - \left[\frac{x_{i+1} - x_k}{\delta} \right] \right\} R_{A_{i+1}, C_j}$$

$$= \left\{ \frac{x_{i+1} - x_k}{\delta} \right\} R_{A_i, C_j} + \left\{ 1 - \left[\frac{x_{i+1} - x_k}{\delta} \right] \right\} R_{A_{i+1}, C_j}$$

$$= \mu_{A_i}(x_k) R_{A_i, C_j} + \left[1 - \mu_{A_i}(x_k) \right] R_{A_{i+1}, C_j} \qquad (11.11)$$

We therefore have

$$R_{B_k, C_j} = \mu_{A_i}(x_k) R_{A_i, C_j} + \mu_{A_{i+1}}(x_k) R_{A_{i+1}, C_j} \qquad (11.12)$$

and similarly,

$$R_{B_{k+1}, C_j} = \mu_{A_i}(x_{k+1}) R_{A_i, C_j} + \mu_{A_{i+1}}(x_{k+1}) R_{A_{i+1}, C_j} \qquad (11.13)$$

11.4 Generating an Optimal Control Signal Using a Partially Trained Model

In practice, it may take a long time for the online identification scheme to update all of the rules in the original fuzzy model used in the FMPC. To achieve satisfactory control performance during the training period, the controller should take more account of the rules in the partially trained model that have been updated (which more accurately predict the behaviour of the plant) instead of those that have not been updated (which still provide only generic descriptions of the behaviour of the plant).

11.4.1 Taking the Amount of Training into Account

The basic FDM scheme can be modified to take account of how much the rules currently being used by the adaptive fuzzy FRM-based predictive control (AFMPC) controller have been trained (Wu and Dexter, 2008). The modification is based on a variant of the method of height defuzzification with a threshold. A weight is put on the membership values of the fuzzy optimal control signal dependent upon how much the associated rules have been trained.

As discussed in Section 10.1.2.4, after a significant amount of training the elements of the F matrix will converge to the value $1/(1 - \lambda)$ if the RRSK fuzzy identification scheme is used to estimate the rule confidences. The current value of a particular element $F(t)$ can therefore be used to indicate the degree to which the associated rule confidences $\hat{R}(t)$ have been trained. If the control signal has a relatively high possibility of being optimal

$$(\text{that is } \mu(u(n)) \geq \alpha), \ \mu_1(u_i(n)) = w_1(u_i(n)) \cdot \mu(u_i(n))$$

where α is an application-dependent weighting threshold and the weight $w_1(u_i(n))$ is given by

$$w_1(u_i(n)) = \frac{1}{2} \left\{ \frac{1 + (1 - \lambda) [F(n) - 2F(0)]}{1 - F(0)(1 - \lambda)} \right\} \qquad (11.14)$$

Note that the value of the element of the matrix $F(n)$ will remain at $F(0)$ if the rule confidences have not been updated and the weight $w_1(u_i(n)) = 0.5$. This means only half-weight is given to control signals which have not been trained, whereas the weight $w_1(u_i(n)) = 1.0$ when the value of $F(n)$ has converged to $1/(1 - \lambda)$ and full weight is given to the membership grade of the optimal control signal.

If the candidate value of the control signal has a small possibility of being optimal (that is, $\mu(u(n)) < \alpha$) $\mu_2(u_i(n)) = w_2(u_i(n)) \cdot \mu(u_i(n))$ where the weight $w_2(u_i(n))$ is given by

$$w_2(u_i(n)) = \frac{1}{2}\left[\frac{1 - (1-\lambda)F(n)}{1 - F(0)(1-\lambda)}\right] \tag{11.15}$$

Note that the value of $F(n)$ will remain at $F(0)$ if the rule confidences have not been updated and, as before, only half-weight is given to control signals which have not been trained, whereas the weight $w_2(u_i(n)) = 0.0$ when the value of $F(n)$ has converged to $1/(1 - \lambda)$ and zero weight is given to the membership grades of the optimal control signal.

It should be noted that, if the fuzzy relational model has not been trained at all, all the membership grades will have the same value of weight (0.5). Height defuzzification will therefore give the same result as is the case in the original FDM scheme without weighting.

An example of the variation of the weights during the training of the model when $\lambda = 0.5$ and $F(0) = 1$ is shown in Figure 11.3. The top curve shows the variation of w_1 against the number of updates of the rule confidences and the bottom curve shows the variation of w_2 against the number of updates. It can be seen that around 10 updates are required to update a rule confidence fully. Note that more updates are required to fully train the model as the forgetting factor increases.

It should be noted that the scheme has two limitations: the value of forgetting factor λ cannot be equal to unity and the initial value of $F(t)$ must be less than $1/1 - \lambda$.

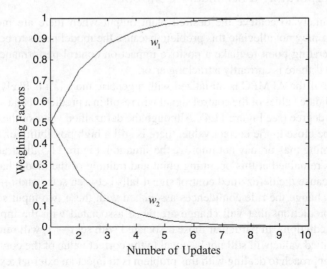

Figure 11.3 Variation of the weights during training.

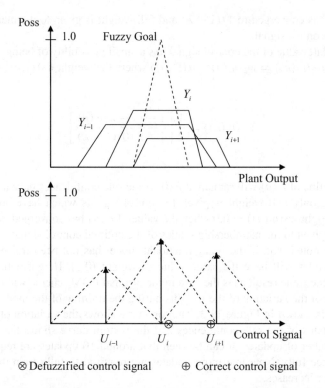

Figure 11.4 Illustration of fuzzy predictions and resulting fuzzy optimal control signal from an untrained AFMPC initialized with a generic model.

11.4.2   Incorporating a Secondary Controller

AFMPC is unlikely to achieve the desired plant output when there are modelling errors. Online learning may not alleviate this problem because the model needs to be trained around the desired operating point to make a positive impact on control performance, and this will not be the case if there is currently a tracking error.

For example, if the AFMPC is initialized with a generic model, it is likely that more than one of the candidate values of the control signal will result in a prediction that satisfies the goal to a significant degree (see Figure 11.4). Although the defuzzified value of the optimal control signal should be close to the correct value, there is still a high possibility of a tracking error and online learning may or may not improve the situation. Figure 11.5 illustrates the situation if the plant has remained at this operating point and training of the model has continued for some time. Because the defuzzified control signal falls between sets i and $i + 1$, the training data will only change the rule confidences associated with these two input sets. This means that the only predictions that will change are those associated with the input values i and $i + 1$. Other predictions, in particular those associated with set $i - 1$, will remain unchanged and the defuzzified value will still not be equal to the correct value of the control signal.

A common approach to dealing with this problem is to inject an external test signal into the plant. The varying control signal allows the online learning to observe the plant in previously untrained states. Unfortunately the goals of this process can be at odds with improving control

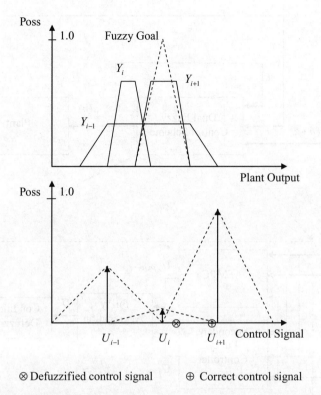

Figure 11.5 Illustration of fuzzy predictions and resulting fuzzy optimal control signal from the AFMPC after significant online training.

performance (Jacobs and Patchell, 1972). Although the final result may be a thoroughly trained model, it requires an extended testing period which may be unacceptable in some applications.

An alternative approach is to incorporate a secondary controller into the control scheme to reduce the tracking errors during the initial period of training but, just as importantly, to generate previously unavailable training data for the online learning scheme.

The two controllers act independently and generate separate fuzzy control signals, which are combined using a fuzzy aggregation scheme to determine the optimal control signal (see Figure 11.6). The influence of a particular controller on the optimal control action is based on a measure of confidence in that controller at the current operating point.

For example, the secondary controller might be based on a fuzzy PI design using generic expert fuzzy rules, as it must be assumed that there is very little prior knowledge of the behaviour of the plant other than whether the plant output is positively or negatively related to the control signal. Here the tracking error and rate of change of plant output are used in the antecedents for the rules in the controller and changes to the control signal are used in the consequents. The fuzzy rules are representative of a human operator's instinct to make small or large increases or decreases in the control signal. Table 11.1 shows the rule-base for a plant with a negatively correlated input and output. To avoid sluggish transient behaviour, the rate of change of the output is taken relative to the setpoint. We therefore have

$$\Delta y_r(n) = \frac{[y(n) - y(n-1)] - [w(n) - w(n-1)]}{h} \tag{11.16}$$

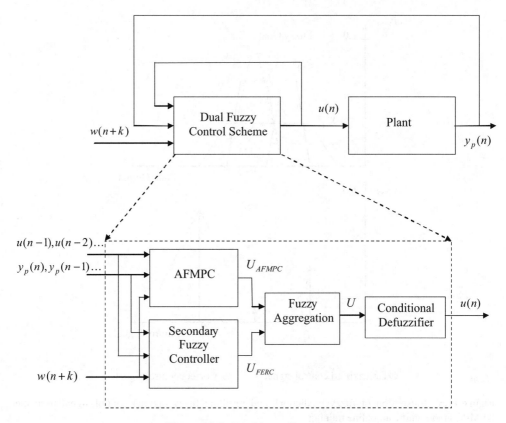

Figure 11.6 Dual controller scheme.

where $\Delta y_r(n)$ is the relative rate of change of the output signal at time n, h is the controller sample time, and $y(n)$ and $w(n)$ are the measured values of the plant output and setpoint, respectively, at time n. The error is described by five fuzzy sets (large negative, negative, zero, positive and large positive) and the relative rate of change by three fuzzy sets (negative, zero and positive). The changes in the control signal are described by five fuzzy sets (negative change, small negative change, no change, small positive change and positive change).

The weights used to aggregate the control signals depend on a measure of confidence in the control signal generated by the AFMPC that is based on how much training has occurred at the current operating point. The fuzzy expert rule controller (FERC) is assigned a constant equivalent value that indicates the amount of training the AFMPC needs for the designer to be as confident in its control decisions as those of the FERC. The value selected will depend on the accuracy and precision of the generic model used to initialize the AFMPC and the quality of the training data to be used to train the AFMPC (as determined by the expected measurement noise and disturbance levels). As the amount of training increases at a particular operating condition, the control signal from the AFMPC will have increasing impact on the control signal for such conditions.

Table 11.1 Rule-base of the expert rule controller.

FERC Rule-base: Change in output signal relative to previous output

		Output Error $e =$ setpoint $-$ plant output				
		Large negative (N+)	Small negative (N−)	Zero (Z)	Small positive (P−)	Large positive (P+)
Relative rate of	Negative	P	SP	SP	NC	SN
change of output	Zero	P	SP	NC	SN	N
signal Δy_r	Positive	SP	NC	SN	SN	N

The possibility of each element of the optimal fuzzy control signal at the nth sampling instant is given by Equation (11.17).

$$Poss(u_{\text{optimal}_i}(n)) = \frac{w_i(n)}{w_i(n) + w_e} Poss(u_{\text{AFMPC}_i}(n)) + \frac{w_e}{w_i(n) + w_e} Poss(u_{\text{FERC}_i}(n)) \quad (11.17)$$

where $Poss(u_{\text{optimal}_i}(n))$, $Poss(u_{\text{AFMPC}_i}(n))$ and $Poss(u_{\text{FERC}_i}(n))$ are the possibilities of the ith elements of the fuzzy optimal control signals generated by the AFMPC and FERC, w_e is the user-defined weighting factor for the FERC and $w_i(n)$ is the weighting factor for element i of the fuzzy control signal generated by the AFMPC.

The weighting factor $w_i(n)$ is given by

$$w_i(n) = \frac{\sum_{k=1}^{is1} \sum_{j=1}^{is2} Poss(Y_j(n)) \times Poss(U_k(n)) \times F_{jk}(n)}{\sum_{k=1}^{is1} \sum_{j=1}^{is2} Poss(Y_j(n)) \times Poss(U_k(n))} \quad (11.18)$$

where $Y(n)$ is the plant output, $U(n)$ is the control signal and $F_{jk}(n)$ is the element of the F matrix (see Section 10.1.2) associated with rules whose antecedent is '$Y(n)$ is Y_j' AND 'U is U_k'. Note that the weighting factor is itself a weighted average of the measures of the amounts of training that have gone into each of the rules used in the calculation of the possibility of the ith element of U_{AFMPC}.

Example 11.1

Consider the use of computer simulation based on a detailed dynamic model of a cooling coil system (see Section 16.5) to test the dual controller scheme. The design of the AFMPC assumes no prior knowledge so the initial values of all of the relational elements of its fuzzy FRM are the same. Recursive fuzzy least-squares estimation with a forgetting factor of unity is used for fuzzy identification (see Section 10.1.1). The tests are conducted under noise-free and disturbance-free conditions.

Figures 11.7 and 11.8 show the behaviour of the plant in response to a square wave variation of the setpoint, when it is controlled by the FERC alone and the AFMPC alone. The closed-loop behaviour of the dual controller scheme is shown in Figure 11.9. The contribution from each controller is displayed along the bottom of the graph.

It can be seen that the FERC is able to eliminate steady-state errors but its transient behaviour is unsatisfactory whereas the transient behaviour of the AFMPC is better but it is unable to remove the steady-state errors.

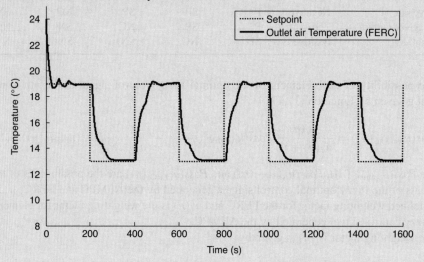

Figure 11.7   Plant output and setpoint while under control of the fuzzy expert rule controller. Reproduced by permission of Andrew Wright.

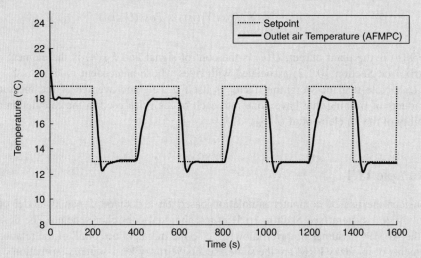

Figure 11.8   Plant output and setpoint while under control of the adaptive fuzzy model-based controller. Reproduced by permission of Andrew Wright.

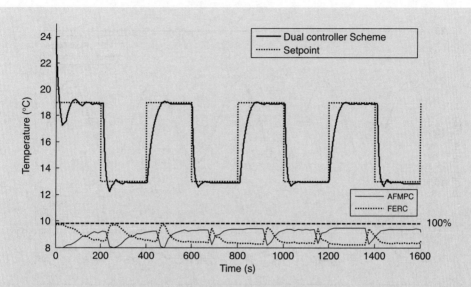

Figure 11.9   Plant output and setpoint while under control of the dual controller scheme. Reproduced by permission of Andrew Wright.

Table 11.2   Comparison of the overall control performances.

	RMSE (°C)	
FERC	AFMPC	Dual controller
1.388	1.286	0.697

The secondary axis, which plots the percentage contribution from each controller, shows that the AFMPC becomes more dominant as it is able to train its model around the desired operating conditions.

The root-mean-square errors for each of the controllers are shown in Table 11.2. The improvements to both the steady-state and transient behaviour, resulting from the use of the dual controller control scheme, are clear.

Example 11.2

A sinusoidally varying setpoint is applied to the simulated cooling coil to assess the control performance at three different frequencies. The periods of the sinusoidal setpoint variations relate to the low, medium and high frequency (750, 420 and 180 seconds, respectively) behaviour of the plant in terms of its open-loop dynamics. The resulting closed-loop behaviour is shown in Figures 11.10–11.12, respectively. Table 11.3 compares the overall control performances of the single AFMPC and dual controller schemes.

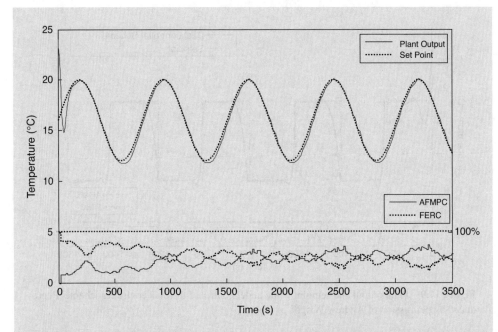

Figure 11.10 Plant output with sinusoidally varying setpoint ($T = 750\,$s) under control of the dual controller scheme. Reproduced by permission of Andrew Wright.

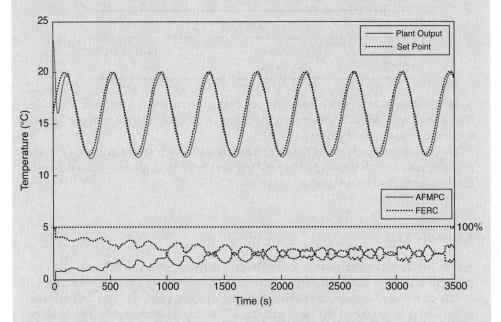

Figure 11.11 Plant output with sinusoidally varying setpoint ($T = 420\,$s) under control of the dual controller scheme. Reproduced by permission of Andrew Wright.

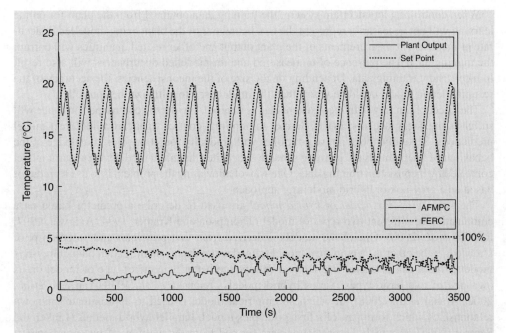

Figure 11.12 Plant output with sinusoidally varying setpoint ($T = 180\,\text{s}$) under control of the dual controller scheme. Reproduced by permission of Andrew Wright.

Table 11.3 Comparison of the overall control performances.

	RMSE (°C)	
Period (sec)	AFMPC control	Dual controller control
750	0.354	0.253
420	0.581	0.586
180	1.738	1.469

The results further demonstrate the improvement resulting from the use of the dual controller scheme.

11.4.3 Combining the Fuzzy Predictions Generated by More than One Model

Another approach is to use two model-based controllers in parallel – one based on a generic model of the plant, the other based on the learnt model – and to aggregate the two fuzzy control signals to create the optimal control signal.

When controlling an uncertain system, the training data obtained from the plant for online learning will not necessarily represent the true behaviour of the plant being modelled. Inaccurate or inconsistent measurements of the plant output and other related quantities will corrupt the training data. The presence of unmeasured and un-modelled disturbances will also result in inconsistent training data. Depending on the size of the inconsistencies, the accuracy of the resulting learnt model could be worse than the model used to initialize the controller.

The effect of training with inconsistent data is twofold. Firstly, the control performance will suffer as the fuzzy decision maker derives the optimal control signal from incorrect model predictions. Secondly, the adaptive controller will gradually forget its prior knowledge of the behaviour of the plant, as the learning scheme overwrites the initial model with a learnt model corrupted by inconsistent training data. One way of alleviating the problem is to use a predictor based on a grey-box or hybrid modelling approach.

The terms *grey-box model* or *hybrid model* are used to describe a predictor based on a combination of different two types of model (Thompson and Kramer, 1994; Agarwal, 1997; Romijn *et al.*, 2008). One of the models is derived from first principles and based on prior knowledge; the other is an empirical model trained using data from the actual plant. Grey-box models, which can be more accurate and extrapolate better than either type of model on its own, are of two basic types. Series hybrid models (Babuska *et al.*, 1999; Van Lith *et al.*, 2002; Arahal *et al.*, 2008), in which an empirical model is used to approximate unknown relationships which form part of a first-principles model. Parallel hybrid models (Linker and Seginer, 2004), in which the predictions are formed from the addition of two components, one generated by the first-principles model and the other from the empirical model. The empirical model is normally used to compensate for the prediction errors associated with the first-principles model.

The use of two or more fuzzy models in a parallel hybrid structure can improve the robustness of an adaptive control scheme that is trained using normal operating data. A simple approach is to use a generic fuzzy model trained offline using simulation data (see Section 5.5.1) to predict future plant outputs if none of the rules in the learnt model are fired (De Barros and Dexter, 2007). Alternatively, the predictions of the learnt fuzzy model could be combined with those of a fuzzy model based on expert rules using fuzzy IF-THEN rules to relate a weighted combination of the two predictions to the overall degree of firing of the rules in each of the two fuzzy models (Wang *et al.*, 2002). A more sophisticated (but more complex) approach is to use a three-level hierarchical rule-base (Yager, 2006) of the type described in Section 10.3.2. The fuzzy model at the top level of the hierarchy consists of high-priority rules. The learning scheme creates new rules in the fuzzy model at the middle level of the hierarchy, and deletes them if they can be merged with similar rules. Rules are transferred from this model to the model at the top level of the hierarchy only if they are seen to be persistent exceptions. The fuzzy model at the bottom of the hierarchy consists of fixed default and generic rules.

11.5 Dealing with the Effects of Disturbances

A number of AFPC schemes have been proposed but they usually ignore the effect of disturbances on the online training and the control performance (Mahouf *et al.*, 2001; Arauzo-Bravo *et al.*, 2004). Some schemes use adaptive IMC to deal with unmeasured disturbances, but they usually assume that the only disturbances are output disturbances (Xie and Rad, 2000) or that the disturbances are constant or have zero mean (Demircan *et al.*, 1999).

11.5.1 Adaptive Feedforward Control Based on an Inaccurate Disturbance Measurement

Taking account of disturbances becomes much more complicated when the system is non-linear, and the disturbances act at the input of the plant and affect the relationship between the control signal and the plant output (Seborg *et al.*, 1989). Adjustments to the model used by the controller are necessary to account for the effect of input disturbances of this type. There are, however, two additional sources of information required to achieve this: a reliable measurement of the disturbance and a knowledge of the plant's behaviour in response to the disturbance, neither of which can be assumed to be available when dealing with an uncertain system. In many applications, all that is available is a relatively poor measurement of the disturbance or an estimate derived from the measurement of some related quantity or quantities, in which case information about the plant's behaviour under the influence of the disturbance must be learnt online.

The dual controller scheme (see Section 11.4.2) can be extended to include feedforward based on an uncertain measurement of the disturbance (see Figure 11.13). Given sufficient training, any bias or noise on the disturbance measurement will be reflected in the fuzzy predictions generated by the model and taken into account by the fuzzy decision-maker when it calculates the fuzzy optimal control signal.

The aggregation of the fuzzy control signals generated by the AFMPC and FERC still depends on the amount of training, but the additional input to the model introduces another dimension to the measure of the amount of training stored in the F matrix. As no prior knowledge of the effect of the disturbance can be assumed, the generic model used to initialize the AFMPC must be identified offline using the mean value of the disturbance. As a result, during the online learning phase predictions from the controller's model for values of the disturbance far from its mean value will not be as reliable as those close to the mean value, for the same amount of online training. This can be taken into account in the fuzzy aggregation by initializing the F matrix with lower values for rules associated with values of the disturbance far from its mean value.

The weight associated with the AFMPC in the fuzzy aggregation is now given by

$$w_i(n) = \frac{\displaystyle\sum_{k=1}^{is1}\sum_{j=1}^{is2}\sum_{l=1}^{is3} Poss(Y_j(n)) \times Poss(U_k) \times Poss(D_l(n)) \times F_{jkl}(n)}{\displaystyle\sum_{k=1}^{is1}\sum_{j=1}^{is2}\sum_{l=1}^{is3} Poss(Y_j(n)) \times Poss(U_k) \times Poss(D_l(n))} \qquad (11.19)$$

where $Poss(D_l(n))$ is the possibility of D_l given the disturbance measurement at the nth sample time.

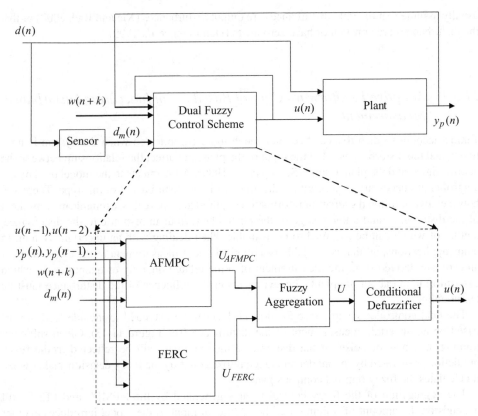

Figure 11.13   Feedforward dual controller control structure.

Example 11.3

Tests are performed to compare the ability of the dual controller scheme with feedforward to reject disturbances with that of the FIMC scheme based on a generic model. A detailed non-linear dynamic simulation of a cooling coil is again used to assess the control performance. The disturbance measurement has a bias of 10% and is corrupted by band-limited noise, which has a standard deviation of $\sim$15% of the disturbance's universe of discourse. To reduce the extra computational demands, the measured disturbance is described by only five fuzzy sets whose membership functions are triangular and evenly-spaced with a partition of unity.

The disturbance is a sinusoidally varying airflow rate that has increasing amplitude. Figure 11.14 shows the disturbance signal applied to the plant at a frequency of 1/30 rad/s. The design mass airflow rate for the plant is approximately 6 kg/s.

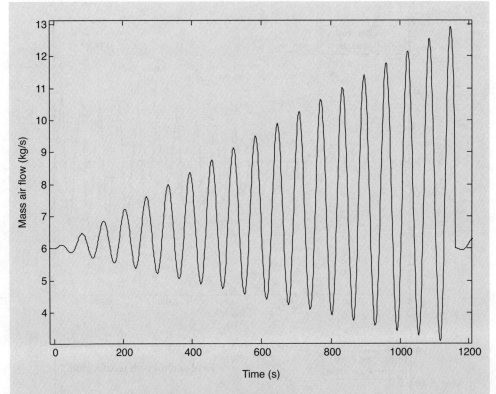

Figure 11.14 Disturbance applied to the mass airflow of the plant for the test at a frequency of 1/30 rad/s. Reproduced by permission of Andrew Wright.

Figure 11.15 compares the variations in the plant output for the non-adaptive FIMC (see Section 9.2) and the adaptive dual controller with feedforward control schemes, resulting from disturbances at three different frequencies. The setpoint is held constant at 15°C. The results shown for the dual controller scheme with feedforward are for the fourth cycle of the disturbance. The values of the forgetting factor and the conditional defuzzification threshold are unity, and the sampling interval of the controllers is 15 seconds.

The results show that dual controller scheme with feedforward handles the disturbance better overall, although the advantages of the feedforward control scheme are more pronounced at higher frequencies and at larger amplitudes of the disturbance. The feedforward aspect allows the controller to pre-emptively account for the effect of the disturbance as opposed to reacting to it, as is the case with the FIMC. The FIMC at higher frequencies and larger amplitudes of the disturbance fails to react quickly enough and, when it does react, it is hampered by the fact that its model is a poor representation of the plant under such conditions.

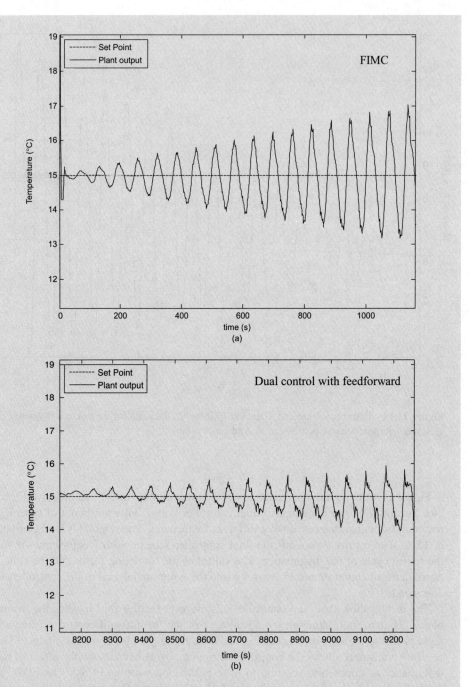

Figure 11.15 FIMC and AFMPC with feedforward under the influence of the increasing amplitude disturbance at frequencies: (a, b) 1/30 rad/s; (c, d) 1/60 rad/s; and (e, f) 1/120 rad/s. Reproduced by permission of Andrew Wright.

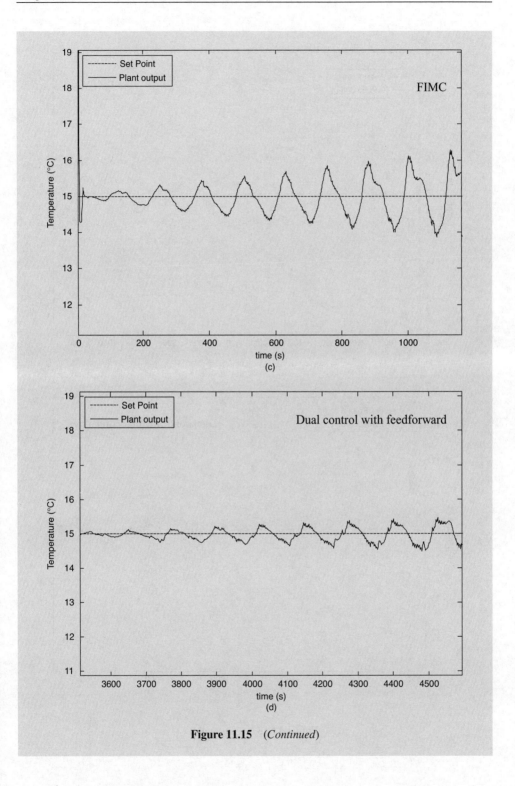

Figure 11.15 (*Continued*)

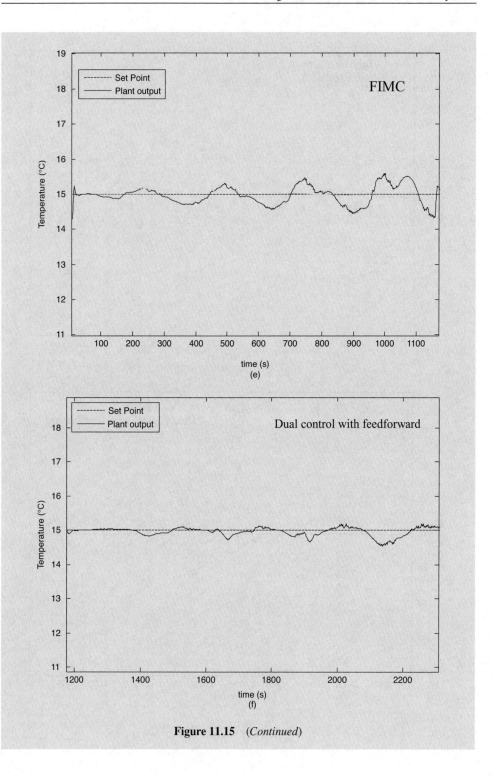

Figure 11.15 (*Continued*)

11.6 Summary

The issues associated with applying an adaptive version of the fuzzy FRM-based control scheme to information-poor systems have been considered. A brief introduction has been given to robust adaptive fuzzy control and an adaptive version of the fuzzy FRM-based predictive controller has been described. Ways of incorporating prior knowledge have been explained and a scheme for initializing the adaptive controllers using a generic FRM has been suggested. Three methods of generating an optimal control signal when the model is only partially trained (weighted fuzzy decision-making, secondary controller and dual model schemes) have been described. Results have been presented that show that the feedforward version of the adaptive controller can reject the effects of a disturbance when the disturbance measurement is inaccurate.

References

Agarwal, M. (1997) Combining neural and conventional paradigms for modelling, prediction and control. *International Journal of System Science*, **28**(1), 65–81.

Arahal, M.R., Cirre, C.M. and Berenguel, M. (2008) Serial grey-box model of a stratified thermal tank for hierarchical control of a solar plant. *Solar Energy*, **82**(5), 441–451.

Arauzo-Bravo, M.J., Cano-Izquierdo, J.M., Gomez-Sanchez, E, *et al.* (2004) Automatization of a penicillin production process with soft sensors and an adaptive controller based on neuro fuzzy systems. *Control Engineering Practice*, **12**, 1073–1090.

Babuska, R., Verbruggen, H.B. and van Can, H.J.L. (1999) Fuzzy modeling of enzymatic penicillin-G conversion. *Engineering Applications of Artificial Intelligence*, **12**(1), 79–92.

Carmona, P., Castro, J.L. and Zurita, J.M. (2002) Contradiction sensitive fuzzy model-based control. *International Journal of Approximate Reasoning*, **30**, 107–129.

De Barros, J.-C. and Dexter, A.L. (2007) Evolving Fuzzy Model-based Adaptive Control. Proceedings of IEEE International Conference on Fuzzy Systems, London, UK.

Demircan, M., Camurdan, M.C. and Postlethwaite, B.E. (1999) On-line learning fuzzy relational model based dynamic matrix control of an open-loop unstable process. *Transactions of Institute of Chemical Engineers*, **77**, Part A, 421–428.

Eliasi, H., Davilu, H. and Menhaj, M.B. (2007) Adaptive fuzzy model based predictive control of nuclear steam generators. *Nuclear Engineering and Design*, **237**, 668–676.

Ibarrola, J.J., Pinzolas, M. and Cano, J.M. (2005) A neurofuzzy scheme to on-line identification in adaptive-predictive control. *Neural Computing and Applications*, **15**, 41–48.

Jacobs, O.L.R. and Patchell, J.W. (1972) Caution and probing in stochastic control. *International Journal of Control* **16**(1), 189–199.

Linker, R. and Seginer, I. (2004) Greenhouse temperature modelling: a comparison between sigmoid neural networks and hybrid models. *Mathematics and Computers in Simulation*, **65**(1), 19–29.

Liu, Y.-J., Tong, S.-C. and Wang, W. (2009) Adaptive fuzzy output tracking control for a class of uncertain nonlinear systems. *Fuzzy Sets and Systems*, **160**, 2727–2754.

Mahouf, M., Kandiah, S. and Linkens, D.A. (2001) Adaptive estimation for fuzzy TSK model-based predictive control. *Transactions of Institute of Measurement & Control*, **23**(1), 31–50.

Ordonez, R. and Passino, K.M. (1999) Stable multi-input multi-output adaptive fuzzy/neural control. *IEEE Transactions on Fuzzy Systems*, **7**(3), 345–353.

Ordonez, R., Spooner, J.T. and Passino, K.M. (2006) Experimental studies in nonlinear discrete-time adaptive prediction and control. *IEEE Transactions on Fuzzy Systems*, **14**(2), 275–286.

Romijn, R., Ozkan, L., Weiland, S., *et al.* (2008) A grey-box modelling approach for the reduction of nonlinear systems. *Journal of Process Control*, **18**, 906–914.

Seborg, D.E., Edgar, T.F. and Mellichamp, D.A. (1989) *Process Dynamics and Control*, Wiley, New York.

Spooner, J.T. and Passino, K.M. (1996) Stable adaptive control using fuzzy systems and neural networks. *IEEE Transactions on Fuzzy Systems*, **4**(3), 339–359.

Thompson, M. and Kramer, M.A. (1994) Modeling chemical processes using prior knowledge and neural networks. *American Institute of Chemical Engineers Journal on Process Systems Engineering*, **40**(8), 1328–1340.

Tong, S. and Li, Y. (2009) Observer-based fuzzy adaptive control for strict-feedback nonlinear systems. *Fuzzy Sets and Systems*, **160**, 1749–1764.

Tong, S., He, X-L. and Zhang, H. (2009a) A combined backstepping and small-gain approach to robust adaptive fuzzy output feedback control. *IEEE Transactions on Fuzzy Systems*, **17**(5), 1059–1069.

Tong, S, Li, C. and Li, Y. (2009b) Fuzzy adaptive observer backstepping control for MIMO nonlinear systems. *Fuzzy Sets and Systems*, **160**, 2755–2775.

Tong, S., He, X., Li, Y. and Zhang, H. (2010) Adaptive fuzzy backstepping robust control for uncertain nonlinear systems based on small gain approach. *Fuzzy Sets and Systems*, **161**, 771–796.

Van Lith, P.F., Betlem, B.H.L. and Roffel, B. (2002) A structured modelling approach for dynamic hybrid fuzzy-first principles models. *Journal of Process Control*, **12**(5), 605–615.

Wai, R-J. and Yang, Z-W (2008) Adaptive fuzzy neural network control design via a T-S fuzzy model for a robotic manipulator including actuator dynamics. *IEEE Transactions on System, Man and Cybernetics, Part B*, **38**(5), 1326–1346.

Wang, Y., Rong, G. and Wang, S. (2002) Hybrid fuzzy modelling of chemical processes. *Fuzzy Sets and Systems*, **130**(2), 265–275.

Wu, Y. and Dexter, A.L. (2008) Adaptive Fuzzy Model-based Predictive Control using Fuzzy Decision-making. Proceedings of UKACC 2008, Manchester.

Xie, W.F. and Rad, A.B. (2000) Fuzzy adaptive internal model control. *IEEE Transactions on Industrial Electronics*, **47**(1), 193–202.

Yager, R. (2006) Learning methods for intelligent evolving systems. IEEE International Conference on Evolving Fuzzy Systems, pp. 3–7.

Yang, Y. and Ren, J. (2003) Adaptive fuzzy robust tracking controller design via small gain approach and its application. *IEEE Transactions on Fuzzy Systems*, **11**(6), 783–795.

Yang, Y. and Feng, G. (2004) A combined backstepping and small-gain approach to robust adaptive fuzzy control for strict feedback nonlinear systems. *Transactions of IEEE on Systems, Man, and Cybernetics, Part A*, **34**(3), 406–420.

Zheng, F., Wang, Q.-G. and Lee, T.H. (2004) Adaptive and robust controller design for uncertain nonlinear systems via fuzzy modelling approach. *Transactions of IEEE on Systems, Man, and Cybernetics, Part B*, **34**(1), 166–178.

Zhou, S., Feng, G. and Feng, C-B. (2005) Robust control for a class of uncertain nonlinear systems: adaptive fuzzy approach based on backstepping. *Fuzzy Sets and Systems*, **151**, 1–20.

12

Adaptive Model-Free Control of Information-Poor Systems

12.1 Introduction to Model-Free Adaptive Control of Non-Linear Systems

The model-based (or indirect) approach to adaptive control described in the previous chapter has two main weaknesses. The first is that the goal of the online adaption is to reduce the modelling errors (the differences between the actual and predicted plant output) and not the tracking errors (the differences between the plant output and the setpoint) (Wu and Dexter, 2008). The second is the difficulty of identifying a plant model online from normal operating data, which are usually incomplete and may be corrupted by unmeasured disturbances and measurement noise (see Section 11.4.3). Model-free (or direct) adaptive control algorithms have just one goal: to minimize the tracking errors.

Several fuzzy model-free control schemes (Abonyi *et al.*, 1999; Ordonez and Passino, 1999; Tan and Dexter, 2000; Angelov, 2004; Pomares *et al.*, 2004; Wai and Chen, 2006; Wai and Lee, 2008) have been proposed but most are concerned with the problem of controlling systems whose behaviour is non-linear; relatively few take explicit account of the uncertainties (Yang, 2005).

Schemes that are based on feedback error learning use a fixed feedback controller and an adaptive feedforward controller, which is trained using the control signal generated by a feedback controller as its target (Kawato and Gomi, 1992). The aim is to train the feedforward controller to act as an inverse controller so that the feedback control signal is eliminated. Feedback error learning schemes have been proposed that use a feedforward controller implemented using an FRM with a defuzzified output, and either a PI, PD or PID control algorithm as the feedback controller (Tan and Dexter, 1999, 2000; Santos and Dexter, 2001).

12.2 Fuzzy FRM-Based Direct Adaptive Control

A detailed block diagram of the control scheme is shown in Figure 12.1. The self-learning fuzzy control scheme has six major components: a reference model, a proportional feedback

---

Monitoring and Control of Information-Poor Systems: An Approach based on Fuzzy Relational Models, First Edition.
Arthur L. Dexter.
© 2012 John Wiley & Sons, Ltd. Published 2012 by John Wiley & Sons, Ltd.

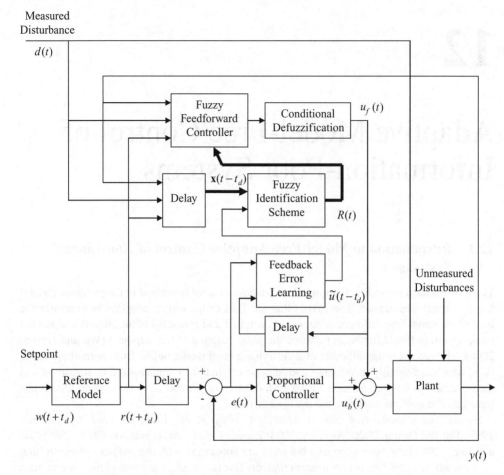

Figure 12.1 The fuzzy FRM-based model-free adaptive controller.

controller, an online fuzzy identification scheme, a feedback error learning scheme, a fuzzy feedforward controller based on a fuzzy FRM and a conditional defuzzification scheme.

The reference model is used to make the setpoint trajectory achievable. Determination of the reference model parameters is an important issue. If the time constant or order of the reference model is much smaller than that of an open-loop plant then large control signal may be generated. The proportional controller, which has a very small gain, reduces the effects of unmeasured disturbances and improves the control performance during the initial training phase (Ruan *et al.*, 2007), but it is not essential. The control signal driving the plant is the sum of the control signal from the feedback controller $u_b(t)$ and the control signal from the feedforward controller $u_f(t)$. The value of the output of the feedback controller should ideally tend to zero when the feedforward controller has been fully trained and the tracking error is zero. As a first-order plus dead-time model of the plant is assumed, the feedforward control signal $u_f(t)$ depends on the current input vector $\mathbf{x}(t) = [r(t + t_d), y(t)]$, where t_d is the effective time delay of the plant. The feedback error learning scheme calculates a target value

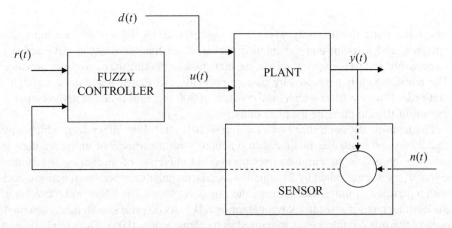

Figure 12.2 Block diagram of the control system with a noisy measurement of the plant output.

of the feedforward control signal, which is used by the online fuzzy identification scheme to update the rule confidences $R(t)$ of the fuzzy FRM-based feedforward controller. The target value, which depends on the tracking error $e(t) = r(t - t_d) - y(t)$ resulting from the previous incorrect control signal $u_f(t - t_d)$, is given by

$$\tilde{u}_f(t - t_d) = u_f(t - t_d) + \gamma e(t) \tag{12.1}$$

When the learning rate γ is unity, this is the value of the feedforward control signal that would have resulted in no tracking error. The fuzzy identification scheme uses the recursive form of the RSK algorithm with forgetting (see Section 10.1.2.2). The values of $R(t)$ are updated using the delayed input vector $\mathbf{x}(t - t_d)$ and the target value of the control action $\tilde{u}_f(t - t_d)$ generated by the feedback error learning scheme.

12.3   Behaviour in the Presence of a Noisy Measurement of the Plant Output

A block diagram of the control system with a noisy measurement of the plant output is shown in Figure 12.2. In this case, the disturbance is assumed to be constant.

Example 12.1

The control of a Hammerstein model of a non-linear dynamic system is used to evaluate the performance of the controller. The Laplace transform transfer function relating the controlled output $y(t)$ to the control signal $u(t)$ is given by:

$$Y(s) = L(f(u(t))) \left[\frac{0.899}{0.9d + 0.8} \right] \left[\frac{e^{-10s}}{1 + 200s} \right], \tag{12.2}$$

where the static non-linearity $f(u(t)) = \frac{1}{3.434}\ln(30\,u(t)+1)$ and d is an input disturbance. The setpoint and the input disturbance are held constant at 0.85 and 0.1, respectively, so that the sensor noise is the main factor affecting the control performance. The noise, which is generated by passing normally distributed white noise through a first-order low-pass filter with a time constant of 300 seconds, is added to the output of the sensor used to measure the plant output.

The input to the controller is $\mathbf{x}(t) = [r(t+10), y(t)]$. Five fuzzy sets, which have equally spaced triangular membership functions with a partition of unity, are used to describe the two input variables over normalized universes of discourse. The control signal $u_f(t)$ is represented by 21 equally spaced triangular membership functions, each with a partition of unity. The value of the forgetting factor λ is 0.999 and the value of the learning rate γ for feedback error learning is 0.5. As no prior knowledge is assumed, each of the rule confidences is initialized to the same value (0.05). The sample time of the controller $h = 10$ s and the simulation is run for a total of 1.2×10^5 samples. Height defuzzification is used for the first 1.0×10^5 samples while the controller is trained, and conditional defuzzification is used for the remainder of the time.

The effect of the sensor noise on the possibility distribution of the control signal is examined first. The resulting fuzzy control signals are plotted in Figure 12.3 for three values of the standard deviation of the measurement noise. It can be seen that the possibility values have an approximately normal distribution and the support for the fuzzy control signal increases as the sensor noise level is increased.

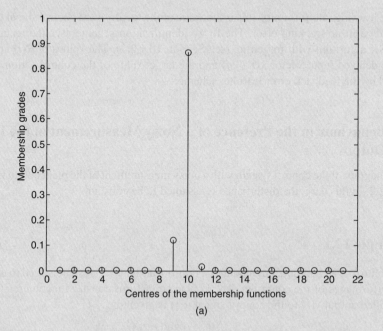

(a)

Figure 12.3 Fuzzy control signal in the presence of output sensor noise where σ equals (a) 0.009; (b) 0.045; and (c) 0.08.

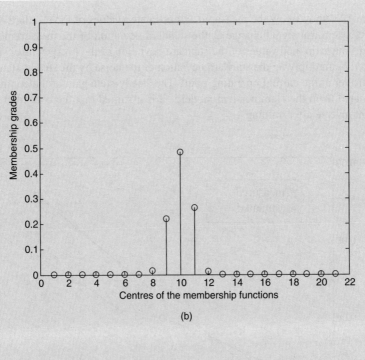

(b)

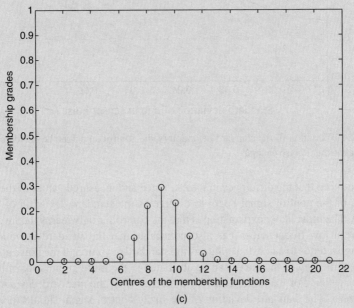

(c)

Figure 12.3   (*Continued*)

Figure 12.4 shows how the measured and estimated values of the standard deviation of the fuzzy control signal change as the standard deviation of the measurement noise is varied. The estimated values of the standard deviation of the fuzzy control signal are calculated by multiplying the standard deviation of the noise by the steady-state gain of the controller at the current operating point. The steady-state gain of the controller can be estimated from the Hammerstein model if it is assumed that the controller behaves as a plant inverse after training.

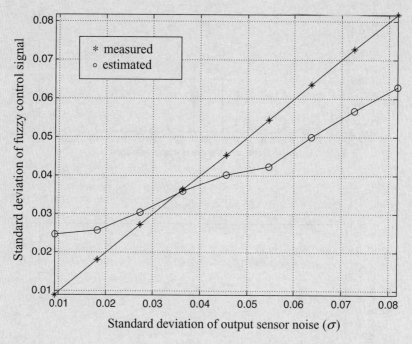

Figure 12.4 Variation of the standard deviation of the control signal as the standard deviation of the measurement noise changes.

It can be seen that the difference in the estimated and measured values of the standard deviation of the control signal begin to diverge as the standard deviation of the noise increases. The most likely explanation is that the controller only learns the inverse plant behaviour at low frequencies. It is also noticeable that the standard deviation of the control signal tends to a constant value as the standard deviation of the noise approaches zero. The flattening of the curve can be attributed to the finite granularity of the FRM-based controller. For example, consider the case when the standard deviation of the noise is zero. The standard deviation of the fuzzy control signal should then be zero but the observed standard deviation of the fuzzy control signal is 0.02. The feedforward controller will only generate a fuzzy singleton if the required value of the control signal coincides with the apex of the membership function for the controller output. Otherwise a fuzzy doubleton will be generated. As the spacing between the apexes of the membership functions of the controller output is 0.05, the standard deviation of the control signal will have a value between 0.0 and 0.025.

Although the measured values of the standard deviation of the fuzzy control signal deviate from the estimated values, it can be seen that the possibility distribution of the fuzzy control signal does reflect the uncertainties resulting from the measurement noise.

If the fuzzy control signal is representative of the noise level on the plant output sensor, the use of conditional defuzzification should avoid unnecessary changes to the control signal and reduce the control activity. The root-mean-square values of the tracking error (RMSE) and the mean absolute values of the changes to the control signal (MADU) are plotted against the value of the threshold used for conditional defuzzification in Figure 12.5. It can be seen that the control activity is highest when the threshold value is 1.0 and the controller is defuzzifying at every sampling instant, which is equivalent to using height defuzzification. The results show that the control activity reduces as the threshold for conditional defuzzification is reduced, but at the expense of larger tracking errors. However, the percentage decrease in the control activity is large (the maximum change is 81%) compared to the percentage increase in the RMSE of the tracking error (the maximum change is 9%). These percentage changes demonstrate that conditional defuzzification can reduce the control activity when there is uncertainty associated with the measurement of the plant output, without significantly compromising the control performance.

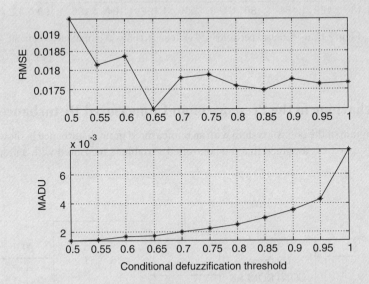

Figure 12.5   Effect of the conditional defuzzification threshold on the control performance.

The behaviour of the closed-loop control system, when the threshold value for conditional defuzzification is 0.9, is shown in Figure 12.6. It can be seen that the actual plant output follows the setpoint reasonably closely, even though the controller frequently keeps the value of the control signal constant for long periods of time.

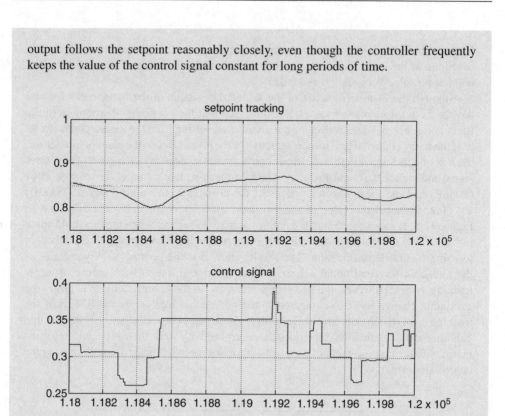

Figure 12.6   Control performance with uncertainty in output measurement.

12.4   Behaviour in the Presence of an Unmeasured Disturbance

A block diagram of the control system with an unmeasured input disturbance is shown in Figure 12.7. In this case, it is assumed that the controlled variable is measured with an ideal sensor.

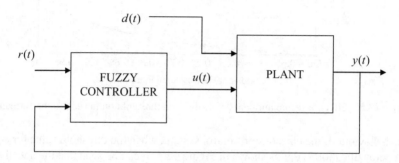

Figure 12.7   Block diagram of the control system with an unmeasured input disturbance.

Example 12.2

The control of a Hammerstein model of a non-linear dynamic system described in Example 12.1 is used to evaluate the performance of the controller.

The unmeasured input disturbance d varies randomly and has a mean of 0.1. The probability distribution of the random component is generated by passing the output of a normally distributed white noise generator through a first-order low-pass filter with a time constant of 300 seconds.

The effect of the unmeasured disturbance on the possibility distribution of the control signal is examined first. The value of the unmeasured input disturbance d is held constant at its average value after training is completed. The resulting fuzzy control signals are shown in Figure 12.8 for three values of the standard deviation of the random disturbance. It can be seen that the width of the distribution of the fuzzy control signal increases as the standard deviation increases. In all three cases the modal value of the fuzzy control signal remains at the same position, coinciding with the apex of the tenth membership function in the universe of discourse of the control signal. This suggests that the controller was able to learn the value of the control signal required to reject the mean value of the unmeasured disturbance.

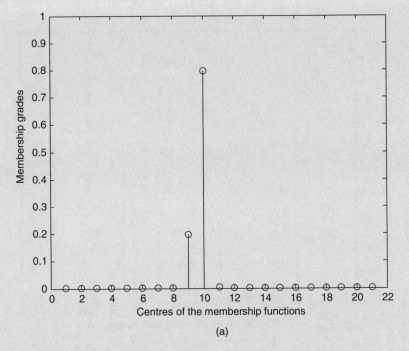

(a)

Figure 12.8 Fuzzy control signal generated in the presence of a random disturbance where σ equals (a) 0.009; (b) 0.045; and (c) 0.08.

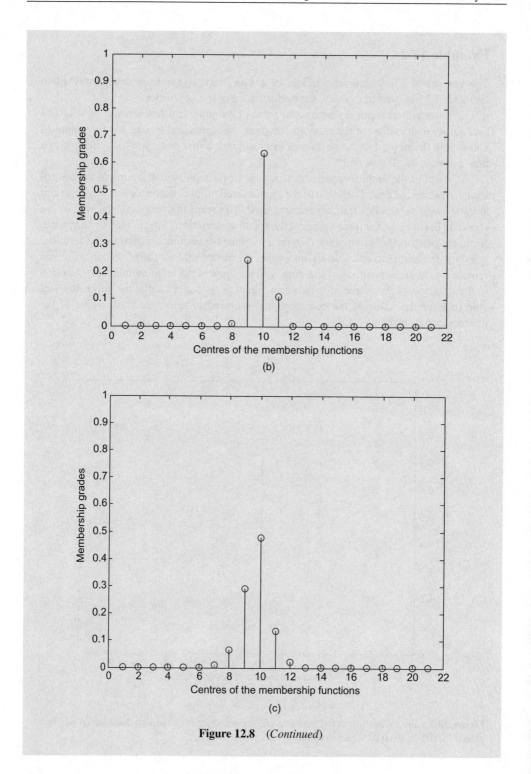

Figure 12.8   (*Continued*)

The measured and estimated values of the standard deviation of the fuzzy control signal are plotted against the standard deviation of the disturbance in Figure 12.9. It can be seen that the standard deviation of the fuzzy control signal increases as the standard deviation of the disturbance increases and, once again, the possibility distribution of the fuzzy control signal reflects the resulting uncertainties.

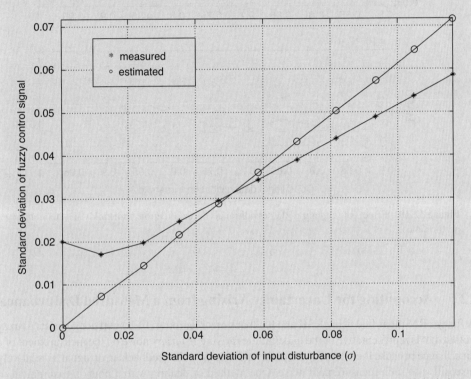

Figure 12.9   Variation of the standard deviation of the control signal as the standard deviation of the disturbance changes.

The effect of conditional defuzzification on the RMSE values of the tracking error and the mean absolute values of the changes to the control signal are shown in Figure 12.10 when the standard deviation of the disturbance is 0.045. It can be seen that there is a decrease in the value of MADU and an increase in the value of RMSE as the threshold value α for conditional defuzzification is reduced from 1.0 to 0.5. However, the overall percentage decrease in the control activity is 86% whereas the overall percentage increase in the tracking error is only 26%. As before, it can be concluded that the use of conditional defuzzification can reduce the control activity significantly while maintaining the tracking error within reasonable bounds.

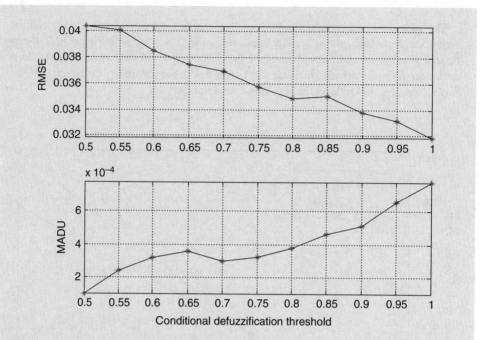

Figure 12.10   Effect of changing the conditional defuzzification threshold on the control performance.

12.5   Accounting for Uncertainty Arising from a Measured Disturbance

A fuzzy FRM identified using RSK and feedback error learning (FEL) will not generate a fuzzy output that is representative of the uncertainties arising from any noise on the measurement of a disturbance because the output of the plant, and therefore the feedback error signal, is relatively insensitive to such measurement noise. One method of dealing with a noisy measurement of the disturbance is to use a dual model FEL control scheme.

A block diagram of the dual model learning scheme for a plant with no transport delay is shown in Figure 12.11. The FEL scheme is used to identify a fuzzy FRM of the relationship between measured value of the disturbance and the state of the system x, and the control signal required to ensure that the average value of the plant output is equal to its setpoint. The crisp output u_{PM} of this model (the *primary model*), together with the measured value of the disturbance z, is then used to identify a second fuzzy FRM (the *ancillary model*) whose fuzzy output U gives an indication of the uncertainty associated with the control signal, resulting from the noisy measurement of the disturbance.

The low-pass filters are used to remove any high-frequency components of the measurement noise, which are outside of the bandwidth of the closed-loop control system, from the disturbance measurement. The defuzzified output of the primary model of the controller is used to control the plant during training. After training has been completed, the plant is controlled by conditionally defuzzifying the fuzzy output produced by the ancillary model of the controller.

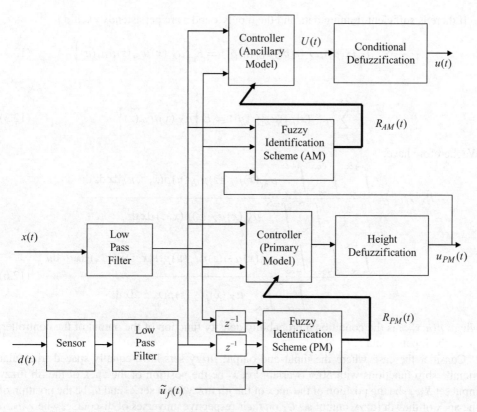

Figure 12.11 Dual model online learning scheme.

The training is terminated when the mean value of the controlled variable follows the setpoint changes within a specified tolerance.

It can be shown that the fuzzy output produced by the ancillary model of the controller is related to the measurement noise. Let d be the true value of the disturbance and $z = d + n$ be the measured value of the disturbance, where n is the associated measurement noise. Consider the case where the output of the controller is given by $u = f(x, d)$ and the measurement of the state x is ideal.

If the non-recursive form of the RSK scheme (see Section 5.3) is used to identify the FRM from u, x and z (the measured value of d), the estimated rule confidences are given by,

$$\hat{R}_{X_i, Z_j, U_k} = \frac{\dfrac{1}{N} \sum_{p=1}^{N} \mu_{X_i}(x[p]) \mu_{Z_j}(z[p]) \mu_{U_k}(u[p])}{\dfrac{1}{N} \sum_{p=1}^{N} \mu_{X_i}(x[p]) \mu_{Z_j}(z[p])} \tag{12.3}$$

where X_i are the fuzzy sets used to describe x, Z_j are the fuzzy sets used to describe z, U_k are the fuzzy sets used to describe u, N is the total number of training datasets and $\mu_{U_k}(u)$ is the membership function defining the kth fuzzy output set U_k.

If there is sufficient training data and the inputs x and z are persistently exciting,

$$\frac{1}{N}\sum_{p=1}^{N}\mu_{X_i}(x[p])\mu_{Z_j}(z[p])\mu_{U_k}(u[p]) = E\left\{\mu_{X_i}(x)\mu_{Z_j}(z)\mu_{U_k}(u)\right\} \tag{12.4}$$

and

$$\frac{1}{N}\sum_{p=1}^{N}\mu_{X_i}(x[p])\mu_{Z_j}(z[p]) = E\left\{\mu_{X_i}(x)\mu_{Z_j}(z)\right\} \tag{12.5}$$

We therefore have:

$$
\hat{R}_{X_i,Z_j,U_k} \cong \frac{\displaystyle\int_{-\infty}^{+\infty}\int_{-\infty}^{+\infty}\int_{-\infty}^{+\infty}\mu_{X_i}(x)\mu_{Z_j}(z)\mu_{U_k}(u)p(x,z,u)\,dx\,dz\,du}{\displaystyle\int_{-\infty}^{+\infty}\int_{-\infty}^{+\infty}\mu_{X_i}(x)\mu_{Z_j}(z)p(x,z)\,dx\,dz}
$$

$$
= \frac{\displaystyle\int_{-\infty}^{+\infty}\int_{-\infty}^{+\infty}\int_{-\infty}^{+\infty}\mu_{X_i}(x)\mu_{Z_j}(z)\mu_{U_k}(u)p(u|x,z)p(x,z)\,dx\,dz\,du}{\displaystyle\int_{-\infty}^{+\infty}\int_{-\infty}^{+\infty}\mu_{X_i}(x)\mu_{Z_j}(z)p(x,z)\,dx\,dz} \tag{12.6}
$$

where $p(u|x,z)$ is the conditional probability density function of the output of the controller, given the inputs x and z.

Consider the case where the input and output fuzzy sets have equally spaced triangular membership functions with 50% overlap. Let x_i be the position of the apex of the ith fuzzy input set X_i, z_j be the position of the apex of the jth fuzzy input set Z_j and u_k be the position of the apex of the kth fuzzy output set U_k on their respective universes of discourse, where u_{k-1}, u_k and u_{k+1} are the positions of the apexes of the $(k-1)$th, kth and $(k+1)$th output sets on the output universe of discourse. Then,

$$
\hat{R}_{X_i,Z_j,U_k} \cong \frac{\displaystyle\int_{u_{k-1}}^{u_{k+1}}\int_{z_{j-1}}^{z_{j+1}}\int_{x_{i-1}}^{x_{i+1}}\mu_{X_i}(x)\mu_{Z_j}(z)\mu_{U_k}(u)p(u|x,z)p(x,z)\,dx\,dz\,du}{\displaystyle\int_{z_{j-1}}^{z_{j+1}}\int_{x_{i-1}}^{x_{i+1}}\mu_{X_i}(x)\mu_{Z_j}(z)p(x,z)\,dx\,dz} \tag{12.7}
$$

If there are a very large number of fuzzy input and output sets, the width of each set will be very narrow and the conditional probability density function $p(u|x,z)$ may be assumed to be approximately constant when u is in the range $[u_{k-1}, u_{k+1}]$, x is in the range $[x_{i-1}, x_{i+1}]$ and z is in the range $[z_{j-1}, z_{j+1}]$.

Therefore

$$
\hat{R}_{X_i,Z_j,U_k} \cong \frac{p(u_k|x_i,z_j)\displaystyle\int_{u_{k-1}}^{u_{k+1}}\mu_{U_k}(u)\int_{z_{j-1}}^{z_{j+1}}\int_{x_{i-1}}^{x_{i+1}}\mu_{X_i}(x)\mu_{Z_j}(z)p(x,z)\,dx\,dz\,du}{\displaystyle\int_{z_{j-1}}^{z_{j+1}}\int_{x_{i-1}}^{x_{i+1}}\mu_{X_i}(x)\mu_{Z_j}(z)p(x,z)\,dx\,dz}
$$

$$
= p(u_k|x_i,z_j)\int_{u_{k-1}}^{u_{k+1}}\mu_{U_k}(u)\,du \tag{12.8}
$$

if

$$\int_{z_{j-1}}^{z_{j+1}} \int_{x_{i-1}}^{x_{i+1}} \mu_{X_i}(x)\mu_{Z_j}(z)p(x, z)\,dx\,dz \neq 0$$

Assuming that the base of the triangular membership function defining the output references sets is $2a$, it can be easily shown that

$$\hat{R}_{X_i, Z_j, U_k} \cong p(u_k | x_i, z_j)\,[a] = \Pr(u_k < u \leq u_{k+1} | x_i, z_j) \qquad (12.9)$$

However, $n = z - d$ and $p(n) = p(z - d) = p_n(z - d) = p_n(z - g(u, x))$, where $p_n(n)$ is the probability density function of the measurement noise and $d = g(u, x)$ is obtained from $u = f(x, d)$. We therefore have

$$\hat{R}_{X_i, Z_j, U_k} \cong p(u_k | x_i, z_j)\,[a] = p_n(z_j - g(u_k, x_i))\,[a] \qquad (12.10)$$

Example 12.3

The control of a Hammerstein model of a non-linear dynamic system described in Example 12.1 is again used to evaluate the performance of the controller. However, in this example, there is no transport delay associated with the dynamic behaviour of the plant.

Five, eleven and twenty-one equally-spaced fuzzy sets with triangular membership functions having a partition of unity are used to describe the plant output, the measured disturbance and the control signal, respectively, all over normalized universes of discourse from 0 to 1. The value of the forgetting factors (λ) used to identify the primary model and ancillary model are 0.99 and 0.995, respectively. A learning rate (γ) of 0.1 is used in the feedback error learning scheme. All of the rule confidences in both the *primary* and *ancillary* models are initially set to zero. The disturbance varies sinusoidally and has a period of 100000 seconds, a mean value of 0.1 and an amplitude of 0.1. The sensor noise is normally distributed and has a mean of 0.01. The sampling time of the controller is 10 seconds and the setpoint is held constant at 0.85. The threshold for conditional defuzzification is 0.5. The time constants of the second-order low-pass filters used to remove high-frequency measurement noise are both 1000 s.

Figure 12.12 shows the behaviour of the control system during training and testing, when the standard deviation of the measurement noise was approximately 5.0. Training is terminated after 10^6 samples.

Figure 12.13 compares the control performance of the dual model control scheme with that of a fuzzy FRM-based model-free adaptive controller (see Section 12.2) for different levels of noise on the output of the sensor measuring the disturbance.

The results show that there are no significant differences in the performance in terms of the tracking errors. However, use of the dual model control scheme results in a significant reduction in the control activity whenever measurement noise is present.

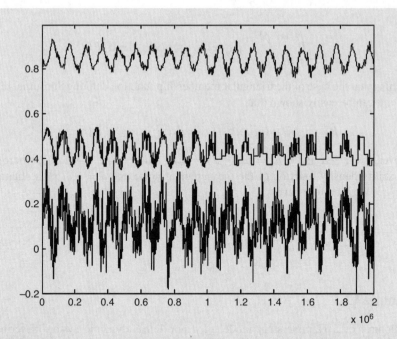

Figure 12.12 Example of the variations in the plant output (upper plot), the measured disturbance (lower plot) and the control signal (middle plot) during the training and testing periods.

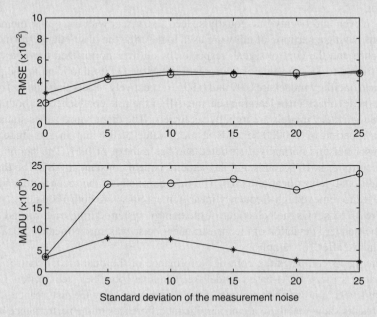

Figure 12.13 Comparison of the performance of the single (o) and dual model (*) control schemes.

12.6   Summary

This chapter has explained how a model-free adaptive control scheme suitable for use in information-poor systems can be designed. A brief introduction to model-free adaptive control has been given, and a fuzzy FRM-based direct adaptive control scheme that is based on feedback error learning has been described. The ability of the control scheme to account for noise on the measurement of the controlled variable, and a random unmeasured input disturbance, has been examined. The use of feedforward to reject a measured disturbance when the measurement of the disturbance has both random and systematic errors has also been investigated. A dual model control scheme has been described in which an *ancillary model* of the controller is identified using the output from a *primary model* of the relationship between the true value of the disturbance and the control signal. Simulation results have been presented which show that the control scheme is able to represent the uncertainties in the fuzzy control signal it produces, and demonstrated that conditional defuzzification can be used to reduce the control activity without compromising the tracking error to any great extent.

References

Abonyi, J., Andersen, H., Nagy, L. and Szeifert, F. (1999) Inverse fuzzy-process-model based direct adaptive control. *Mathematics and Computers in Simulation*, **51**, 119–132.

Angelov, P. (2004) A fuzzy controller with evolving structure. *Information Sciences*, **161**, 21–35.

Kawato, M. and Gomi, H. (1992) A computational model of four regions of the cerebellum based on feedback-error learning. *Biological Cybernetics*, **68**, 95–103.

Ordonez, R. and Passino, K.M. (1999) Stable multi-input multi-output adaptive fuzzy/neural control. *IEEE Transactions on Fuzzy Systems*, **7**(3), 345–353.

Pomares, H., Rojas, I., Gonzalez, J., *et al.* (2004) Online global learning in direct fuzzy controllers. *IEEE Transactions on Fuzzy Systems*, **12**(2), 218–229.

Ruan, X., Ding, M., Gong, D. and Qiao, J. (2007) On-line adaptive control for inverted pendulum balancing based on feedback-error-learning. *Neurocomputing*, **70**(4), 770–776.

Santos, M. and Dexter, A.L., (2001) Temperature control in a liquid helium cryostat using a self-learning neurofuzzy controller. *Proceedings of IEE: Control Theory and Applications*, **148**(3), 233–238.

Tan, W.W. and Dexter, A.L. (1999) Self-learning neurofuzzy control of a liquid helium cryostat. *Control Engineering Practice*, **7**(10), 1209–1220.

Tan, W.W. and Dexter, A.L. (2000) A self-learning fuzzy controller for embedded applications. *Automatica*, **36**(8), 1189–1198.

Wai, R-J. and Lee, J-D. (2008) Adaptive fuzzy-neural-network control for Maglev transportation system. *IEEE Transactions on Neural Networks*, **19**(1), 54–70.

Wai, R-J. and Chen, P-C. (2006) Robust neural-fuzzy-network control for robot manipulator including actuator dynamics. *IEEE Transactions on Industrial Electronics*, **53**(4), 1328–1349.

Wu, Y. and Dexter, A.L. (2008) Adaptive Fuzzy Model-based Predictive Control using Fuzzy Decision-making. Proceedings of UKACC 2008, Manchester.

Yang, Y. (2005) Direct robust adaptive fuzzy control (DRAFC) for uncertain nonlinear systems using small gain theory. *Fuzzy Sets and Systems*, **151**, 79–97.

13

Fault Diagnosis in Information-Poor Systems

13.1 Introduction to Fault Detection and Isolation in Non-Linear Uncertain Systems

Safety and reliability are a growing concern in modern complex processes. Process monitoring and the early diagnosis of faulty operation is increasingly important (Isermann, 2005a). Faults are non-permitted deviations of a process variable from its normal value that may lead to a failure of the system. They can be divided into three types: process (or component) faults, sensor faults and actuator faults. Some faults cause sudden changes in the plant behaviour (abrupt faults); others result in a gradual degradation of the performance (incipient faults).

An excellent summary of the terms used in fault detection and diagnosis is given by Isermann and Balle (1997). Fault diagnosis can be broken down into three stages. The first stage (*fault detection*) is the detection of a fault and the time of its occurrence. The second stage (*fault isolation*) involves determining the class of fault and its location in the process. The third stage (*fault identification*) involves finding the type, magnitude and cause of the fault. A system that is capable of detecting and isolating faults is often referred to as a fault detection and isolation (FDI) system.

There are three basic approaches to fault diagnosis (Frank *et al.*, 2000): hardware redundancy, analytical redundancy based on signal processing and model-based analytical redundancy. Methods based on hardware redundancy identify faults by comparing the behaviour of multiple sensors, actuators and components. A major disadvantage of this approach is the extra cost of the redundant equipment. Methods based on analytical redundancy compare the measured variables with expected values or with estimates generated by mathematical models of the plant.

A response is required if a significant fault has been diagnosed (*fault accommodation*). Either the faulty component or components must be repaired or replaced (*preventative maintenance*) (Muenchhof *et al.*, 2009) or the mode of operation of the process must change (*fault tolerant control*) usually by reconfiguring (Balle *et al.*, 1998) or redesigning the associated control

Monitoring and Control of Information-Poor Systems: An Approach based on Fuzzy Relational Models, First Edition.
Arthur L. Dexter.
© 2012 John Wiley & Sons, Ltd. Published 2012 by John Wiley & Sons, Ltd.

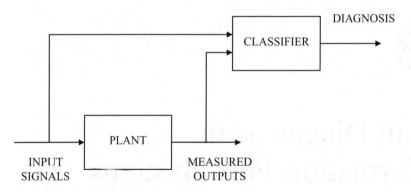

Figure 13.1   The direct approach to fault diagnosis.

scheme (Balle *et al.*, 1998; Liu and Dexter, 2001; Ichtev *et al.*, 2002; Mendonca *et al.*, 2006, 2007, 2008).

There are two basic methods for diagnosing faults: direct methods (Patton *et al.*, 2000; Reaz *et al.*, 2006) and indirect (or model-based) methods (Isermann, 2005b). Direct methods use a classifier that has been taught to recognize the presence of a fault from the observed behaviour of the process (see Figure 13.1). The main disadvantage of this approach is the difficulty of training the classifier to recognize the symptoms of each of the possible faults at all possible operating conditions. Indirect methods generate residuals by comparing the observed behaviour with that of different mathematical models of the process: one for fault-free operation and one for each of the possible faults (see Figure 13.2). In many cases, this will simplify the design of the classifier.

13.1.1   Model-Based Methods for Non-Linear Systems

Some type of non-linear reference model must be used to represent the fault-free and faulty behaviour of the process (Frank *et al.*, 2001). Fault diagnosis schemes based on T-S fuzzy models (Balle, 1999; Diao and Passino, 2004; Kowal and Korbicz, 2007; Twiddle *et al.*, 2008; Mendonca *et al.*, 2009), neurofuzzy models (Tan and Huo, 2005; Uppal *et al.*, 2006; Mok and Chan, 2008) and fuzzy relational models (FRM) (Amann *et al.*, 2001) have been proposed. Some schemes use the model(s) to estimate the transient behaviour of the measured variable(s) in dynamic systems (Balle and Spreitzer, 1998; Balle *et al.*, 1998; Amann *et al.*, 2001; Diao and Passino, 2004; Mendonca *et al.*, 2005; Uppal *et al.*, 2006; Korbicz and Kowal, 2007). Others extract steady-state information from the measurements and use a reference model of the steady-state behaviour (Dexter and Ngo, 2001; Tan and Huo, 2005; Twiddle *et al.*, 2008). The training data are easier to generate and less training data are required when a steady-state model is used, particularly if the model is identified using data obtained from a computer simulation (Tan and Huo, 2005).

It should be noted that dynamic reference models, which are based on a one-step-ahead prediction, so-called series-parallel models (Balle and Spreitzer, 1998; Balle, 1999), cannot use previous values of the measured outputs as inputs to a model that is designed to estimate the fault-free behaviour, after a fault has occurred (Korbicz and Kowal, 2007). One solution

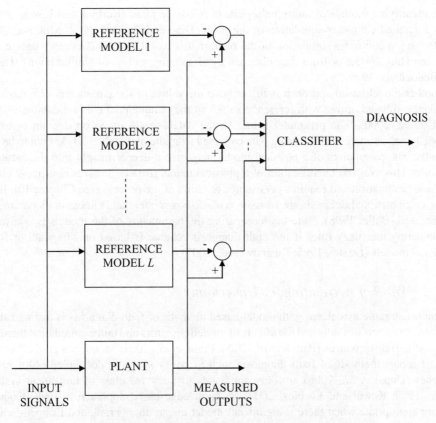

Figure 13.2   The indirect approach to fault diagnosis.

to this problem is to avoid the use of autoregressive inputs in the model (Korbicz and Kowal, 2007). Another is to use several reference models to describe the behaviour of the plant, with and without faults (Diao and Passino, 2004; Mendonca *et al.*, 2009). Another is to use only previously estimated values of the plant output as inputs to the model, the so-called parallel model (Frank and Koppen-Seliger, 1997b; Amann *et al.*, 2001).

Some schemes use only a reference model of fault-free behaviour and multiple measurements and estimates to differentiate between different faults (Balle *et al.*, 1998). In such schemes, the diagnosis depends on which residuals are close to zero and the sign of the residuals (Balle and Spreitzer, 1998). Others use a single measurement and multiple reference models of fault-free and faulty behaviour (possibly including that resulting from the simultaneous occurrence of several different faults) (Dexter and Ngo, 2001; Diao and Passino, 2004; Uppal *et al.*, 2006) or multiple measurements and reference models of fault-free and faulty behaviour (Mendonca *et al.*, 2005, 2009). An expert supervisor can be used to decide on the minimum combination of reference models that are required in a multi-model scheme (Diao and Passino, 2004). In these schemes, the diagnosis depends on which of the models generate residuals that are close to zero.

The reference models can be identified from data obtained from the actual plant (Balle *et al.*, 1998; Ichtev *et al.*, 2002; Mendonca *et al.*, 2008), although this is seldom an option when identifying models of faulty behaviour in full-scale plant (Korbicz and Kowal, 2007); from a detailed computer simulation of the plant (Diao and Passino, 2004; Mok and Chan, 2008); from a computer simulation of the plant that is based on manufacturer's design data (Tan and Huo, 2005); or from data obtained by simulating a class of similar plants (Dexter and Benouarets, 1996).

Another model-based approach is to compare the values of the parameters of a model of the plant estimated online with reference values of the parameters held in a database (Balle and Fuessel, 2000). The parameter values in the database are based on domain or expert knowledge or estimated offline using data collected from the plant when it is known to be free of faults. The parameters of a physical model may give a deeper insight into the nature of the faults. However the identification of a physical model usually involves non-linear online parameter estimation and requires persistent excitation of the process (see Chapter 10). If the online computational demands are an issue, a black-box model that is linear in its parameters can be used (Balle, 1998). Note that comparing the parameters of the models is equivalent to comparing the fuzzy rules if the fault diagnosis scheme is based on Mamdani or fuzzy relational models (Dexter, 1995; Rusinov et al., 2007).

13.1.2   Ways of Accounting for Uncertainty

An important issue associated with model-based methods of fault diagnosis is distinguishing between the effects of faults and the effects of modelling errors and unmeasured (and therefore un-modelled) disturbances (Patton et al., 2000; Frank et al., 2001; Mendonca et al., 2009).

Most robust methods of fault diagnosis such as parity space or decoupled observer approaches (Uppal et al., 2006) are only applicable to a narrow class of non-linear systems (Balle, 1999; Kowal and Korbicz, 2007). A knowledge-based approach to fault diagnosis is more appropriate when there is significant model uncertainty (Frank and Koppen-Seliger, 1997a, b; Frank et al., 2000). Several methods of dealing with uncertainty have been suggested.

An adaptive (i.e. variable) threshold can be used, which takes the size of the uncertainties into account when analysing the residuals. This allows a value of the threshold to be selected that will not cause the scheme to be insensitive to faults, nor result in frequent false alarms. The threshold can be identified from fault-free data (Balle, 1999; Tan and Huo, 2005) or derived from prior knowledge of the modelling errors (Korbicz and Kowal, 2007).

As shown in Figure 13.3, fuzzy rules can be used to analyse the residuals (Ulieru and Isermann, 1993). The use of expert fuzzy rules (Balle and Spreitzer, 1998; Isermann, 1998; Rizzo and Gabriella Xibilia, 2002; Reaz et al., 2006; Mendonca et al., 2008; Twiddle et al., 2008), fuzzy rules identified from data (Balle and Fuessel, 2000; Mok and Chan, 2008) and fuzzy fault trees (Ulieru, 1994) have all been suggested.

Alternatively, an interval fuzzy model can be used as the reference model so that a confidence interval can be generated for the predicted plant output (Oblak et al., 2007). A fault is detected if the measured output of the plant is outside of this interval.

To reduce the rate of false alarms, decisions are often based on more than one set of observations so that an alarm is raised only if the same diagnosis is generated on several occasions (Diao and Passino, 2004; Mendonca et al., 2006, 2008; Rusinov et al., 2007), a moving average window is used to filter the residuals (Ichtev et al., 2002) or evidence is accumulated using the Dempster-Shafer theory of evidence (see Section 3.2.2) (Dexter and Benouarets, 1997; Rakar et al., 1999; Jones et al., 2002).

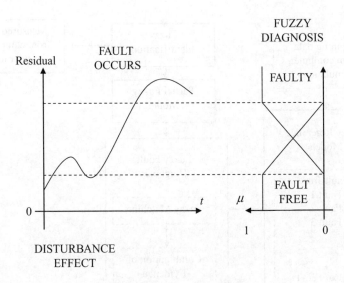

Figure 13.3 Fuzzy threshold.

Fuzzy decision factors can also be used to analyse the residuals. Each fault has an associated fuzzy decision factor, which is the t-norm of the degrees of membership in 'residual is zero' for those residuals associated with that particular fault (Mendonca *et al.*, 2008, 2009). An alarm is raised only if the value of the decision factor of one of the faults is above a specified threshold and the decision factors of all of the other faults are less than the threshold, on a given number of consecutive occasions.

13.2 A Fuzzy FRM-Based Fault Diagnosis Scheme

The fuzzy FRM-based fault diagnosis scheme is a model-based approach that is based on online parameter estimation and rule comparison. The method of diagnosis accounts for any ambiguity which may result from fault-free and faulty operation or operation with different faults, having similar symptoms at some operating conditions.

The scheme uses fuzzy FRMs as reference models to represent the fault-free and faulty behaviour of the system (Dexter and Benouarets, 1997). The reference models are obtained from data generated by simulating a similar plant or a typical plant of the same type. Each fuzzy reference model is a qualitative description of the relationship between input-output variables. One of the models describes the correct operation of the system and each of the other models describes the behaviour of the system in the presence of a particular fault or group of faults. An overview of the fault diagnosis scheme is shown in Figure 13.4. The measured data are pre-processed to remove any high-frequency noise and, when static fuzzy reference models are to be used, a transient detector is used to determine whether the system is sufficiently close to steady state.

A partial fuzzy model which describes the current behaviour of the system is identified online from the measured data. The term partial model is used since the model only describes the behaviour of the system at the current operating point and will have a large number of rules whose rule confidences are zero. The reference models describe the behaviour of the system

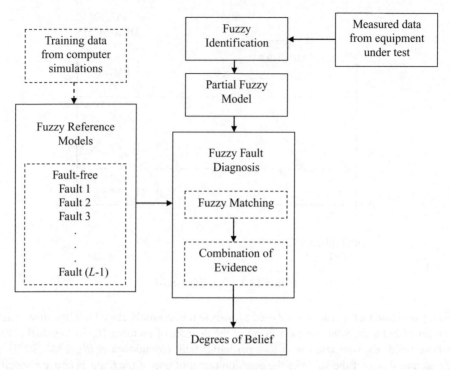

Figure 13.4 Overview of the fault diagnosis scheme.

at all possible operating points and will have many more rules with non-zero rule confidences. A fuzzy matching scheme is used to compare the rules of the partial fuzzy model with the rules in the fuzzy reference models that have the same antecedents as rules with non-zero rule confidences in the partial model. The reference models are also compared to each other to determine any ambiguity which may arise if fault-free and faulty operation or different faults have similar symptoms at a particular operating point.

13.2.1 Measuring the Similarity of FRMs

The degree of similarity of fuzzy FRMs that have the same structure can be evaluated in the same way as the similarity of conventional fuzzy sets (see Section 6.2.3) if the models are themselves considered as level-2 fuzzy sets (Sosnowski and Pedrycz, 1992) with discrete membership functions defined by the rule confidences of their rules (Dexter, 1995). Thus, a measure of the degree of similarity between two fuzzy models M_i and M_j is given by

$$Sim_{M_iM_j} = \frac{\sum_{n=1}^{N} MIN\{R_{M_i}(n), R_{M_j}(n)\}}{\sum_{n=1}^{N} R_{M_i}(n)} \tag{13.1}$$

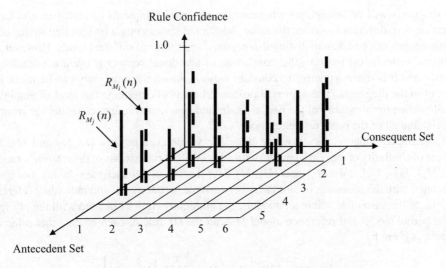

Figure 13.5 Fuzzy similarity of two fuzzy models.

where $R_{M_i}(n)$ and $R_{M_j}(n)$ are the rule confidences of the nth rule in the fuzzy models M_i and M_j, respectively, and N is the number of rules in each of the models.

The value $Sim_{M_iM_j}$ can be regarded as a measure of the extent to which the symptoms of the behaviour in the operating state represented by fuzzy model M_i are similar to those of behaviour in the state represented by the fuzzy model M_j (see Figure 13.5).

The degrees of similarity of the rules of the partial fuzzy model M_P and the rules of the L fuzzy reference models are therefore given by

$$Sim_{M_PM_i} = \frac{\sum_{n=1}^{N} MIN\{R_P(n), R_{M_i}(n)\}}{\sum_{n=1}^{N} R_P(n)}, i = 1, 2, \ldots L \qquad (13.2)$$

where $R_{M_P}(n)$ is the confidence in the nth rule in the partial model denoted M_P.

In practice, the symptoms of the behaviour described by the partial model may be similar to those described by more than one of the reference models. The same approach can be used to calculate the degree of similarity of the partial model and any two or more of the reference models (Dexter and Benouarets, 1997). For example, the degree of similarity of the partial model and all of the reference models $M_1, M_2, \ldots, M_L$ is given by

$$Sim_{M_PM_1M_2\ldots M_L} = \frac{\sum_{n=1}^{N} MIN\{R_{M_P}, R_{M_1}(n), R_{M_2}(n), \ldots, R_{M_L}(n)\}}{\sum_{n=1}^{N} R_{M_p}(n)} \qquad (13.3)$$

The diagnosis will be ambiguous whenever the observed symptoms of fault-free and faulty operation or of different faults are the same. Additional sensors might be installed so that other measurements can be taken to distinguish between symptoms of different faults. However, for economic or technical reasons, the installation of additional sensors is often not feasible in practice and it is more attractive to consider ways in which the ambiguity can be taken into account in the diagnosis. The approach proposed here is to determine the level of ambiguity by calculating the similarity of the partial model and a particular reference model (or group of models) and all of the other reference models.

For example, suppose that there are three fuzzy reference models M_1, M_2 and M_3. The degrees of similarity of the partial model with all possible combinations of the reference models $\{M_1, M_2\}$, $\{M_1, M_3\}$, $\{M_2, M_3\}$ and $\{M_1, M_2, M_3\}$ determine whether there is any ambiguity associated with the measures of similarity between the partial model and individual reference models. In the case of M reference models, the ambiguity Amb associated with the nth rules of the partial model and reference model M_i, and the nth rule of each of the other reference models, is given by

$$Amb_{M_i}(n) = MIN\left\{R_{MP}, R_{M_i}, \underset{j=1, j\neq i}{\overset{L}{MAX}}\left[R_{M_j}(n)\right]\right\} \tag{13.4}$$

Similarly, the ambiguity associated with the nth rule of two reference models $\{M_i, M_j\}$ and the nth rule of all of the other reference models is given by

$$Amb_{\{M_i, M_j\}}(n) = MIN\left\{R_{MP}, R_{M_i}, R_{M_j}, \underset{l=1, j\neq i\neq j}{\overset{L}{MAX}}\left[R_{M_l}(n)\right]\right\} \tag{13.5}$$

and so on.

13.2.2   Accumulating Evidence of Fault-Free or Faulty Operation

The similarity measures Sim and the levels of ambiguity Amb are used to generate the strength of evidence (or normalized basic assignment) that the system is in any one of its possible states. Here 'possible state' can mean operation with no faults or operation in the presence of an individual fault or any one of a group of faults. The basic assignments are defined with respect to a frame of discernment Φ based on a finite number of propositions, where a basic assignment of value of 0 indicates no evidence and a value of 1 indicates complete evidence in support of a particular proposition. When there are L reference models $M_1, M_2, \ldots, M_L$ (where M_1 is the model of the correct operation, and the other $L-1$ models define the operation of the system in the presence of each of $L-1$ different faults), the possible states that must be considered are fault-free operation, operation in the presence of each of the $R–1$ single faults, operation in the presence of one of every combination of possible faults and fault-free operation.

If the reference models do not define the behaviour in the presence of every one of the faults that can possibly occur, the partial model may exhibit symptoms of an unrecognized fault. In this case, a proposition $\{M_U\}$ must also be included in Φ to account for situations where the symptoms described by the partial model differ from those described by any of the reference models. In this case, the frame of discernment is

$$\Phi = \{\{M_1\}, \{M_2\}, \ldots \{M_L\}, \{M_1, M_2\}, \{M_1, M_3\}, \ldots \{M_{L-1}, M_L\}, \ldots \{M_1, M_2, \ldots M_L\}, \{M_U\}\}$$

(13.6)

The strength of evidence that the system is in the state associated with model M_i is given by the degree to which the partial model P is only similar to model M_i (Dexter and Benouarets, 1997). This value is calculated by subtracting the total ambiguity Λ_{M_i} associated with reference model M_i from the degree of similarity between the partial model and that reference model. The strength of evidence $m(\{M_i\})$ is therefore given by

$$m(\{M_i\}) = \frac{\sum_{n=1}^{N} \{MIN\left[R_{M_P}(n), R_{M_i}(n) - Amb_{M_i}(n)\right]\}}{\sum_{n=1}^{N} R_{M_P}(n)}$$

(13.7)

or

$$m(\{M_i\}) = Sim_{M_P M_i} - \Lambda_{M_i}$$

(13.8)

where Λ_{M_i} is the total ambiguity associated with reference model M_i, defined

$$\Lambda_{M_i} = \frac{\sum_{n=1}^{N} Amb_{M_i}(n)}{\sum_{n=1}^{N} R_{M_P}(n)}$$

(13.9)

The degree to which the partial model is only similar to both M_i and M_j is used to estimate the strength of evidence that the system is in either the state associated with model M_i or the state associated with model M_j. Thus,

$$m(\{M_i, M_j\}) = \frac{\sum_{n=1}^{N} \left\{MIN\left[R_{M_P}(n), R_{M_i}(n), R_{M_j}\right] - Amb_{\{M_i, M_j\}}(n)\right\}}{\sum_{n=1}^{N} R_{M_P}(n)}$$

(13.10)

The strength of evidence that the system is in the state associated with model M_i, the state associated with model M_j or the state associated with model M_k is similarly given by

$$m(\{M_i, M_j, M_k\}) = \frac{\sum_{n=1}^{N} \left\{MIN\left[R_{M_P}(n), R_{M_i}(n), R_{M_j}, R_{M_k}\right] - Amb_{\{M_i, M_j M_k\}}(n)\right\}}{\sum_{n=1}^{N} R_{M_P}(n)}$$

(13.11)

and so on.

The strength of evidence that the system is in an unrecognized state M_U is given by

$$m(\{M_U\}) = \frac{\sum_{n=1}^{N} \left\{ R_{M_P}(n) - MIN\left[R_{M_P}(n), \; MAX_{j=1}^{L} R_{M_j}(n) \right] \right\}}{\sum_{n=1}^{N} R_{M_P}(n)} \qquad (13.12)$$

The evidence associated with each proposition Q_i in the frame of discernment can therefore be generated:

$$m(\{M_1\}), m(\{M_2\}), \ldots m(\{M_L\}), m(\{M_1, M_2\}), m(\{M_1, M_3\}), \ldots, m(\{M_{L-1}, M_L\}),$$
$$\ldots, m(\{M_1, M_2, \ldots M_L\}), m(\{M_U\}) \}$$
$$(13.13)$$

This method of calculating the basic assignments ensures that

$$0 \le m(Q_i) \le 1, \quad \forall Q_i \in \Phi \quad \text{and} \quad \sum_{Q_i \in \Phi} m(Q_i) \le 1$$

New values for the basic assignments must be computed every time the partial model is identified and the degrees of belief in each of the propositions are updated whenever new evidence is obtained. If there is a sufficiently long time interval between the current and last identification of the partial model, the evidence can be assumed to come from independent sources and Dempster's rule can be used to combine the new evidence with evidence collected previously. Alarms are set whenever the belief in the presence of an individual fault reaches a user-defined threshold, but the current values of belief in each of the propositions are also displayed to allow the plant operator to make the final diagnosis.

   Although the reference models can be generated offline using data collected from computer simulations of typical systems, and fuzzy matching is not (in itself) computationally very demanding, there are some problems associated with using the approach for fault diagnosis in large-scale systems. Processing times lengthen rapidly as the number of possible faults (and therefore the number of reference models) increases, and when the number of inputs to each reference model becomes large. The processing overhead associated with generating the evidence and calculating the degrees of belief is particularly sensitive to the number of states in which it is assumed the system could be operating.

   To avoid the online computational complexity associated with combining evidence when a large number of possible faults is considered, the diagnosis can be restricted to propositions associated with individual models (Gertler and Anderson, 1992), to propositions associated with individual models and pairs of models, or to individual models and propositions associated with groups of faults which are known to have similar symptoms. In such cases, $\sum_{Q_i \in \Phi} m(Q_i)$ is strictly less than 1 and the remaining unassigned evidence is associated with the proposition $\{M_1, M_2, \cdots M_L\}$.

   The degree of belief that the system is in a particular operating state is calculated from the combined strengths of evidence using evidence theory (see Section 3.2.2). Belief in a single operating state, which is referred to as unambiguous belief, is equal to the strength of evidence that the system is in that operating state.

The results of the fuzzy model-based diagnosis are belief values in the range 0% (no belief) to 100% (complete belief) in each of the possible operating states of the subsystem under test. The possible operating states may involve a single fault condition or states where one of two or more faults are present. Where faults exhibit similar symptoms at some operating points, evidence collected at a number of operating conditions must be combined to generate a more precise diagnosis.

13.2.3   Generating Robust Generic Models of Faulty Operation

A major source of uncertainty is bias (systematic error) on the output of sensors arising as a result of poor calibration, long-term sensor drift or the use of single point sensors to represent spatial averages (see Chapter 17). One approach is to assume that sensor errors cannot be eliminated and to base the diagnosis on robust generic reference models that take sensor bias into account (Ngo and Dexter, 1999).

Worst-case positive and negative offsets are added to the training data produced by the simulations to account for bias on the output sensor. Fuzzy FRMs are first identified that describe the behaviour of the individual designs when the sensor has its largest possible positive bias and its largest possible negative bias. The robust generic models are produced by taking the maximum value of each rule confidence for each of the individual fuzzy models and then post-processing the resulting values to ensure that the rule confidences of rules with the same antecedent form a unimodal set of values over the output universe of discourse (see Section 5.5.1). This method of generating the reference models guarantees that the diagnosis will generate no unambiguous evidence of a fault if: the equipment under test is fault-free and is a member of the class of designs used to generate the training data; the test data are collected at operating conditions used to generate the training data; and the actual sensor bias is less than the sensor offset included in the training data.

A large number of non-zero values of belief may be generated even when the number of possible faults is small (e.g. with only 5 possible faults, the diagnosis can generate up to 64 non-zero values of belief) and a decision must be taken as to which, if any, alarm messages should be displayed. Since diagnosis based on robust generic reference models is unlikely to generate erroneous values of belief, the largest non-zero value of belief in a single-fault condition (including belief in an unrecognized fault) determines the alarm message. If all the single-state beliefs are zero, the largest non-zero value of belief in one of two fault conditions determines the alarm message, and so on.

It should be noted that the fault sensitivity of fuzzy model-based diagnosis based on robust generic reference models is determined by both the size of the class that the generic reference models represent and the magnitude of the sensor offsets superimposed on the training data.

13.2.4   Multi-Step Fault Diagnosis

The precision of the results can be improved if the number of reference models used in the diagnosis is reduced by first eliminating some of the faults that could have occurred (Dexter and

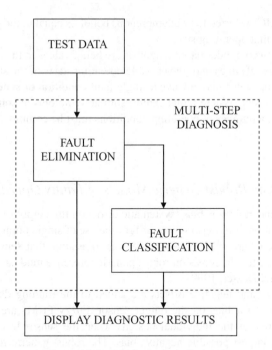

Figure 13.6 Multi-step fault diagnosis.

Ngo, 2001). There will also be a significant reduction in the processing overhead associated with combining evidence if fewer references models are used in each diagnosis. A fault can be eliminated if a comparison of the observed behaviour with the behaviour when that particular fault has occurred and when no faults are present generates no unambiguous belief in its presence. Note that there are similarities between this approach and the use of a model-based fault detector to check for faulty behaviour before diagnosis is initiated to isolate the particular type of fault.

Figure 13.6 is a block diagram of the multi-step fault diagnosis scheme. The diagnostic process is divided into two phases: fault elimination and fault classification.

Multiple diagnoses are performed in the fault-elimination phase (see Figure 13.7), each based on the use of only two reference models: the fault-free reference model and one of the reference models describing faulty behaviour. Only one of three conclusions is possible: unambiguous belief in fault-free operation; unambiguous belief in the presence of the fault; or it is unclear whether the fault is, or is not, present. The subsystem under test is assumed to be fault-free if the result of all diagnoses is unambiguous belief in fault-free operation. Faulty operation is assumed if one or more of the diagnoses generate unambiguous beliefs in the presence of a fault. The diagnosis is inconclusive if at least one of the diagnoses generates an ambiguous result and no diagnosis generates unambiguous belief in the presence of a fault.

The fault classification phase (see Figure 13.8) is entered whenever the fault elimination phase indicates that two or more faults may be present. In this phase, the test data are re-evaluated to isolate the particular fault that is present. This time only the reference

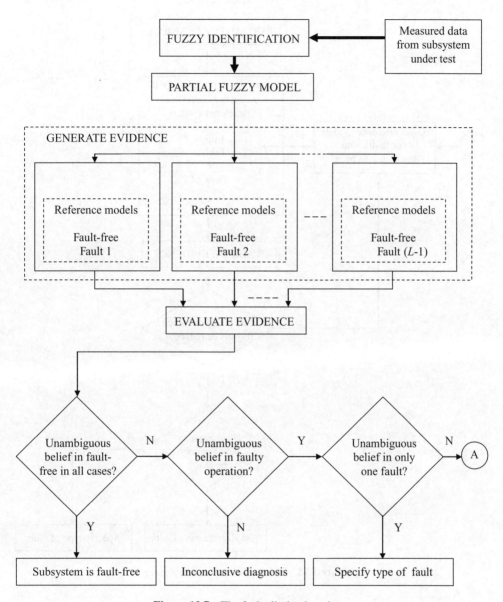

Figure 13.7   The fault elimination phase.

models associated with faults, for which the fault elimination phase generated some unambiguous belief, are used. If this process still generates unambiguous belief in more than one fault, it is repeated until either a single non-zero value of unambiguous belief is generated or the same number of reference models would be used in the next step of the diagnosis. If the latter occurs, a list is produced of all of the faults in which there is still some unambiguous belief.

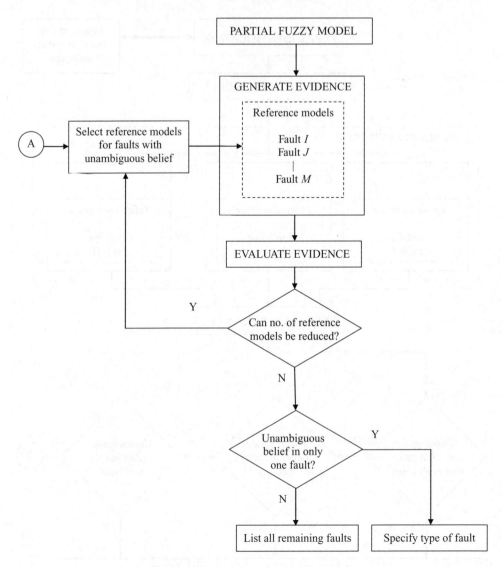

Figure 13.8   The fault classification phase.

13.3 Summary

This chapter has shown how FRMs can be used to diagnose faults in information-poor systems without generating false alarms. An introduction has been given to fault detection and isolation (FDI) in non-linear uncertain systems and a fuzzy FRM-based approach to FDI has been described. Methods for measuring the similarity of FRMs, accumulating evidence of fault-free or faulty operation and generating robust generic models of fault-free and faulty operation have been described. A multi-step method of fault diagnosis has also been proposed.

References

Amann, P., Perronne, J.M., Gissinger, G.L. and Frank, P.M. (2001) Identification of fuzzy relational models for fault detection. *Control Engineering Practice*, **9**, 555–562.

Balle, P. (1998) Fuzzy model-based symptom generation and fault diagnosis for nonlinear processes. IEEE Conference on Fuzzy Systems, **1**, 945–950.

Balle, P. (1999) Fuzzy-model-based parity equations for fault isolation. *Control Engineering Practice*, **7**, 261–270.

Balle, P. and Spreitzer, K. (1998) A multi-model approach for detection and isolation of sensor and process faults for a heat exchanger. *IEEE Conference on Fuzzy Systems*, **2**, 1476–1481.

Balle, P. and Fuessel, D. (2000) Closed-loop fault diagnosis based on a nonlinear process model and automatic rule generation. *Engineering Applications of Artificial Intelligence*, **13**, 695–704.

Balle, P., Fischer, M., Nelles, O. and Isermann, R. (1998) Integrated control, diagnosis and reconfiguration of a heat exchanger. *IEEE Control Systems Magazine*, **18**(3), 52–63.

Dexter, A.L. (1995) Fuzzy model-based fault diagnosis using fuzzy matching. *Proceedings of IEE, Part D*, **142**(6), 545–550.

Dexter, A.L. and Benouarets, M. (1996) A generic approach to identifying faults in HVAC plant. *ASHRAE Transactions*, **102**(Pt. 1), 567–573.

Dexter, A.L. and Benouarets, M. (1997) Model-based fault diagnosis using fuzzy matching. *IEEE Transactions on Systems, Man, and Cybernetics - Part A*, **27**(5), 673–682.

Dexter, A.L. and Ngo, D. (2001) Fault diagnosis in HVAC systems: a multi-step fuzzy model-based approach. *International Journal of HVAC&R Research*, **7**(1), 83–102.

Diao, Y. and Passino, K.M. (2004) Fault diagnosis for a turbine engine. *Control Engineering Practice*, **12**(9), 1151–1165.

Frank, P.M. and Koppen-Seliger, B. (1997a) Fuzzy logic and neural network applications to fault diagnosis. *International Journal of Approximate Reasoning*, **16**, 67–88.

Frank, P.M. and Koppen-Seliger, B. (1997b) New developments using AI in fault diagnosis. *Engineering Applications of Artificial Intelligence*, **10**(1), 3–14.

Frank, P.M., Ding, S.X. and Marcu, T. (2000) Model-based fault diagnosis in technical processes. *Transactions of IMC*, **22**(1), 57–101.

Frank, P.M., Alcorta Garcia, E. and Koppen-Seliger, B. (2001) Modelling for fault detection and isolation versus modelling for control. *Mathematical and Computer Modelling of Dynamical systems*, **7**(1), 1–46.

Gertler, J.J. and Anderson, K.C., An evidential reasoning extension to quantitative model-based failure diagnosis. *IEEE Transactions on Systems, Man, and Cybernetics*, **22**(2), 1992.

Ichtev, A., Hellendorn, J., Babuska, R., and Mollov, S. (2002) Fault-tolerant model-based predictive control using multiple T-S fuzzy models. *IEEE Conference on Fuzzy Systems*, **1**, 346–351.

Isermann, R. (1998) On fuzzy logic applications for automatic control, supervision, and fault diagnosis. *IEEE Transactions on Systems, Man and Cybernetics, Part A*, **28**(2), 221–235.

Isermann, R. (2005a) *Fault-diagnosis Systems: An Introduction from Fault Detection to Fault Tolerance*, Springer, Berlin, Heidelberg, New York.

Isermann, R. (2005b) Model-based fault detection and diagnosis – status and applications. *Annual Reviews in Control*, **29**, 71–85.

Isermann, R. and Balle, P. (1997) Trends in the application of model-based fault detection and diagnosis of technical processes. *Control Engineering Practice*, **5**(5), 709–719.

Jones, R.W., Lowe, A. and Harrison, M.J. (2002) A framework for intelligent medical diagnosis using the theory of evidence. *Knowledge-based Systems*, **15**, 77–84.

Korbicz, J. and Kowal, M. (2007) Neuro-fuzzy networks and their application to fault detection of dynamical systems. *Engineering Applications of Artificial Intelligence*, **20**, 609–617.

Kowal, M. and Korbicz, J. (2007) Fault detection under fuzzy model uncertainty. *International Journal of Automation and Computing*, **4**(2), 117–124.

Liu, X-F and Dexter, A.L. (2001) Fault tolerant supervisory control of VAV air-conditioning systems. *Energy and Buildings*, **33**, 379–389.

Mendonca, L.F., Sousa, J.M.C. and Sa da Costa, J.M.G. (2005) Fault detection and isolation of industrial processes using optimized fuzzy models. *IEEE Conference on Fuzzy Systems*, 851–856.

Mendonca, L.F., Vieira, S.M., Sousa, J.M.C. and Sa da Costa, J.M.G. (2006) Fault accommodation using fuzzy predictive control. *IEEE Conference on Fuzzy Systems*, 1535–1542.

Mendonca, L.F., Sousa, J.M.C. and Sa da Costa, J.M.G. (2007) Fault tolerant control of a three tank benchmark using fuzzy predictive control. *Lecture Notes on AI*, 4529, pp. 732–742.

Mendonca, L.F., Sousa, J.M.C. and Sa da Costa, J.M.G. (2008) Fault accommodation of an experimental three tank system using fuzzy predictive control. *IEEE Conference on Fuzzy Systems*, 1619–1625.

Mendonca, L.F., Sousa, J.M.C. and Sa da Costa, J.M.G. (2009) An architecture for fault detection and isolation based on fuzzy methods. *Expert Systems with Artificial Intelligence*, 36(2), 1092–1104.

Mok, H.T. and Chan, C.W. (2008) Online fault detection and isolation of nonlinear systems based on neurofuzzy networks. *Engineering Applications of Artificial Intelligence*, 21, 171–181.

Muenchhof, M., Beck, M. and Isermann, R. (2009) Fault-tolerant actuators and drives – structures, fault detection principles and application. *Annual Reviews in Control*, 33(2), 136–148.

Ngo, D. and Dexter, A.L. (1999) A robust model-based approach to diagnosing faults in air-handling units. *Transactions of ASHRAE*, 105(Pt. 1), 1078–1086.

Oblak, S., Skrjanc, I. and Blazic, S. (2007) Fault detection for nonlinear systems with uncertain parameters based on the interval fuzzy model. *Engineering Applications of Artificial Intelligence*, 20, 503–510.

Patton, R.J., Chen, J. and Benkhedda, H. (2000) A study on neuro-fuzzy systems for fault diagnosis. *International Journal of System Science*, 31(11), 1441–1448.

Rakar, A., Juricic, D. and Balle, P. (1999) Transferable belief model in fault diagnosis. *Engineering Applications of Artificial Intelligence*, 12, 555–567.

Reaz, M.B.I., Choong, F., Sulaiman, M.S. and Mohd-Yasin, F. (2006) Design and implementation of a power quality disturbance classifier: an AI approach. *Journal of Intelligent and Fuzzy Systems*, 17, 623–631.

Rizzo, A. and Gabriella Xibilia, M. (2002) An innovative intelligent system for sensor validation in Tokomak machines. *IEEE Transactions on Control Systems Technology*, 10(3), 421–431.

Rusinov, L.A., Rudakova, I.V. and Kurkina, V.V. (2007) Real-time diagnostics of technological processes and field equipment. *Chemometrics and Intelligent Laboratory Systems*, 88, 18–25.

Sosnowski, Z.A. and Pedrycz, W. (1992) *Fuzzy Logic for the Management of Uncertainty* (eds L. Zadeh and J. Kacprzyk), John Wiley & Sons, Inc., New York, pp. 501–505.

Tan, W-W. and Huo, H. (2005) A generic neurofuzzy model-based approach for detecting faults in induction motors. *IEEE Transactions on Industrial Electronics*, 52(5), 1420–1427.

Twiddle, J.A., Jones, N.B. and Spurgeon, S.K. (2008) Fuzzy model-based condition monitoring of a dry vacuum pump via time and frequency analysis of exhaust pressure signal. *Proceedings of Institute of Mechanical Engineering*, 222(2), 287–293.

Ulieru, M. (1994) Diagnosis by approximate reasoning on dynamic fuzzy fault trees. *IEEE Conference on Fuzzy Systems*, 3, 2051–2056.

Ulieru, M. and Isermann, R. (1993) Design of a fuzzy logic based diagnostic model for technical processes. *Fuzzy Sets and Systems*, 58, 249–271.

Uppal, F.J., Patton, R.J. and Witczak, M. (2006) A neuro-fuzzy multiple-model observer approach to robust fault diagnosis based on the DAMADICS benchmark problem. *Control Engineering Practice*, 14(6), 699–717.

Part IV

Some example applications

14

Control of Thermal Comfort

The thermal comfort of the occupants of a building depends on many factors including metabolic rates, clothing, air temperature, mean radiant temperature and air velocity and humidity (Tse and So, 2000; Mirinejad et al., 2008). In most buildings, however, only temperature and humidity can be controlled (Schumacher et al., 1998). Indeed, in many European buildings, over a wide range of humidity, only zone temperature is controlled (Kummert et al., 1997). In such cases, the control objective is then to maintain the zone temperature within a pre-defined range (Lute and van Paassen, 1995; Hagras et al., 2008).

The predicted mean vote (PMV) is often used as an overall index of the global thermal comfort conditions (Calvino et al., 2004, 2010). PMV refers to "the mean value of the votes of a large group of persons on a 7-point thermal sensation scale (see Table 14.1) based on the heat balance of the human body" (Fanger, 1972). PMV depends on an individual's metabolic rate and the thermal resistance of his/her clothing as well as the air temperature, the mean radiant temperature, the air velocity and the relative humidity of the air. A neutral thermal balance is achieved when the internal heat production in the body is equal to the loss of heat to the environment. Neutrality of thermal sensation, which can be assumed to correspond to the desired thermal comfort condition, is given by a PMV of zero with a tolerance of ±0.5.

Predicted percentage of dissatisfied people (PPD) is also used to evaluate the general long-term thermal comfort (Argiriou et al., 2004). PMV and PPD are related through a one-to-one correspondence. A PPD of 5%, which is the lowest value achievable in practice, is equivalent to a PMV of zero.

Because there is considerable uncertainty associated with calculating some of the factors affecting PMV, there are major difficulties in computing both the PMV and the PPD (Bruant et al., 2001).

In many cases, some of the factors affecting PMV are fixed (Dounis and Manolakis, 2001; Gouda et al., 2001) and PMV is assumed to be only a function of the air temperature, mean radiant temperature and relative humidity, a function of the air temperature and velocity (Farzaneh and Tootoonchi, 2009), a function of the air temperature and relative humidity (Xi et al., 2007; Soyguder and Alli, 2009) or even simply a function of air temperature alone (Lute and van Paassen, 1995).

Monitoring and Control of Information-Poor Systems: An Approach based on Fuzzy Relational Models, First Edition.
Arthur L. Dexter.
© 2012 John Wiley & Sons, Ltd. Published 2012 by John Wiley & Sons, Ltd.

Table 14.1   Seven-point thermal sensation scale.

PMV	Thermal sensation
+3	Hot
+2	Warm
+1	Slightly warm
0	Neutral
−1	Slightly cool
−2	Cool
−3	Cold

14.1   Main Sources of Uncertainty and Practical Considerations

Human comfort is a complex concept because it is determined by three interrelated factors: thermal comfort, visual comfort and indoor air quality. Individual preferences mean that comfort must be treated as a concept with uncertainty and, ideally, all three factors must be taken into account (Dounis *et al.*, 2011).

Thermal comfort is a highly subjective issue and it is difficult, perhaps impossible, to have agreement among even a small number of individuals sharing the same working or living environment. As a result, there is uncertainty associated with the control objective in a comfort control system.

In practice, the problem is also multi-objective as there are conflicting requirements (high occupant satisfaction, low energy consumption and low maintenance costs). Although thermal comfort can have a significant effect on the productivity of the occupants of a modern energy-efficient building, the economic cost of low productivity is difficult to quantify precisely (Federspiel, 2000). Striking the correct balance between the different control objectives is therefore extremely difficult and there is always uncertainty associated with the definition of the overall control objective.

The thermal conditions (temperatures and airflows) vary spatially in the different parts of the occupied zones of a building but the control system must take decisions based on measurements from a finite (usually relatively small) number of sensors. The measurement uncertainties are even greater in naturally ventilated buildings, in which the external conditions (e.g. the magnitude and direction of the wind) can have a significant influence on the air flowing around the interior of the building (Eftekhari and Marjanovic, 2003).

The thermal conditions inside the building are subject to time-varying disturbances (e.g. the amount of solar radiation falling on the different facades of the building, the internal heat gains generated by the occupants and their equipment, and changes in the rate at which outside air is flowing in and out of the building). In practice, few of the disturbances can be estimated accurately from the available measurements. As a result, disturbances are another major source of uncertainty in a thermal comfort control system.

The comfort control problem is further complicated by non-linear behaviour of the air-conditioning equipment (see Chapter 16) which is difficult to model accurately as the associated thermo-fluid processes are spatially distributed and complex, there is frequently a lack of design data and it is difficult to obtain representative training data from the actual system.

Although a modern building control system consists of a network of embedded micro-computers (so-called outstations) with supervisory software running on one or more personal computers, the computing power available for the implementation of individual controllers is usually very limited.

Another important issue is that commissioning typically takes place over a relatively short period of a few weeks, so that control schemes that require extensive manual tuning are unlikely to be of any commercial interest.

14.2 Review of Approaches Suggested for Dealing with the Uncertainty

Several different ways of handling the uncertainties associated with controlling the comfort conditions inside of a building have been proposed (Dounis and Caraiscos, 2009).

Adaptive controllers attempt to reduce the model uncertainty by learning about the behaviour of the system online. For example, adaptive predictive controllers have been used to control the globe (or operating) temperature inside a room heated by an underfloor radiant panel heating system (Chen, 2002), to reduce the use of artificial lighting in a control scheme designed to optimize the integration of visual comfort, thermal comfort and energy consumption (Kurian *et al.*, 2008) and to control the start-up of heating plant in non-residential buildings (Dexter, 1981). The main drawback of adaptive model-based control of thermal comfort (So *et al.*, 1997) is that many of the proposed algorithms calculate overly precise values of the control signal and are therefore computationally demanding (Sousa *et al.*, 1997).

The design of robust controllers takes account of the effect of bounded uncertainties in such a way as to guarantee robust stability and robust performance. For example, an uncertainty polytope can be used to describe the uncertainties associated with the gain and time constant of an air-conditioned building, and a robust air temperature controller can be designed using optimization over linear matrix inequalities (Xu *et al.*, 2010). The main disadvantage of robust control techniques is the amount of online computing power required and the possibility that, if the uncertainty bounds are chosen too conservatively, a suitable control strategy may not exist. Fortunately, the control performance needed to maintain acceptable thermal comfort is usually less demanding than that required in process control applications, and fuzzy control of the air temperature is an attractive option.

The main drawback of comfort control schemes which use a fuzzy controller based on fixed expert rules (Funakoshi and Matsuo, 1995; Kolokotsa *et al.*, 2001; Gouda, 2005; Kolokotsa *et al.*, 2006; Ahmed *et al.*, 2007; Chiou *et al.*, 2009; Kristl *et al.*, 2009) is the time and effort required to acquire a correct, complete and consistent set of rules. This is especially challenging in low-energy buildings as expert knowledge is difficult to find (Yu and Dexter, 2009). The quality of the rules might be improved by using data obtained from an offline computer simulation of the building to identify the rules (Yu and Dexter, 2010) or to fine tune the membership functions of the fuzzy sets used in the expert rules (Dounis *et al.*, 2011).

14.3 Design of the Fuzzy FRM-Based Control System

The scheme for controlling thermal comfort in an air-conditioned building is based on a combination of fuzzy modelling and predictive control techniques. A block diagram of the scheme is shown in Figure 14.1. The controller consists of three parts: (1) a fuzzy model of the

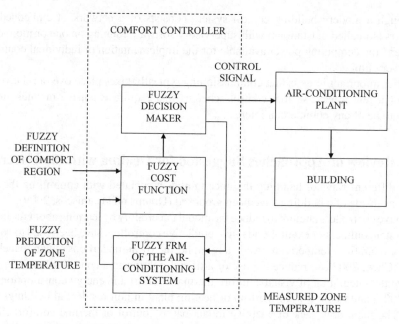

Figure 14.1   Fuzzy model-based zone temperature control.

process; (2) two fuzzy cost functions to be optimized, the comfort cost $J_C(U)$ and energy cost $J_E(U)$; and (3) a fuzzy decision-making scheme that determines the optimum fuzzy control signal U_{OPT} which minimizes the combined cost function:

$$J(U) = J_C(U) \textbf{ AND } J_E(U).$$

14.3.1   The Fuzzy FRM

A fuzzy relational model with five inputs and one output is used in the control scheme (see Figure 14.2). The structure of this non-linear first-order plus time-delay auto-regressive model is given by:

$$\hat{T}_Z(n+1) = T_Z(n) \circ U(n - T_D) \circ Q_s(n) \circ T_a(n) \circ p(n) \circ R \qquad (14.1)$$

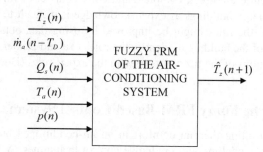

Figure 14.2   Inputs and output of the fuzzy FRM.

where $\hat{T}_Z(n+1)$ and $T_Z(n)$ are the predicted and current values of the zone temperature, respectively; $U(n-T_D)$ is the delayed control signal (in this case, the setpoint for the pressure-independent variable air volume terminal boxes supplying cold air to the zone at a rate $\dot{m}_a$); $Q_s(n)$ is an estimate of the current solar gains; $T_a(n)$ is the current ambient temperature; $p(n)$ is the Boolean output of a presence sensor that indicates whether the zone is currently occupied; $\circ$ is the fuzzy composition operator (here, sum-product is used); T_D is the overall time delay associated with the control action (expressed as an integer number of sampling intervals); and R is the fuzzy relational array.

The reference sets, all of which have triangular membership functions, are equally spaced over the universes of discourse and have a partition of unity (see Figure 14.3).

The fuzzy model is identified using the RSK fuzzy identification scheme and training data obtained from a simple first-order linear dynamic model of the building and plant, whose parameters are based on available design information (see Figure 14.4). S_r is the amount of solar radiation falling on the external surfaces of the building.

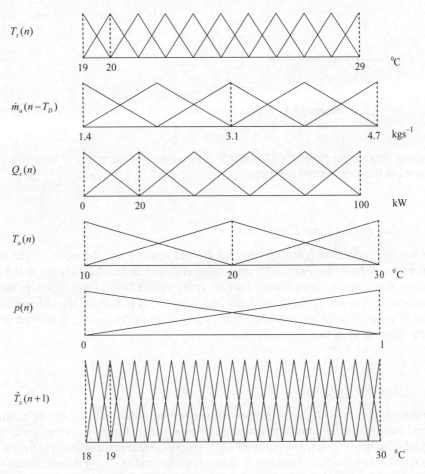

Figure 14.3 Fuzzy reference sets used by the fuzzy FRM.

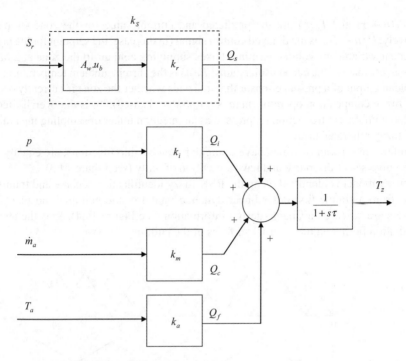

Figure 14.4 Simple linear thermal model of the zone.

Training data are generated for all possible combinations of the values of the five inputs at the apexes of the membership functions.

14.3.2   The Fuzzy Cost Functions

There are two fuzzy cost functions: one for thermal comfort and the other for the energy consumption. A fuzzy measure of thermal comfort $J_C(U)$ is found by matching the fuzzy prediction of the zone air temperature, for each fuzzy value of the control signal, to the zone air temperature setpoint. A fuzzy proximity measure is used to indicate the closeness of the two fuzzy values. The fuzzy measure of the energy consumption $J_E(U)$ is derived directly from the fuzzy value of the control signal.

14.3.3   The Fuzzy Goals

The fuzzy comfort goal G_C has a trapezoidal membership function to reflect the acceptable comfort band around the setpoint temperature (Figure 14.5a). The fuzzy energy goal G_E has a triangular membership function, centred on an energy consumption of 0% (Figure 14.5b).

The relative importance of maintaining thermal comfort and of saving energy is determined by the supports of the fuzzy goals. If after exhaustively searching through all fuzzy values

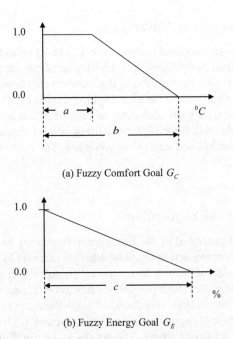

(a) Fuzzy Comfort Goal G_C

(b) Fuzzy Energy Goal G_E

Figure 14.5   Fuzzy goals for comfort and energy.

of U the controller can find no value of the control signal which, to some extent, satisfies both goals simultaneously, the goals are automatically relaxed by lengthening the bases of the membership functions until they are achievable. The results shown in Table 14.2 give some idea of how changing the support of the goals affects the energy consumption and control activity. The values of energy consumption and control activity are normalized with respect to the values obtained in Case 3. The conditional defuzzification threshold is 0.8 in all cases.

It can be seen that the energy consumption decreases as the support of the fuzzy energy goal is reduced and that increasing the support of the fuzzy comfort goal decreases both the energy consumption and the control activity.

Table 14.2   Comparison of the normalized energy consumption and control activity when the supports of the fuzzy goals are varied.

Case	a (°C)	b (°C)	c (% maximum airflow rate)	Energy consumption (%)	Control activity (%)
1	1.0	1.5	36	81.1	100.0
2	1.0	1.5	48	93.5	100.0
3	1.0	1.5	120	100.0	100.0
4	1.0	2.0	120	83.1	64.9

14.3.4   The Fuzzy Decision-Maker

The fuzzy cost functions are compared to their respective fuzzy goals to determine the extent to which the goals are satisfied (see Section 6.2). The degree of similarity between the fuzzy cost function and fuzzy goal is taken as the ratio of the common area between the two membership functions to the total area underneath the membership function of the fuzzy cost function. The two area ratios produced by a particular value of control action are multiplied (**AND**ed) to produce a discrete membership function for the fuzzy control signal U_{OPT} that is optimal in terms of both thermal comfort and energy consumption. The fuzzy optimal control signal is then conditionally defuzzified.

14.3.5   The Conditional Defuzzifier

The fuzzy control signal generated by the fuzzy decision-making process is compared to the current value of optimal control action to see whether the setpoint of the flow actuator should be changed. If the possibility of the new fuzzy optimal control signal given the current value of the control signal is greater than some pre-defined threshold, the setpoint remains unchanged (see Section 7.4.2). Otherwise, U_{OPT} is defuzzified using the mean of maxima defuzzification and a new value of the control signal is sent to the flow actuator. The choice of the threshold determines the level of actuator activity. The results shown in Table 14.3 give some idea of the effect on the energy consumption and control activity of changing the threshold in this application. The values of energy consumption and control activity are normalized with respect to the values obtained in Case 1. The support for the comfort and energy goals are fixed ($a = 1°C$, $b = 1.5°C$ and $c = 48\%$).

It can be seen that reducing the defuzzification threshold results in lower values of both the energy consumption and the control activity. The size of the reduction in the control activity is however much greater.

14.4   Performance of the Thermal Comfort Controller

A detailed computer simulation of one zone of a multi-storey office building is used to assess the performance of the controllers. The external conditions are typical of those occurring on a midsummer day. The varying external air temperature and solar radiation are shown in Figures 14.6 and 14.7, respectively.

Table 14.3   Comparison of the energy consumption and control activity when the conditional defuzzification threshold is varied.

Case	Defuzzification Threshold	Energy Consumption (%)	Control Activity (%)
1	0.8	100.0	100.0
2	0.6	97.1	74.1
3	0.4	85.0	48.1

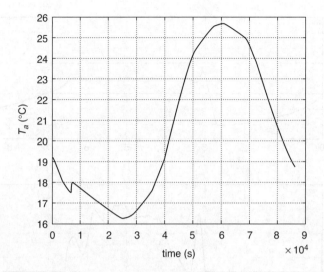

Figure 14.6 Diurnal change in the outside temperature variation. Reproduced by permission of Richard Thompson.

A constant internal gain of 26.55 kW is assumed during the occupancy period, which is from 7.30 to 18.30. The controller varies the mass airflow rate into the zone to maintain the comfort conditions in the zone.

The performance of the fuzzy FRM-based thermal comfort controller is evaluated using PI control of the zone temperature as a benchmark. The proportional gain, the integral time and the setpoint of the PI controller are 0.05, 10 s and 24°C, respectively. Figure 14.8 shows the closed-loop behaviour when the PI controller adjusts the airflow rate.

The simulation results presented in Figures 14.9 and 14.10 show the performance of the fuzzy FRM-based comfort control scheme when it is used to control the zone in two different

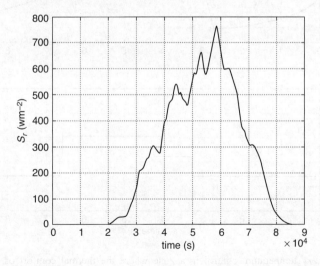

Figure 14.7 Diurnal change in the solar radiation. Reproduced by permission of Richard Thompson.

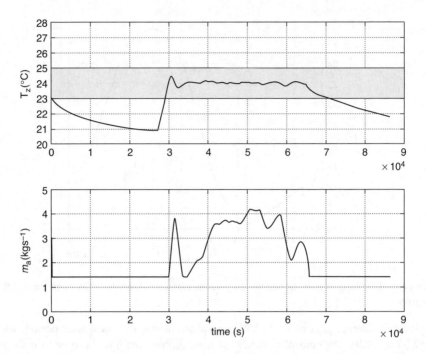

Figure 14.8   PI control of the zone temperature. Reproduced by permission of Richard Thompson.

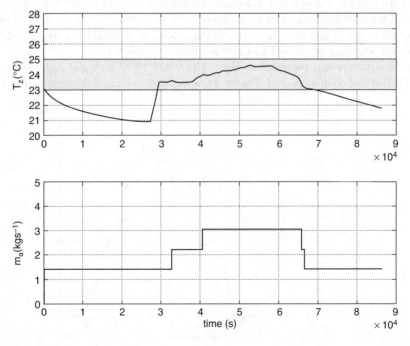

Figure 14.9   Fuzzy temperature control in a zone where the thermal comfort of the occupants is important ($b = 1$; $c = 10$).

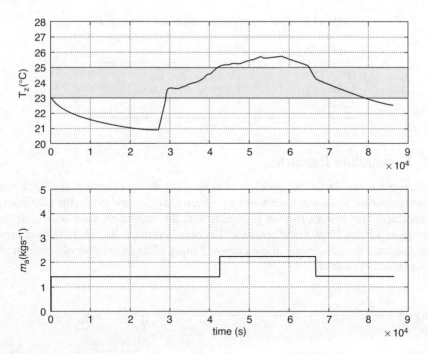

Figure 14.10 Fuzzy temperature control in a zone where energy savings are important ($b = 2$; $c = 4$).

modes. The comfort region is the shaded area between 23 and 25°C. In both cases, the setpoint of the controller is set to the *neutral temperature* of the zone, which is defined as the temperature at which people are neither too hot nor too cold. A neutral temperature of 24°C is assumed here. The threshold for conditional defuzzification is set at 0.8.

The first mode (see Figure 14.9) is one in which the thermal comfort of the occupants of the zone is a priority, such as would be the case if the zone was being used as a conference room of a modern office building. Compared to conventional PI control using a constant setpoint of 24°C, the energy consumption is reduced by 6% and the control activity by 74%.

In the second mode (see Figure 14.10), the energy consumption is the major concern. This scenario could occur if the zone were employed as a school classroom, for example, where thermal comfort is not necessarily a great priority but where conserving energy certainly might be. Compared to conventional PI control using a constant setpoint of 24°C, the energy consumption is reduced by 22% and the control activity by 83%.

If the zone is unoccupied at lunchtime, the comfort goal can be relaxed in the middle of the day (see Table 14.4) and greater energy savings can be made. Figure 14.11 shows the closed-loop behaviour when the goals are varied in this way and the neutral temperature is assumed to be 21.5°C. It can be seen that increasing the support of the fuzzy comfort goal and decreasing the support of the fuzzy energy goal results in a higher zone temperature and a lower airflow rate (and therefore greater energy savings) during the lunch period.

Table 14.4 Time variation of comfort and energy goals.

Time of Day	a (°C)	b (°C)	c (%)
0900–1200	1.0	1.5	192
1201–1400	1.0	2.0	48
1401–1700	1.0	1.5	192

14.5 Concluding Remarks

This chapter has considered the problem of controlling thermal comfort in a large building. The most important sources of uncertainty have been identified and the design constraints have been specified. Previous approaches to dealing with the uncertainty have been reviewed and the design of a fuzzy FRM-based controller has been described. Results have been presented that demonstrate the advantages of the proposed fuzzy FRM-based controller in comparison to conventional PI control.

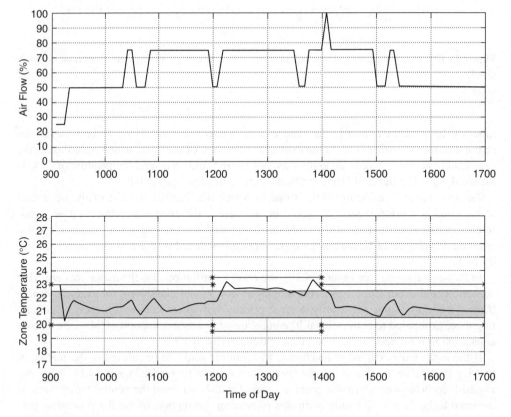

Figure 14.11 Closed-loop behaviour with variable definition of the goals. Reproduced by permission of Richard Thompson.

References

Ahmed, S.S., Majid, M.S., Novia, H. and Rahman, H.A. (2007) Fuzzy logic based energy saving technique for a central air conditioning system. *Energy*, **32**, 1222–1234.

Argiriou, A.A., Bellas-Velidis, I., Kummert, M. and Andre, P. (2004) A neural network controller for hydronic heating systems of solar buildings. *Neural Networks*, **17**(3), 427–440.

Bruant, M., Guarracino, G. and Michel, P. (2001) Design and tuning of a fuzzy controller for indoor air quality and thermal comfort management. *International Journal of Sustainable Energy*, **21**(2–3), 81–109.

Calvino, F., La Gennusa, M., Morale, M., *et al.* (2004) The control of indoor thermal comfort conditions: introducing a fuzzy adaptive controller. *Energy and Buildings*, **36**(2), 97–102.

Calvino, F., La Gennusa, M., Morale, M., *et al.* (2010) Comparing different control strategies for indoor thermal comfort aimed at the evaluation of the energy cost of quality of building. *Applied Thermal Engineering*, **30**(16), 2386–2395.

Chen, T.Y. (2002) Application of adaptive predictive control to a floor heating system with a large thermal lag. *Energy and Buildings*, **34**(1), 45–51.

Chiou, C.B., Chiou, C.H., Chu, C.M. and Lin, S.L. (2009) The application of fuzzy control on energy saving for multi-unit room air-conditioners. *Applied Thermal Engineering*, **29**, 310–316.

Dexter, A.L. (1981) Self-tuning optimum start control of heating plant. *Automatica*, **17**(3), 483–492.

Dounis, A.I. and Manolakis, D.E. (2001) Design of a fuzzy system for living space thermal-comfort regulation. *Applied Energy*, **6**(2), 119–144.

Dounis, A.J. and Caraiscos, C. (2009) Advanced control systems engineering for energy and comfort management in a building environment – a review. *Renewable and Sustainable Energy Reviews*, **13**(6), 1246–1261.

Dounis, A.I., Tiropanis, P., Argiriou, A. and Diamantis, A. (2011) Intelligent control system for reconciliation of the energy savings with comfort in buildings using soft computing techniques. *Energy and Buildings*, **43**(1), 66–74.

Eftekhari, M.M. and Marjanovic, L.D. (2003) Application of fuzzy control in naturally ventilated buildings for summer conditions. *Energy and Buildings*, **35**(7), 645–655.

Fanger, P. O. (1972) *Thermal Comfort Analysis and Application In Environmental Design*, New York, McGraw Hill.

Farzaneh, Y. and Tootoonchi, A.A. (2009) Controlling automobile thermal comfort using optimized fuzzy controller. *Applied Thermal Engineering*, **28**(14–15), 1906–1917.

Federspiel, C.C. (2000) Predicting the frequency and cost of hot and cold complaints in buildings. *International Journal of HVAC&R Research*, **6**(4), 289–305.

Funakoshi, S. and Matsuo, K. (1995) PMV-based train air conditioning control system. *ASHRAE Transactions*, **101**(Pt. 1), 423–430.

Gouda, M.M. (2005) Fuzzy ventilation control for zone temperature and relative humidity. *Proceedings of 2005 American Control Conference*, **1**, 507–512.

Gouda, M.M., Danaher, S. and Underwood, C.P. (2001) Thermal comfort based fuzzy logic controller. *Building Services Engineering Research and Technology*, **22**(4), 237–253.

Hagras, H., Packham, I., Vanderstockt, Y., *et al.* (2008) An intelligent agent based approach for energy management in commercial buildings. *IEEE Conference on Fuzzy Systems*, pp. 156–162.

Kolokotsa, D., Tsiavos, D., Stavrakakis, G.S., *et al.* (2001) Advanced fuzzy logic controllers design and evaluation for buildings' occupants thermal-visual comfort and indoor air quality satisfaction. *Energy and Buildings*, **33**(6), 531–543.

Kolokotsa, D., Saridakis, G., Pouliezos, A. and Stavrakakis, G.S. (2006) Design and installation of an advanced EIB fuzzy indoor comfort controller using Matlab. *Energy and Buildings*, **38**(9), 1084–1092.

Kristl, Z., Kosir, M., Trobec Lah, M. and Krainer, A. (2009) Fuzzy control system for thermal and visual comfort in building. *Renewable Energy*, **33**(4), 694–702.

Kummert, M., Andre, P. and Nicolas, J. (1997) Optimized thermal zone controller for integration within a Building Energy Management System. CD-ROM of Proceedings of Clima 2000, Brussels, Paper No. 329.

Kurian, C.P., Aithal, R.S., Bhat, J. and George, V.I. (2008) Robust control and optimisation of energy consumption in daylight-artificial light integrated schemes. *Lighting Research & Technology*, **40**(1), 7–24.

Lute, P. and van Paassen, D. (1995) Optimal indoor temperature control using a predictor. *IEEE Control Systems Magazine*, **15**(4), 4–10.

Mirinejad, H., Sadati, S.H., Ghasemian, M. and Torab, H. (2008) Control techniques in heating, ventilating and air conditioning (HVAC) systems. *Journal of Computer Science*, **4**(9), 777–783.

Schumacher, B., Bachmann, G. and Guebeli, M. (1998) Economiser tx2. Proceedings of UKACC International Conference on Control 1998, pp. 1711–1716.

So, A.T.P., Chan, W.L. and Tse, W.L. (1997) Self-Learning fuzzy air handling system controller. *Proceedings of CIBSE A: Building Services Engineering Research & Technology*, **18**(2), 99–108.

Sousa, J.M., Babuska, R. and Verbruggen, H.B. (1997) Fuzzy predictive control applied to an air-conditioning system. *Control Engineering Practice*, **5**(10), 1395–1406.

Soyguder, S. and Alli, H. (2009) An expert system for the humidity and temperature control in HVAC systems using ANFIS and optimization with fuzzy modelling approach. *Energy and Buildings*, **41**(8), 814–822.

Tse, W. L. and So, A.T.P. (2000) Implementation of comfort-based air-handling unit control algorithms. *Transactions of ASHRAE*, **106**(Pt. 1).

Xi, X-C., Poo, A-N. and Chou, S-K. (2007) Support vector regression model predictive control on a HVAC plant. *Control Engineering Practice*, **15**, 897–908.

Xu, X., Wang, S. and Huang, G. (2010) Robust MPC for temperature control of air-conditioning systems concerning on constraints and multitype uncertainties. *Building Services Research & Technology*, **31**(1), 39–55.

Yu, Z. and Dexter, A.L. (2009) Simulation-based predictive control of low-energy building systems using two-stage optimisation. Proc. IBPSA Building Simulation 2009, paper 659.

Yu, Z. and Dexter, A. (2010) Hierarchical fuzzy control of low energy building systems. *Solar Energy*, **84**(4), 538–548.

15

Identification of Faults in Air-Conditioning Systems

Early detection of the faults in air-conditioning systems can prevent energy wastage and avoid occupant discomfort (Qin and Wang, 2005). When faced with high levels of uncertainty however, there is a real risk of incorrect diagnosis and the cost of failing to diagnose a fault must be weighed against the cost of having to respond to a false alarm (Dodier *et al.*, 1998). The plant operator may even turn off the fault diagnosis system if there are too many false alarms (Visier *et al.*, 1997). One of the main requirements of any fault diagnosis scheme is therefore that it should generate very few false alarms.

15.1   Main Sources of Uncertainty and Practical Considerations

The problems associated with isolating and identifying faults in air-conditioning systems are more severe than those that occur in most process control applications. The behaviour of heating, ventilating and air-conditioning (HVAC) plants and buildings is more difficult to predict. Accurate mathematical models cannot be produced because most HVAC designs are unique and financial considerations restrict the amount of time and effort that can be put into deriving a model (Lo *et al.*, 2007). Detailed design information is seldom available (Haves *et al.*, 1996a) and measured data from the actual plant are often a poor indicator of the overall behaviour, since test signals (Haves *et al.*, 1996b) cannot usually be injected during normal operation and buildings are subject to seasonal disturbances. The prediction of faulty behaviour is even more problematic because the deliberate insertion of most faults will result in an unacceptable increase in either energy costs or occupant discomfort and, even if this is not the case, many faults cannot be introduced in a realistic manner (Haves, 1997). Another problem is that many variables cannot be measured accurately (see Chapter 17) and some measurements are not available. For example, air and water flow rates are measured in relatively few places in many HVAC systems. This is a particular problem in fault diagnosis since the presence of some faults may be very difficult to detect using the available measurements; with a limited number of measurements, several faults may have similar symptoms.

Monitoring and Control of Information-Poor Systems: An Approach based on Fuzzy Relational Models, First Edition. Arthur L. Dexter.
© 2012 John Wiley & Sons, Ltd. Published 2012 by John Wiley & Sons, Ltd.

For example, the air temperature drop across a cooling coil is not very sensitive to a reduction in the water flow rate caused by fouling of the tubes of the coil, and any observed change might also be a result of drift in the sensor measuring the chilled water supply temperature.

A further problem is that variables which cannot be measured directly are often only crudely estimated. For example, the use of single-point air temperature sensors to indicate average values over the entire cross-section of a large duct can result in biased estimates of the average air temperature (Robinson, 1999). The behaviour of HVAC equipment may also be highly non-linear. For example, an incorrectly sized damper will have a non-linear installed characteristic (Federspiel, 1997). In addition, the behaviour of the plant will vary as its mode of operation changes. For example, the relationship between zone air temperature and the position of the valve of the re-heating coil in a terminal box will be very different to the relationship between zone air temperature and the position of the damper controlling the air flow through the terminal box. There are also constraints on the operation of most of the equipment. For example, there will be a lower limit imposed on the position of the fresh air dampers; the supply air temperature must not be allowed to drop below a specified value. Problems also arise because, in most cases, the design intent is poorly specified. For example, maintaining thermal comfort levels does not usually equate to tight control of zone air temperature (see Chapter 14). The importance of closely controlling intermediate variables, such as the supply air temperature, is usually unknown. It is therefore difficult to quantify the economic cost of operating an air-conditioning system in the presence of faults that do not cause catastrophic failure but result in poor thermal comfort or over-active control (Breuker and Braun, 1998).

There are also a number of important practical issues that constrain the design of the FDI system. The number of application-dependent parameters must be kept to a minimum and the use of application-specific thresholds should be avoided. No fault diagnosis scheme is entirely generic and all need to be commissioned to some extent. Manual tuning usually requires specialist knowledge and can be extremely time consuming. However, many diagnosis schemes require careful selection of a large number of application-dependent parameters.

Measurement errors must be taken into account unless any sensor faults can first be detected and eliminated (Wang and Wang, 1999; Du et al., 2009). Most fault diagnosis and control schemes assume that any sensor faults have been eliminated during commissioning (Sun et al., 2010). Unless the diagnosis is based on commissioning data, data must be collected at a number of operating conditions if unambiguous results are to be obtained and false alarms are to be avoided (see Chapter 13). For example, the symptoms of a leaky damper and a stuck-slightly-open damper are very similar when the damper is closed, but they are very different when the damper is fully open. Clearly, the time required for the diagnosis scheme to produce an unambiguous result will depend on the richness of the test data in terms of the faults to be diagnosed. For example, it will take an extremely long time to unambiguously diagnose that a damper is leaking if the damper is never fully closed. During commissioning, data can be collected at a number of different test conditions that are chosen according to the types of fault to be diagnosed (Ngo and Dexter, 1998a). A final diagnosis is then produced by combining evidence obtained at one test condition with evidence obtained at the other test conditions.

Diagnosing faults such as valve leakage (Habbi et al., 2009) or the fouling of a coil (Zhou and Dexter, 2009), which develop slowly over a long period of time, is particularly difficult as the effect of the uncertainties may be greater than any observed short-term changes in the behaviour of the plant.

Several methods have been suggested for reducing the probability of a false alarm without compromising the ability of the FDI scheme to identify faults of an acceptable size. Some fault diagnosis schemes use thresholds that are determined heuristically (Glass *et al.*, 1995; Qin and Wang, 2005). Other schemes base the value of the threshold on statistical analysis (Rossi and Braun, 1997; Wang and Xiao, 2004; Wang and Qin, 2005) or hypothesis testing (Dodier *et al.*, 1998).

The fuzzy model-based approach (Dexter and Benouarets, 1997) requires no explicit alarm thresholds to be selected as it uses generic reference models that have very few application-dependent parameters to account for the uncertainties. It also produces the final diagnosis by combining evidence generated from data collected at a number of operating points.

15.2   Design of a Fuzzy FRM-Based Monitoring System for a Cooling Coil Subsystem

The design is based on the multi-step fault diagnosis scheme described in Chapter 13. Generic fuzzy reference models are identified using data obtained from computer simulations, which are based on information taken from manufacturers' data sheets and consultants' design specifications (Ngo and Dexter, 1998b). The models are identified from steady-state training data obtained by simulating nine cooling coil subsystems (Ngo and Dexter, 1999) using a detailed component model originally developed for the evaluation of control algorithms and strategies (Haves *et al.*, 1998). The coils are designed for chilled water supply temperatures in the range 5–9°C and airflow rates in the range 1.0–5.0 kg/s. Each subsystem is designed for the same inlet (24.0°C and 48% RH) and outlet conditions (13°C and 90% RH). The training data are generated for fault-free operation and for operation with one of the following faults present: 1 mm of water-side fouling, a valve leakage of approximately 10%, valve stuck fully closed, valve stuck midway and valve stuck fully open. The training data are collected at a representative number of part load conditions over the full operating range. The values of the rule confidences are estimated from the data using a max-product fuzzy identification scheme (Xu and Lu, 1987).

All of the fuzzy models have the same structure and consist of rules that are based on predefined fuzzy sets with triangular membership functions. The models have four inputs: the valve control signal, the normalized dry bulb temperature and relative humidity of the air entering the coil and the normalized air mass flow rate. The output of the models is the normalized air temperature difference across the coil. Five fuzzy sets are used to describe the control signal, the inlet air temperature and the temperature difference; three fuzzy sets are used to describe the mass airflow rate and the relative humidity.

The results presented in Section 15.3 are generated from test data produced by simulating a commissioning test on the cooling coil subsystem designed for a chilled water supply temperature of 7.0°C and an airflow rate of 3.0 kg/s. Steady-state data are collected when the mixing-valve is 0%, 10%, 40%, 70% and 100% fully open. The temperature and relative humidity of the inlet air are held constant at 22.0°C and 50%, respectively, during the simulated commissioning test. To ensure test conditions are well defined and to avoid overly ambiguous diagnosis, the experiments are performed on one of the cooling coil subsystem designs used to generate the generic references models. It should be noted that the training data take no account of sensor bias, as none is present in the simulated test data.

The results presented in Section 15.4 are generated from test data collected during commissioning tests on a real air-conditioning system in a commercial office building. Robust generic reference models, which allow for offsets on both the coil inlet and supply air temperature sensors of up to ±1 K, are used for the diagnosis.

15.3   Diagnosis of Known Faults in a Simulated Cooling Coil Subsystem

15.3.1   Fault-Free Operation

Table 15.1 shows the results obtained after the fault elimination phase. All of the diagnoses indicate that the subsystem under test is fault-free. The highest values of unambiguous belief in fault-free operation are generated when the diagnosis is based on the fault-free and stuck reference models, since the symptoms of fault-free and stuck operation are dissimilar at most operating conditions. Lower values of belief in fault-free operation are generated when the fouled and leaky reference models are used. In this case, the fault elimination phase indicates that the subsystem is fault-free and there is no need to proceed to the fault classification phase.

It should be noted that single-step diagnosis based on all six reference models generates a significantly lower value of unambiguous belief in fault-free operation (4.5%).

15.3.2   Leaky Valve

Table 15.2 shows the results obtained after the fault elimination phase when there is 10% leakage through the fully closed valve. The results indicate fault-free operation except for the case when the diagnosis is based on the fault-free and leaky valve reference models. Since there is no unambiguous belief in the presence of any other fault, the fault elimination phase indicates that the subsystem under test has a leaky valve, and there is no need to proceed to the fault classification phase.

It should be noted that single-step diagnosis based on all six reference models generates only a slightly lower value of unambiguous belief in a leaky valve (25.9%).

Table 15.1   Results of the *fault elimination* phase (fault-free operation).

Reference models used	Degree of belief (%)
Fault-free Valve-leak (10%)	Bel(fault-free) = 24.7 Bel(valve-leak) = 0.0
Fault-free Fouled-coil (1 mm)	Bel(fault-free) = 18.2 Bel(fouled-coil) = 0.0
Fault-free Valve-stuck-closed (0%)	Bel(fault-free) = 100.0 Bel(valve-stuck-closed) = 0.0
Fault-free Valve-stuck-midway (50%)	Bel(fault-free) = 90.6 Bel(valve-stuck-midway) = 0.0
Fault-free Valve-stuck-open (100%)	Bel(fault-free) = 98.6 Bel(valve-stuck-open) = 0.0

Table 15.2 Results of the *fault elimination* phase (leaky valve).

Reference models used	Degree of belief (%)
Fault-free	Bel(fault-free) = 0.0
Valve-leak (10%)	Bel(valve-leak) = 27.8
Fault-free	Bel(fault-free) = 18.2
Fouled-coil (1 mm)	Bel(fouled-coil) = 0.0
Fault-free	Bel(fault-free) = 100.0
Valve-stuck-closed (0%)	Bel(valve-stuck-closed) = 0.0
Fault-free	Bel(fault-free) = 84.9
Valve-stuck-midway (50%)	Bel(valve-stuck-midway) = 0.0
Fault-free	Bel(fault-free) = 97.9
Valve-stuck-open (100%)	Bel(valve-stuck-open) = 0.0

Note that some of the degrees of beliefs in fault-free operation generated during the fault elimination phase are much larger than the degrees of beliefs in the fault. The large values of belief in fault-free operation arise because, at many operating points, the symptoms of the leakage fault are similar to the symptoms of fault-free operation but dissimilar to those of the fault described by the reference model. For the same reason, a fault detector based on a single reference model describing fault-free operation is unable to determine whether the system is genuinely fault-free if the symptoms of fault-free and faulty operation are similar at some operating points; it is only able to indicate the presence of a fault.

15.3.3 Fouled Coil

Table 15.3 shows the results obtained after the fault elimination phase when 1 mm of water-side fouling is simulated in the tubes of the cooling coil. The result of the diagnosis based on the fault-free and leaky valve reference models and the fault-free and valve stuck open

Table 15.3 Results of the *fault elimination* phase (fouled coil).

Reference models used	Degree of belief (%)
Fault-free	Bel(fault-free) = 24.7
Valve-leak (10%)	Bel(valve-leak) = 0.0
Fault-free	Bel(fault-free) = 0.0
Fouled-coil (1 mm)	Bel(fouled-coil) = 41.9
Fault-free	Bel(fault-free) = 92.8
Valve-stuck-closed (0%)	Bel(valve-stuck-closed) = 4.8
Fault-free	Bel(fault-free) = 78.7
Valve-stuck-midway (50%)	Bel(valve-stuck-midway) = 7.7
Fault-free	Bel(fault-free) = 98.9
Valve-stuck-open (100%)	Bel(valve-stuck-open) = 0.0

Table 15.4   Results of the *fault classification* phase (fouled coil).

Reference models used	Degree of belief (%)
Fouled-coil (1mm)	Bel(fouled-coil) = 83.5
Valve-stuck-closed	Bel(valve-stuck-closed) = 0.0
Valve-stuck-midway	Bel(valve-stuck-midway) = 0.0

reference models is unambiguous belief in fault-free operation. The diagnosis based on the fault-free and coil-fouled reference models indicates the presence of the fault. There is some unambiguous belief in both fault-free and faulty operation when the diagnosis is based on the fault-free and valve-stuck-closed or valve-stuck-midway reference models.

Since there is more than one non-zero value of unambiguous belief in a fault, the test data are re-evaluated using the fouled-coil, the valve-stuck-fully-closed and the valve-stuck-midway reference models. The result of this diagnosis (see Table 15.4) is that the cooling coil is fouled.

It should be noted that the unambiguous value of belief in a fouled coil is significantly higher than the 34.7% generated when single-step diagnosis based on all six of the reference models is used.

15.3.4   Valve Stuck in the Fully Closed Position

Table 15.5 shows the results obtained after the fault elimination phase when the valve is stuck fully closed. The result of the diagnosis based on the fault-free and valve-stuck-fully-closed reference models generates a high value of unambiguous belief in the valve being stuck in the fully closed position. The diagnosis generates unambiguous belief in fault-free operation when it is based on either the fault-free and leaky valve reference models or the fault-free and valve-stuck-fully-open reference models. However, there is some unambiguous belief in both fault-free and faulty operation when the fault-free and valve-stuck-midway reference models are used for the diagnosis. The belief in both fault-free and faulty operation arises because

Table 15.5   Results of the *fault elimination* phase (valve stuck fully closed).

Reference models used	Degree of belief (%)
Fault-free	Bel(fault-free) = 23.9
Valve-leak (10%)	Bel(valve-leak) = 0.0
Fault-free	Bel(fault-free) = 1.6
Fouled-coil (1 mm)	Bel(fouled-coil) = 70.2
Fault-free	Bel(fault-free) = 0.0
Valve–stuck-closed (0%)	Bel(valve-stuck-closed) = 98.2
Fault-free	Bel(fault-free) = 58.8
Valve-stuck-midway (50%)	Bel(valve-stuck-midway) = 28.2
Fault-free	Bel(fault-free) = 93.8
Valve-stuck-open (100%)	Bel(valve-stuck-open) = 0.0

Table 15.6   Results of the *fault classification* phase (valve stuck fully closed).

Reference models used	Degree of belief (%)
Fouled-coil (1 mm)	Bel(fouled-coil) = 0.0
Valve-stuck-closed	Bel(valve-stuck-closed) = 93.6
Valve-stuck-midway	Bel(valve-stuck-midway) = 0.0

the behaviour of the coil when the valve is stuck closed is similar to the behaviour when the valve is stuck midway at high loads and similar to fault-free operation at low loads. Diagnosis based on the fault-free and fouled-coil reference models generates unambiguous belief in a fouled coil because the behaviour when the valve is stuck closed has some similarity to that of a fouled coil, but is entirely dissimilar to that of a fault-free coil, at high loads.

Table 15.6 shows the result of the fault classification phase when the diagnosis is based on the fouled coil, valve-stuck-fully-closed and valve-stuck-midway reference models. The diagnosis generates a single high value of unambiguous belief in the valve being stuck in the fully closed position. In this case, single-step diagnosis generates the same value of unambiguous belief as multi-step diagnosis.

15.3.5   Valve Stuck in the Midway Position

Table 15.7 shows the results obtained after the fault elimination phase when the valve is stuck midway. None of the diagnoses generate unambiguous belief in fault-free operation. The diagnosis based on the fault-free and leaky valve reference models generates a very small belief in a leaky valve. The diagnosis based on the fault-free and valve-stuck-midway reference models generates significant belief that the valve is stuck midway. Some unambiguous belief in both fault-free and faulty operation is generated when the diagnosis is based on the fault-free and fouled-coil reference models, the fault-free and valve-stuck-fully-closed reference models or the fault-free and valve-stuck-fully-open reference models.

Table 15.7   Results of the *fault elimination* phase (valve stuck midway).

Reference models used	Degree of belief (%)
Fault-free	Bel(fault-free) = 0.0
Valve-leak (10%)	Bel(valve-leak) = 1.7
Fault-free	Bel(fault-free) = 10.5
Fouled-coil (1 mm)	Bel(fouled-coil) = 31.1
Fault-free	Bel(fault-free) = 90.7
Valve-stuck-closed (0%)	Bel(valve-stuck-closed) = 6.6
Fault-free	Bel(fault-free) = 0.0
Valve-stuck-midway (50%)	Bel(valve-stuck-midway) = 88.8
Fault-free	Bel(fault-free) = 26.9
Valve-stuck-open (100%)	Bel(valve-stuck-open) = 68.6

Table 15.8 Results of the *fault classification* phase (valve stuck midway).

Reference models used	Degree of belief (%)
Leaky-valve (10%)	Bel(valve-leak) = 0.0
Fouled-coil (1 mm)	Bel(fouled-coil) = 0.0
Valve-stuck-closed	Bel(valve-stuck-closed) = 0.0
Valve-stuck-midway	Bel(valve-stuck-midway) = 59.4
Valve-stuck-open	Bel(valve-stuck-open) = 0.0

Table 15.8 shows the result of the fault classification phase when the diagnosis is based on all five of the faulty reference models. The result is now unambiguous belief in the valve being stuck in the midway position. Again, both the single-step and multi-step diagnosis schemes generate identical results.

15.3.6   Valve Stuck in the Fully Open Position

Table 15.9 shows the results of the diagnosis after the fault elimination phase when the valve is stuck fully open. Diagnosis based on the fault-free and valve-leak reference models is unable to produce any unambiguous beliefs (i.e. the result is inconclusive).

Diagnosis based on the fault-free and fouled-coil reference models, and the fault-free and valve-stuck-fully closed reference models, generates unambiguous belief in fault-free operation. This arises because the behaviour of a coil with its valve stuck open is dissimilar to that of both a fouled coil and a coil with its valve stuck fully closed (at all operating conditions), but is similar to that of a fault-free coil at high loads. There is a large unambiguous belief in the presence of a fault when the diagnosis is based on the fault-free and the valve-stuck-fully-open or the valve-stuck-midway reference models.

Table 15.10 shows the results of the fault classification phase when the diagnosis is based on the valve-stuck-midway and valve-stuck-fully-open reference models. The diagnosis now

Table 15.9 Results of the *fault elimination* phase (valve stuck fully open).

Reference models used	Degree of belief (%)
Fault-free Valve-leak (10%)	no unambiguous belief is generated
Fault-free Fouled-coil (1 mm)	Bel(fault-free) = 46.8 Bel(fouled-coil) = 0.0
Fault-free Valve-stuck-closed (0%)	Bel(fault-free) = 99.0 Bel(valve-stuck-closed) = 0.0
Fault-free Valve-stuck-midway (50%)	Bel(fault-free) = 0.5 Bel(valve-stuck-midway) = 98.4
Fault-free Valve-stuck-open (100%)	Bel(fault-free) = 0.0 Bel(valve-stuck-open) = 98.4

Table 15.10 Results of the *fault classification* phase (valve stuck fully open).

Reference models used	Degree of belief (%)
Valve-stuck-midway	Bel(valve-stuck-midway) = 0.0
Valve-stuck-open	Bel(valve-stuck-open) = 80.1

generates only unambiguous belief in the valve being stuck in the fully open position. Once again, single-step and multi-step diagnosis generate identical results.

15.4 Commissioning of Air-Handling Units

The commissioning tests were performed on an air-conditioning system in a commercial office building. Figure 15.1 is a schematic diagram of the HVAC plant.

The system has eight air-handling units of widely differing sizes. The plant room itself acts as a common mixing plenum. The results presented here were obtained during the commissioning of the cooling coil in air-handling unit AHU7. This air-handling unit was chosen for the tests because the associated zone is unoccupied in the evenings and the design specification of the coil is within the class of designs that were used to produce the reference models. Table 15.11 summarizes the technical specification of the cooling coil subsystem.

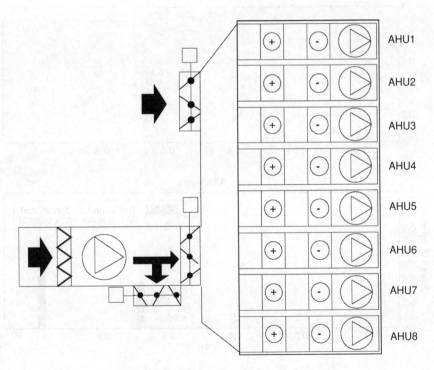

Figure 15.1 The air-conditioning system.

Table 15.11   Specification for the cooling coil subsystem in AHU7
(db: dry bulb; wb: wet bulb).

Type	Copper tube/Aluminium fin
Air ON temperature	24.0°C db/17.0°C wb
Air OFF temperature	11.0°C db/10.3°C wb
Chilled water	5.5°C supply/11.0°C return
Airflow rate	0.96 kg/s
Water flow rate	0.82 l/s

A performance monitoring and automated commissioning tool (Ngo and Dexter, 1998a) was
used to commission the plant remotely. Steady-state test data were collected at 0%, 10%, 40%,
70% and 100% valve opening. The temperature of the cooling coil inlet and supply air and
the relative humidity of the mixed air were all measured directly using single-point sensors.
The cooling coil outlet air temperature was estimated from the measurement of the supply air
temperature by subtracting an amount to allow for the temperature rise across the supply fan.
The design value was used for the supply airflow rate as a flow sensor was not installed.

The observed process characteristic and the final results of the diagnosis are shown in
Figure 15.2. The normalized temperature difference across the cooling coil is plotted against

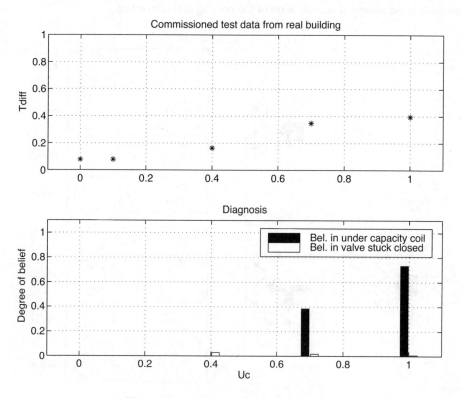

Figure 15.2   Results of the final diagnosis.

Table 15.12 Commissioning results after the *fault elimination* phase.

Reference models used	Degree of belief (%)
Fault-free Valve-leak (10%)	no unambiguous belief is generated
Fault-free Coil-under-capacity	Bel(fault-free) = 2.9 Bel(coil-under-capacity) = 1.0
Fault-free Valve-stuck-closed (0%)	Bel(fault-free) = 73.8 Bel(valve-stuck-closed) = 0.6
Fault-free Valve-stuck-midway (50%)	Bel(fault-free) = 86.2 Bel(valve-stuck-midway) = 0.1
Fault-free Valve-stuck-open (100%)	Bel(fault-free) = 97.2 Bel(valve-stuck-open) = 0.0

the control valve position in the upper graph. The lower plot shows how the final diagnosis changes as the steady-state commissioning data are collected.

Table 15.12 shows the results obtained using all of the commissioning data, after the fault elimination phase. Although some of the values of unambiguous belief are very small, the results indicate that the subsystem is either under capacity or that the cooling coil valve is stuck in the fully closed or midway position.

Table 15.13 shows the results of the fault classification phase when the diagnosis is based on the under-capacity-coil reference model and the valve-stuck-closed and valve-stuck-midway reference models. The possibility of the valve being stuck midway is now eliminated and the diagnosis is repeated using only the under-capacity-coil and valve-stuck-closed reference models. As can be seen in Table 15.14, the belief in the coil being under capacity has increased further, although there is still a small residual belief in the valve being stuck closed.

The results obtained from the diagnosis are plausible since subsequent manual inspection confirmed that (1) the temperature of the chilled water supplied to the coil varied between 7.8 and 9.8°C although the coil had been designed to operate with an inlet chilled water temperature of 5.5°C; and (2) a dead band of between 30 and 40% had been introduced into the valve control signal to avoid rapid switching between heating and cooling modes at low loads. The presence of the dead band resulted in behaviour that resembled that of a valve that is stuck closed at low values of the control signal. It should be noted that the degree of belief in the valve being stuck closed begins to decrease as soon as the control signal rises above 40%.

Table 15.13 Commissioning results after the first pass of the *fault classification* phase.

Reference models used	Degree of belief (%)
Coil-under-capacity Valve-stuck-closed Valve-stuck-midway	Bel(coil-under-capacity) = 62.9 Bel(valve-stuck-closed) = 0.8 Bel(valve-stuck-midway) = 0.0

Table 15.14 Commissioning results after completion of the *fault classification* phase.

Reference models used	Degree of belief (%)
Coil-under-capacity	Bel(coil-under-capacity) = 73.3
Valve-stuck-closed	Bel(valve-stuck-closed) = 0.8

15.5 Concluding Remarks

The problem of identifying faults in air-conditioning systems has been considered. The most important sources of uncertainty have been identified and design constraints have been discussed. Previous approaches to dealing with the uncertainty have been reviewed and the design of a fuzzy FRM-based monitoring scheme for a cooling coil subsystem has been described that is unlikely to generate false alarms in practice. The results obtained from computer simulation have shown that the multi-step scheme is able to generate more precise results than those produced by single-step diagnosis. The results obtained from field trials in a real office building have demonstrated that faults in air-conditioning systems can be successfully diagnosed, using measurements currently available from the building energy management system, if the test data are collected at a number of different operating points and the symptoms of the faults are larger than the effects of any modelling errors and sensor bias.

References

Breuker, M.S. and Braun, J.E. (1998) Common Faults and their Impacts for Rooftop Air Conditioners. *International Journal of HVAC R Research*, **4**(3), 303–318.

Dexter, A.L. and Benouarets, M. (1997) Model-based fault diagnosis using fuzzy matching. *Transactions of IEEE on Systems, Man, and Cybernetics - Part A*, **27**(5), 673–682.

Dodier, R.H., Curtiss, P.S. and Kreider, J.F. (1998) Small-scale On-line Diagnostics for an HVAC System. *Transactions of ASHRAE*, **104**(1), 530–539.

Du, Z., Jin, X. and Yang, Y. (2009) Fault diagnosis for temperature, flow rate and pressure sensors in VAV systems using wavelet neural network. *Applied Energy*, **86**(9), 1624–1631.

Federspiel, C.C. (1997) Flow Control with Electric Actuators. *International Journal of HVAC R Research*, **3**(3), 265–289.

Glass, A.S., Gruber, P., Roos, M. and Todtli, J. (1995) Qualitative Model-based fault Detection in Air-handling Units. *IEEE Control Systems Magazine*, **15**(4), 11–22.

Habbi, H., Kinnaert, M. and Zelmat, M. (2009) A complete procedure for leak detection and diagnosis in a complex heat exchanger using data-driven fuzzy models. *ISA Transactions*, **48**(3), 354–361.

Haves, P. (1997) Fault modelling in component-based HVAC simulation. Proceedings of the 5th IBPSA Conference: Building Simulation '97, Prague, Czech Republic, paper P101.

Haves, P., Salsbury, T.I. and Wright, J.A. (1996a) Condition Monitoring in HVAC Subsystems using First Principles Models. *Transactions of ASHRAE*, **102**(1), 519–527.

Haves, P., Jorgensen, D.R., Salsbury, T.I. and Dexter, A.L. (1996b) Development and Testing of a Prototype Tool for HVAC Control System Commissioning. *Transactions of ASHRAE*, **102**(1), 467–475.

Haves, P., Norford, L.K. and DeSimone, M. (1998) A Standard Simulation Test Bed for the Evaluation of Control Algorithms and Strategies. *Transactions of ASHRAE*, **104**(1), 460–473.

Lo, C.H., Chan, P.T, Wong, Y.K., *et al.* (2007) Fuzzy-genetic algorithm for automatic fault detection in HVAC systems. *Applied Soft Computing*, **7**, 554–560.

Ngo, D. and Dexter, A.L. (1998a) Automatic Commissioning of Air-conditioning Plant. *Proceedings of UKACC International Conference on Control '98*, **2**, 1694–1699.

Ngo, D. and Dexter, A.L. (1998b) Fault diagnosis in air-conditioning systems using generic models of HVAC plant. Proceedings of International Conference on System Simulation in Buildings SSB'98, Liege, Belgium, paper 23.

Ngo, D. and Dexter, A.L. (1999) A robust model-based approach to diagnosing faults in air-handling units. *Transactions of ASHRAE*, **105**(1).

Qin, J. and Wang, S. (2005) A fault detection and diagnosis strategy of VAV air-conditioning systems for improved energy and control performances. *Energy and Buildings*, **37**(10), 1035–1048.

Robinson, K.D. (1999) Mixing Effectiveness of AHU Combination Mixing/Filter Box with and without Filters. *Transactions of ASHRAE*, **105**(1).

Rossi, T.M. and Braun, J.E. (1997) A statistical, rule-based fault detection and diagnostic method for vapor compression air conditioners. *International Journal of HVAC Research*, **3**(1), 19–37.

Sun, Y., Wang, S. and Huang, G. (2010) Online sensor fault diagnosis for robust chiller sequencing control. *International Journal of Thermal Sciences*, **49**(3), 589–602.

Visier, J.C, Corailes, P., Irigoin, M., *et al.* (1997) Fault Detection and Diagnosis Tool for Schools Heating Systems. Proceedings of Clima 2000, Brussels, Belgium, paper 220.

Wang, S. and Wang, J-B (1999) Law-based Sensor Fault Diagnosis and Validation for Building Air-conditioning Systems. *International Journal of HVAC R Research*, **5**(4), 353–380.

Wang, S. and Xiao, F. (2004) Detection and diagnosis of AHU sensor faults using principal component analysis method. *Energy Conversion & Management*, **45**(17), 2667–2686.

Wang, S. and Qin, J. (2005) Sensor fault detection and validation of VAV terminals in air conditioning systems. *Energy Conversion & Management*, **45**(15–16), 2482–2500.

Xu, C-W. and Lu, Y-Z. (1987) Fuzzy Model Identification and Self-learning for Dynamic Systems. *IEEE Transactions on Systems, Man and Cybernetics*, SMC-**17**(4), 683–689.

Zhou, Y. and Dexter, A.L. (2009) Estimating the size of incipient faults in HVAC equipment. *HVAC and R Research*, **15**(1), 151–163.

16

Control of Heat Exchangers

Heat exchangers are used to control fluid temperatures in a wide variety of applications. This is normally achieved by controlling the exit temperature of one of the fluids by varying the inlet temperature or flow rate of another fluid. Accurate prediction of the dynamic and steady-state thermal behaviour of heat exchangers is extremely difficult because of the complex nature of their geometries and the non-linear spatially varying nature of the associated thermo-fluid processes (Pacheco-Vega et al., 2009). Many assumptions must be made (e.g. lumped parameter simplification, one-dimensional heat transfer, constant fluid properties and heat transfer coefficients) when generating mathematical models that are suitable for use in model-based controllers (Maidi et al., 2008).

The example application considered in this chapter is the use of an air-to-water heat exchanger to control the supply air temperature in an air-conditioning system. This is a particularly challenging control problem because of the uncertainties that normally characterize the operation of its air-handling units, for example: uncertain process gain (Huang et al., 2009); uncertain dynamics (Huang and Dexter, 2008; Xu et al., 2010); unmeasured disturbances (Lu et al., 2007); uncertain process interactions (Lygouras et al., 2008); and uncertain non-linear behaviour (He et al., 2005; Zhao et al., 2011).

16.1 Main Sources of Uncertainty and Practical Considerations

Because there is usually a lack of design information and it is difficult to obtain representative training data from the actual plant, the main source of uncertainty is the prediction errors resulting from the use of simplified and inaccurate mathematical models. Another major source of uncertainty is a result of the disturbances acting on the control loop (e.g. variations in the fluid flow rates or the inlet temperatures of the fluids), as none of the disturbances can be measured accurately.

In many applications, the control algorithm should require little (if no) onsite commissioning and it must be suitable for implementation on an embedded microcomputer with very limited processing power (Huang and Dexter, 2008). Fuzzy control is therefore an attractive option in applications of this type (Chiou et al., 2009; Navale and Nelson, 2010b).

Monitoring and Control of Information-Poor Systems: An Approach based on Fuzzy Relational Models, First Edition. Arthur L. Dexter.
© 2012 John Wiley & Sons, Ltd. Published 2012 by John Wiley & Sons, Ltd.

The main disadvantage of fuzzy control schemes, which are based on direct fuzzy control using fixed expert rules (Gouda *et al.*, 2001), is the time and effort required to acquire a correct, complete and consistent set of rules. Recent work has concentrated on the schemes for tuning the expert rules to improve the performance of such controllers (Navale and Nelson, 2010). Fuzzy model-based control schemes (So *et al.*, 1997; Ghaius, 2001) are easier to commission but they often expend significant computer resources on calculating precise values of the optimal control signal (Sousa *et al.*, 1997). It is also important to note that the output of the model must be defuzzified and the cost function must be non-fuzzy when either analytical (Linkens and Kandiah, 1996) or conventional direct search (Postlethwaite, 1996) methods of optimization are used.

An alternative approach is to use fuzzy decision-making to find the optimal control signal (see Section 7.2.2). Fuzzy model-based predictive control schemes of this type can take account of the uncertainty associated with predicting future behaviour, allow a fuzzy control objective to be specified and are computationally undemanding.

16.2   Design of a Fuzzy FRM-Based Predictive Controller

The design of the controller is based on the FMPC scheme described in Section 7.4. A fuzzy model generates a fuzzy description of the one-step-ahead predicted process output from candidate fuzzy values of the current control signal generated by the fuzzy decision-maker, the previous value of the control signal and measured values of the current and previous process output. A fuzzy cost is computed by considering the similarity of the setpoint trajectory and the fuzzy predictions. A fuzzy decision-maker generates the optimum fuzzy control signal by determining the degree to which the fuzzy cost associated with each of the candidate fuzzy values of the control signal satisfies a pre-defined fuzzy goal. A conditional defuzzification scheme determines what (if any) changes are to be made to the position of the actuator.

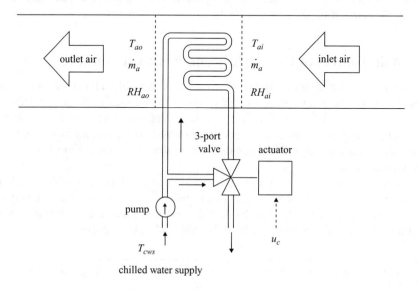

Figure 16.1   Schematic diagram of the cooling coil subsystem.

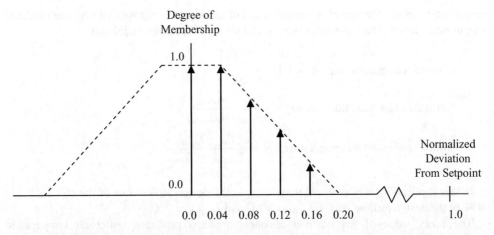

Figure 16.2   The membership function of the fuzzy goal used in this application.

A schematic diagram of the air-to-water heat exchanger used in the example application is shown in Figure 16.1. Inlet air at temperature T_{ai} (°C) and relative humidity RH_{ai} (%) is cooled as it flows at $\dot{m}_a$ (kg/s) through the heat exchanger and emerges as outlet air at temperature T_{ao} (°C) and relative humidity RH_{ao} (%). Chilled water is pumped into the coils of the heat exchanger at a supply temperature T_{cws} (°C). The amount of cooling depends on how much chilled water flows through the coils. The water flow rate is determined by the position of a 3-port mixing valve, which can be adjusted by a motor-driven actuator in response to changes in the value of the control signal u_c.

The FMPC scheme has a single fuzzy goal and a single fuzzy cost function, which is a fuzzy measure of the deviation of the outlet air temperature from its setpoint value. The fuzzy goal has a trapezoidal membership function, as is shown in Figure 16.2. Both the fuzzy cost and the fuzzy goal are defined on normalized discrete universes of discourse.

It is assumed that normalized setpoint deviations of up to 0.04 ($\sim$1.0 K) are totally acceptable, and those in the interval from 0.04 to 0.2 ($\sim$1.0–4.5 K) are partially acceptable but to a decreasing extent.

A non-linear first-order dynamic fuzzy model is used to predict the behaviour of the cooling coil subsystem. As can be seen in Figure 16.3, to keep the number of inputs to a minimum the predictor does not use previous values of the control signal and assumes that there is relatively little variation in the inlet air temperature and relative humidity (assumed to be 25°C and 50%,

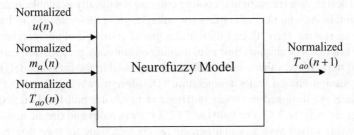

Figure 16.3   Structure of the fuzzy model.

respectively). Since the model is generic, each of the inputs is normalized and a normalized output is produced. The input and output variables are normalized as follows:

$$\text{Normalized actuator control signal } u' = \frac{u_c}{10}$$

$$\text{Normalized air mass flow rate } \dot{m}'_a = \frac{\dot{m}_a - \dot{m}_{a,\min}}{\dot{m}_{a,\max} - \dot{m}_{a,\min}}$$

$$\text{Normalized outlet air temperature } T'_{ao} = \frac{T_{ao} - T_{ao,\min}}{T_{ao,\max} - T_{ao,\min}}$$

In this application, it is assumed that $\dot{m}_{a,\max}$ is the design airflow rate of the coil, $\dot{m}_{a,\min}$ is 40% of the design airflow rate, $T_{ao,\max} = 30°C$ and $T_{ao,\min} = 7°C$.

The fuzzy value of the normalized one-step-ahead predicted outlet air temperature $T'_{ao}(n+1)$ is given by:

$$T'_{ao}(n+1) = u'(n) \circ T'_{ao}(n) \circ \dot{m}'_a(n) \circ R$$

where $T'_{ao}(n)$ is the current value of normalized outlet air temperature, $\dot{m}'_a(n)$ is the current value of normalized air mass flow rate, $u'(n)$ is the normalized actuator control signal, $\circ$ denotes fuzzy compositional inference, and R is the fuzzy relational array.

As can be seen in Figure 16.4, the fuzzy relational model has 11 linearly spaced fuzzy reference sets, with triangular membership functions having 50% overlap, that describe the current value of the normalized outlet air temperature (10% uncertainty in measuring the air temperature is assumed). To reduce the computational demands associated with the fuzzy decision-making process, only five fuzzy reference sets are used to describe the normalized actuator control signal and, to take account of the uncertainties associated with estimating the airflow rate, only three fuzzy reference sets are used to describe the normalized air mass flow rate. However, 26 fuzzy reference sets are used to describe the predicted normalized outlet air temperature so that the model can produce a representative fuzzy output.

A fuzzy model is required that captures the main characteristics of the non-linear dynamic behaviour of the cooling coil subsystem. Because it is difficult and time-consuming to generate training data that cover all possible operating conditions from experiments on an actual cooling coil, the training data are generated by computer simulation using a software package which was originally developed to model the non-linear dynamic behaviour of heating, ventilation and air-conditioning systems (Haves et al., 1998).

As detailed design data for particular cooling coils are not usually available, training data are obtained by simulating a number of cooling coil subsystems of the same type as that used in the air-conditioning system. Here, three different designs of cooling coil are simulated. All three coils have the same design air mass flow rate (training on coils designed for other airflow rates is unnecessary if normalized values of the airflow rate are used in the fuzzy model) but different values of the design chilled water temperature. The design data are shown in Table 16.1. The coiling coils are designed to supply outlet air at 11°C (dry bulb) and 10.5°C (wet bulb) when the inlet air is at 24.0°C (dry bulb) and 17.0°C (wet bulb) and the air mass flow rate is 1 kg/s. The mixing valves have an equal-percentage characteristic for their flow ports, a linear characteristic for their bypass ports and an authority of 0.5.

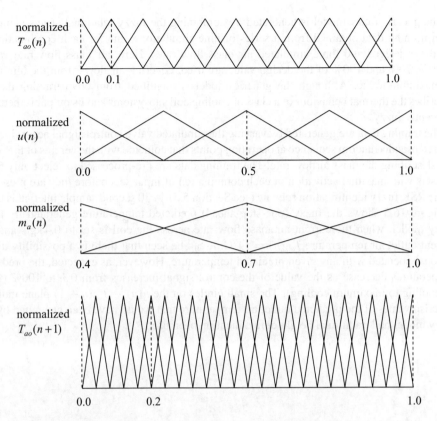

normalized
$T_{ao}(n)$

0.0   0.1   1.0

normalized
$u(n)$

0.0   0.5   1.0

normalized
$m_a(n)$

0.4   0.7   1.0

normalized
$T_{ao}(n+1)$

0.0   0.2   1.0

Figure 16.4   Fuzzy reference sets defined on the normalized input and output spaces.

Table 16.1   Design parameters for each of the simulated cooling coils.

Design parameters	Coil design		
	D1	D2	D3
Chilled water supply temperature (°C)	5.0	7.0	9.0
Coil width (m)	0.61	0.61	0.60
Coil height (m)	0.61	0.61	0.61
No. of circuits	4	4	4
No. of rows of tubes	3	5	10
Tube diameter (m)	0.0127	0.0127	0.0127
Tube thickness (m)	0.00043	0.00043	0.00043
Fins (/m)	472	472	551
Fin spacing (m)	0.0021	0.0021	0.0018
Length (finned sec.) in direction of flow (m)	0.11	0.19	0.38
Flow resistance on air side (0.001 kg m)	0.055	0.089	0.209
Coil water flow resistance (0.001 kg m)	31.2	50.8	98.1
By-pass water flow resistance (0.001 kg m)	31.2	50.8	98.1
Valve capacity (m³/hr @ 1 bar)	6.45	5.05	6.64

The generic fuzzy model is generated by combining the fuzzy relational arrays identified from training data obtained from a series of nine simulations: one for each combination of the three designs and three simulated air mass flow rates. Here, air mass flow rate values of 40%, 70% and 100% of the design value are used, covering air flows from the minimum to maximum value. Although the generic model is identified from crisp training data, it describes the thermal behaviour of a class of cooling coil subsystems and its output is therefore inherently fuzzy.

The training data are generated by varying the simulated valve control signal according to a filtered pseudo-random sequence at operating points that coincide with the centres of the fuzzy sets describing the mass airflow rate. The training data are pre-processed to select only those datasets with maximal activation of each combination of input sets before they are presented to the RSK fuzzy identification scheme (see Section 5.3). A 30 second sample interval is used.

Figure 16.5 shows the fuzzy (one-step-ahead) predicted temperature generated from the fuzzy model, when the current air mass flow rate across the coil is set to 0.4 kg/s and the current outlet air temperature $T_{a0}(n) = 14°C$. It can be seen that there is a possibility distribution associated with any given predicted temperature. However, as expected, the predicted temperatures decrease as the value of the control signal increases from 0% to 100% (from no cooling to maximum cooling). The small circles on the $u(n) - T_{ao}(n+1)$ plane indicate the crisp values of the predicted outlet air temperature that would have been generated by the fuzzy model had traditional height defuzzification been used.

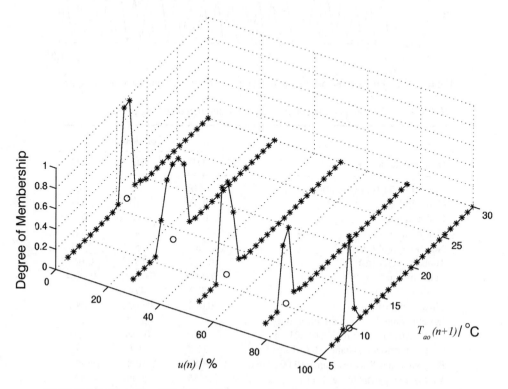

Figure 16.5   Fuzzy predicted air temperature when $\dot{m}_a(n) = 0.4$ kg/s and $T_{a0}(n) = 14°C$.

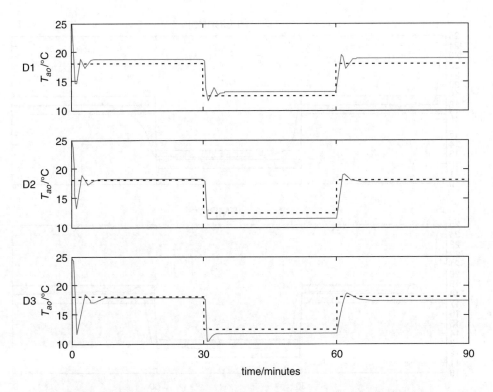

Figure 16.6   Control of different cooling coils using the same FMPC scheme.

Figure 16.6 shows the behaviour of the FMPC when it is used to control each of the simulated coiling coils used to generate the training data. The controller uses the same generic fuzzy model, fuzzy goal, normalization parameters and conditional defuzzification threshold in all three cases. The results demonstrate that the same controller can provide satisfactory control performance when it is used to control chilled-water cooling coils of different designs.

Experiments on a pilot-scale air-handling unit are used to compare the performance of the fuzzy air temperature controller with that of a conventional fixed-parameter proportional-plus-integral (PI) controller. Gain scheduling was not considered because the onsite commissioning of gain-scheduled PI controllers is too time-consuming for them to be used in this type of application. The proportional gain and the integral time constant of the PI controller, which does not have a neutral zone, are $5.0\,°C^{-1}$ and $60\,s$, respectively. The sampling interval of the PI controller is $10\,s$. The fuzzy model-based predictive controller uses a crude estimate of the current airflow rate, which is based on the supply fan control signal as well as a measurement of the current outlet air temperature. The sampling interval is $30\,s$ and the value of the conditional defuzzification threshold is 0.8.

Figures 16.7 and 16.8 show the time variations of the outlet air temperature and the valve control signal in response to step changes in the setpoint at a high airflow rate. Figures 16.9 and 16.10 compare the steady-state and transient performance of the two schemes when they are controlling the outlet air temperature at a low airflow rate. The shading indicates the region

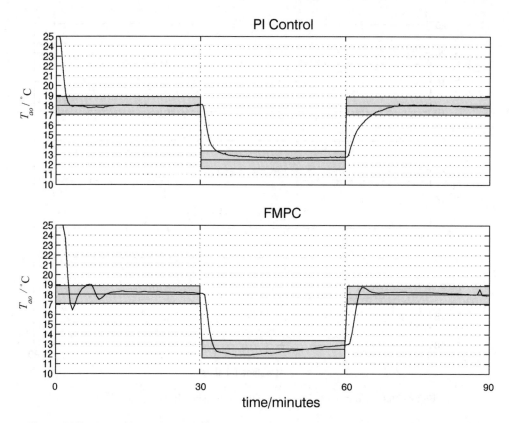

Figure 16.7   Comparison of PI and FMPC control of outlet air temperature at a high airflow rate.

in which the outlet air temperature is considered to be acceptable. The small spikes observed on some of the plots of the outlet air temperature are a result of data collection errors.

At a high airflow rate, the transient behaviour of the PI controller following a step increase in the setpoint is more sluggish than that of the FMPC scheme (see Figure 16.7); the FMPC scheme results in significantly less control activity (see Figure 16.8). As expected however, the steady-state performance of the PI controller is superior to that of the FMPC scheme.

The cycling of the chiller causes the chilled water supply temperature to vary during the experiments. As can be seen in Figure 16.8, the PI controller responds to an increase in chilled water temperature by opening the cooling coil control valve; however, the FMPC scheme allows the outlet air temperature to drift upwards (see Figure 16.7) because it recognizes that the temperature is still within acceptable limits and therefore does not justify changing the position of the control valve.

Controlling the outlet air temperature becomes more difficult at low airflow rates, due to the higher process gain under these conditions. Although the fixed-parameter PI controller produces a sluggish response at high airflow rates (as can be seen in Figure 16.9), the outlet air temperature oscillates around the setpoint at low flow rates. The FMPC controller keeps the outlet air temperature within the desired control limits at the low airflow rate, even though there are far fewer changes in the position of the control valve (see Figure 16.10). It should

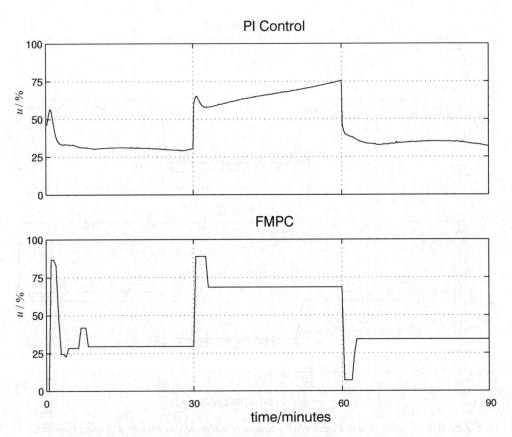

Figure 16.8 Comparison of the control action for PI and FMPC control at a high airflow rate.

also be noted that there is slightly less undershoot following a step down in the setpoint when using the FMPC scheme (see Figure 16.9). The transient behaviour of both control schemes is similar following an increase in setpoint.

Table 16.2 compares the control activity Q associated with each control scheme at both the low and high airflow rates, where

$$Q = \frac{1}{N-1} \sum_{n=2}^{N} |u(n) - u(n-1)|$$

and N is the total number of samples. The values of control activity are normalized with respect to that observed when using the PI control scheme at the low airflow rate.

16.3   Design of a Fuzzy FRM-Based Internal Model Control Scheme

A fuzzy internal model control (FIMC) scheme with an internal model based on a generic fuzzy FRM and a fuzzy model-based predictive controller based on the same fuzzy model

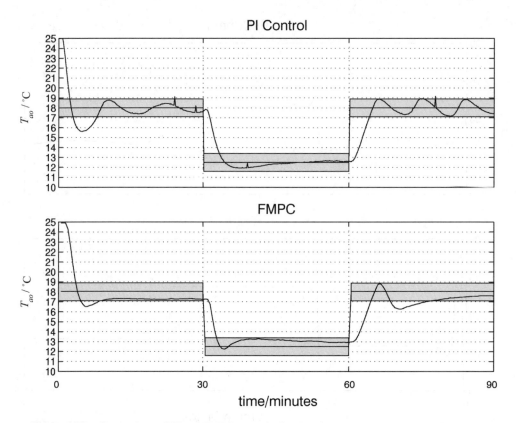

Figure 16.9 Comparison of PI and FMPC control of outlet air temperature at a low airflow rate.

can be used to reject unmeasured disturbances (see Section 9.2). To combat the problems that could arise from modelling errors, the fuzzy predictions (rather than the defuzzified values) generated by the internal model are used to calculate a fuzzy error signal E_{imc}(see Figure 16.11). In this design, the effect of the disturbance on the plant is assumed to be much more significant than the internal model's prediction errors. The defuzzified value of the prediction of the previous plant output, rather than the measured plant output, is used as an input to both the fuzzy model used by the FMPC and the fuzzy internal model of the plant (see Section 9.1.3). The defuzzified values are used because of the additional uncertainties that would be introduced by repeatedly using previous fuzzy predictions as inputs to the models.

As the error signal is fuzzy, the resulting error-adjusted setpoint W_e is also fuzzy. A fuzzy setpoint that reflects the uncertainty in the model's prediction allows the fuzzy decision-making to account for the uncertainty arising from the use of a generic internal model to compensate for the effect of the disturbance.

A detailed computer simulation of the cooling coil subsystem shown in Figure 16.1 is used to assess the performance of the FIMC scheme. The behaviour of the cooling coil is simulated using detailed component models. The value of the air mass flow rate across the coil is varied

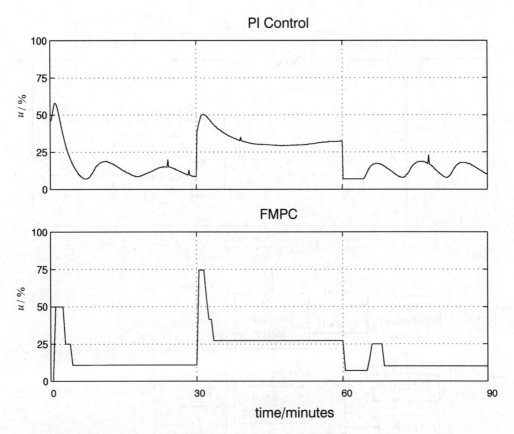

Figure 16.10   Comparison of the control action for PI and FMPC control at a low airflow rate.

to simulate a disturbance on the system. The inlet air temperature is held constant during the tests.

The FIMC uses a generic fuzzy FRM that is identified using training data obtained by simulating nine different designs of the type of cooling coil subsystem under test. The size and structure of the heat exchangers are varied so that training data representative of a class of cooling coils with relatively wide variations in their dynamic and steady-state behaviour can be generated (see Tables 16.3 and 16.4).

Table 16.2   Comparison of the control activities of the two control schemes.

	PI control (%)	FMPC (%)	Reduction in activity (%)
Low flow	100.0	29.3	70.7
High flow	61.6	40.4	21.2

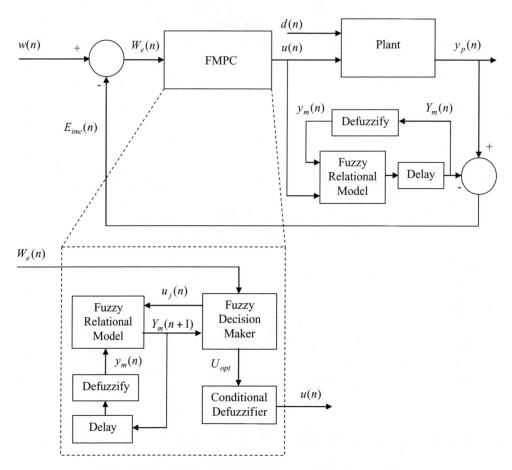

Figure 16.11 FIMC using FMPC with a prediction horizon of one.

The fuzzy FRM predicts the next values of the outlet air temperature from the previous predictions of the outlet air temperature and the control signal determining the position of the 3-port mixing valve. Evenly spaced fuzzy sets with triangular membership functions are used to describe the control signal (11 sets), the previously predicted outlet air temperature (15 sets) and the predicted next value of the outlet air temperature (15 sets).

During the tests, the outlet air setpoint is held constant at 15°C and a sinusoidally varying input disturbance (a varying air mass flow rate) with increasing frequency is applied to the plant under control (see Figure 16.12). Because good low-frequency disturbance rejection requires small incremental steps in the control signal, the conditional defuzzification threshold is set to unity for the FIMC tests. Figure 16.13 shows the observed variations in the outlet air temperature when the FIMC scheme is tested on a simulation of one of the cooling coil designs (D3).

The same tests were repeated using internal model control based on a crisp prediction from the internal model (NIMC). The robustness filter (see Section 9.1.3) used in the NIMC is a second-order low-pass filter with two time constants of 300 s.

Table 16.3   Cooling coil design parameters (designs 1–5).

Design parameters	Design number				
	D1	D2	D3	D4	D5
Airflow rate (kg/s)	1.0	3.0	5.0	1.0	3.0
Inlet water temperature (°C)	5.0	5.0	5.0	7.0	7.0
No. of rows	3	4	3	5	5
No. of circuits	4	13	17	4	13
No. of tubes	16	26	34	16	26
Coil duty (kW)	17.0	52.3	86.8	17.9	53.8
Length of finned section (m)	0.1143	0.1524	0.1143	0.1905	0.1905
Height of finned section (m)	0.6096	0.9906	1.2954	0.6096	0.9906
Width of finned section	0.6100	1.0200	1.3000	0.6100	1.0200
Tube OD (m)	0.0127				
Tube wall thickness (m)	0.00043				
Tube material			Copper		
Fin spacing (/m)	472	394	512	472	551
Fin thickness (m)	0.00024				
Fin material	Aluminium				
Air-side flow resistance (Pa s^2/m^6)	0.0548	0.0077	0.0028	0.0890	0.0137
Coil/bypass water flow resistance	31.198	3.3076	1.7973	50.767	4.0904
Valve type		Equal percentage (coil)/Linear (bypass)			
Valve capacity (m^3/hr@1bar)	6.445	19.795	26.853	5.053	17.800
Valve curvature parameter	20.0				
Valve rangeability	100.0				
Valve leakage	0.0001				

Figure 16.14 shows the observed variations in the outlet air temperature when the NIMC scheme is tested on a simulation of the same cooling coil subsystem (design D3). It can be seen that the outlet air temperature is less sensitive to variations in the airflow rate when the FIMC scheme is used. The improvement in the controller's ability to reject higher-frequency disturbances is most likely a result of the FIMC scheme not requiring a robustness filter to be used.

The overall performance of each of the two controllers is assessed by comparing the root-mean-square value of the tracking error (RMSE) and the control activity, indicated by the mean absolute change in the control signal (MADU) over the test period. The values of the RMSE and the MADU for each controller and for each of the four cooling coil designs are shown in Table 16.5. It can be seen that, in all cases, the performance of the FIMC without a robustness filter is superior to that of the NIMC with a robustness filter. A possible implication of the almost identical values of control activity is that an internal model control scheme using fuzzy predictions from the internal model is much less sensitive to any high-frequency mismatch between the plant and model.

Table 16.4 Cooling coil design parameters (designs 6–9).

Design parameters	Design number			
	D6	D7	D8	D9
Airflow rate (kg/s)	5.0	1.0	3.0	5.0
Inlet water temperature (°C)	7.0	9.0	9.0	9.0
No. of rows	5	10	11	10
No. of circuits	17	4	11	17
No. of tubes	34	16	26	34
Coil duty (kW)	88.8	18.0	54.0	89.8
Length of finned section (m)	0.1905	0.3810	0.4191	0.3810
Height of finned section (m)	1.2954	0.6096	0.9906	1.2954
Width of finned section	1.3000	0.6100	1.0200	1.3000
Tube OD (m)	0.0127			
Tube wall thickness (m)	0.00043			
Tube material		Copper		
Fin spacing (/m)	472	551	512	551
Fin thickness (m)	0.00024			
Fin material		Aluminium		
Air-side flow resistance (Pa s^2/m^6)	0.0042	0.2086	0.0268	0.0095
Coil/bypass water flow resistance	2.929	98.085	14.149	5.757
Valve type		Equal percentage (coil)/Linear (bypass)		
Valve capacity (m^3/hr@1bar)	21.036	3.6349	9.5705	15.003
Valve curvature parameter	20.0			
Valve rangeability	100.0			
Valve leakage	0.0001			

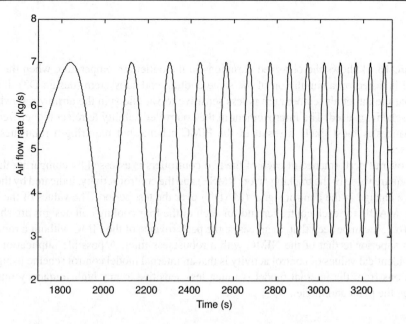

Figure 16.12 Variations in the mass airflow rate. Reproduced by permission of Andrew Wright.

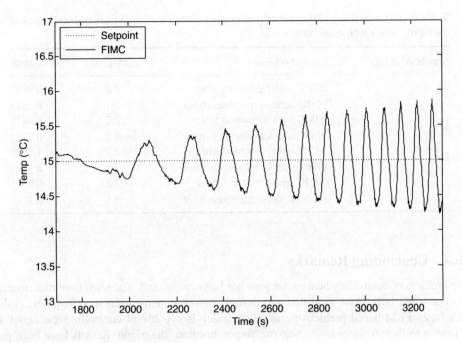

Figure 16.13 Changes in the outlet air temperature from cooling coil design D3 for the FIMC without a robustness filter. Reproduced by permission of Andrew Wright.

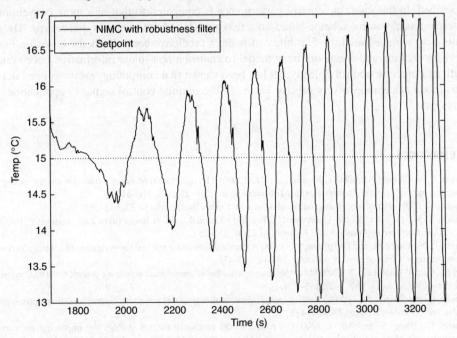

Figure 16.14 Changes in the outlet air temperature from cooling coil design D3 for NIMC control with a robustness filter. Reproduced by permission of Andrew Wright.

Table 16.5 Comparison of the control performance of the FIMC with no robustness filter and the NIMC with a robustness filter.

Simulated design	Control scheme	MADU (%)	RMSE
D3	NIMC with robustness filter	2.2	0.9873
	FIMC without robustness filter	2.1	0.3573
D4	NIMC with robustness filter	2.2	0.9875
	FIMC without robustness filter	2.1	0.3323
D6	NIMC with robustness filter	2.1	1.0024
	FIMC without robustness filter	2.1	0.3582
D7	NIMC with robustness filter	2.2	0.9571
	FIMC without robustness filter	2.1	0.3912

16.4 Concluding Remarks

The problem of controlling heat exchangers has been considered. The most important sources of uncertainty have been identified and design constraints have been specified. The design of a fuzzy FRM-based predictive controller, which is capable of controlling the outlet air temperature from a water-to-air heat exchanger, has been described. Results have been presented that compare the performance of the predictive fuzzy controller to conventional PI control. The design of an internal model control scheme based on a fuzzy FRM has also been described. In this case, the control performance is compared to that of a more conventional internal model control scheme based on a fuzzy model whose output is defuzzified. The results have demonstrated the feasibility of using a predictive control scheme based on fuzzy decision-making and a generic fuzzy model to control a non-linear information-poor system with an imprecise control objective. It has been shown that computing resources need not be wasted on calculating overly precise values of the optimal control signal in applications of this type.

References

Chiou, C.B., Chiou, C.H., Chu, C.M. and Lin, S.L. (2009) The application of fuzzy control on energy saving for multi-unit room air-conditioners. *Applied Thermal Engineering*, **29**(2–3), 310–316.

Ghaius, C. (2001) Fuzzy model and control of a fan-coil. *Energy and Buildings*, **33**, 545–551.

Gouda, M.M., Danaher, S. and Underwood, C.P. (2001) Thermal comfort based fuzzy logic controller. *Building Services Engineering Research and Technology*, **22**(4), 237–253.

Haves, P., Norford, L.K. and DeSimone, M. (1998) A standard simulation test bed for evaluation of control algorithms and strategies. *Transactions of ASHRAE*, **104**(1), 460–473.

He, M., Cai, W-J. and Li, S-Y. (2005) Multiple fuzzy model-based temperature predictive control for HVAC systems. *Information Sciences*, **169**(1–2), 155–174.

Huang, G. and Dexter, A.L. (2008) Realization of robust nonlinear model predictive control by offline optimisation. *Journal of Process Control*, **18**, 431–438.

Huang, G., Wang, S. and Xu, X. (2009) A robust model predictive control strategy for improving the control performance of air-conditioning systems. *Energy Conversion and Management*, **50**(10), 2650–2658.

Linkens, D.A. and Kandiah, S. (1996) Long-range predictive control using fuzzy process models. *Transactions of Institute of Chemical Engineering*, **74** (Part A), 77–88.

Lu, H., Jia, L., Kong, S. and Zhang, Z. (2007) Predictive functional control based on fuzzy T-S model for HVAC systems temperature control. *Journal of Control Theory and Application*, **5**(1), 94–98.

Lygouras, J.N., Kodogiannis, V.S., Pachidis, T., *et al.* (2008) Variable structure TITO fuzzy-logic controller implementation for solar air-conditioning system. *Applied Energy*, **85**, 190–203.

Maidi, A., Diaf, M. and Corriou, J-P. (2008) Optimal linear PI fuzzy controller design of a heat exchanger. *Chemical Engineering and Processing*, **47**(5), 938–945.

Navale, R.L. and Nelson, R.M. (2010a) Use of evolutionary strategies to develop an adaptive fuzzy logic controller for a cooling coil. *Energy and Buildings*, **42**(11), 2213–2218.

Navale, R.L. and Nelson, R.M. (2010b) Use of genetic algorithms to develop an adaptive fuzzy logic controller for a cooling coil. *Energy and Buildings*, **42**(5), 708–716.

Pacheco-Vega, A., Ruiz-Mercado, C., Peters, K. and Vilchiz, L.E. (2009) On-line fuzzy-logic-based temperature control of a concentric-tube heat exchanger facility. *Heat Transfer Engineering*, **30**(14), 1208–1215.

Postlethwaite, B.E. (1996) Building a model-based fuzzy controller. *Fuzzy Sets and Systems*, **79**, 3–13.

So, A.T.P., Chan, W.L., and Tse, W.L. (1997) Self-Learning fuzzy air handling system controller. *Building Services Engineering Research and Technology*, **18**(2), 99–108.

Sousa, J.M., Babuska, R., and Verbruggen, H.B. (1997) Fuzzy predictive control applied to an air-conditioning system. *Control Engineering Practice*, **5**(10), 1395–1406.

Xu, X., Wang, S. and Huang, G. (2010) Robust MPC for temperature control of air-conditioning systems concerning on constraints and multitype uncertainties. *Building Services Research and Technology Research*, **31**(1), 39–55.

Zhao, T., Zhang, J. and Sun, D. (2011) Experimental study on a duty ratio fuzzy control method for fan coils. *Building and Environment*, **46**(2), 527–534.

17

Measurement of Spatially Distributed Quantities

Accurate measurement of the average value of a spatially distributed quantity is notoriously difficult especially if the spatial distribution changes with operating conditions, for example: the average temperature of a fluid in a stratified storage tank (Arahal *et al.*, 2008; Kreuzinger *et al.*, 2008; Belessiotis *et al.*, 2010); the average value of the air temperature in a large office (Jassar *et al.*, 2009, 2011; Liao and Dexter, 2010); or the average value of the flow rate of a fluid flowing through a pipe or duct which has a large cross-sectional area (Tan and Dexter, 2005; Wichman and Braun, 2009; Yu *et al.*, 2011). The main problem with modern electronic sensors is not usually measurement noise, drift or poor calibration that can be eliminated during commissioning, but sensor bias (an offset between the average value of the measurement and the true value of the measured variable).

For example, accurate measurement of the temperature and velocity of the air flowing down a large duct is extremely difficult when there are significant variations in the temperature and velocity over the cross-section of the duct (Carling and Isakson, 1999). Currently available sensors for measuring air temperature and airflow rates in ducts are either inaccurate or expensive. Even commercial multi-point averaging sensors can produce large errors if they are used in certain locations (e.g. immediately downstream of a mixing box; Robinson, 1999). Efficient operation of HVAC systems requires reliable estimates of the average temperature and velocity of air flowing down the ducts. Measurement errors can result in poor control and/or incorrect diagnosis of the causes of poor performance. The presence of these measurement errors is one of the main barriers to the practical application of schemes for optimizing the operation of, and automatically detecting faults in, air-conditioning systems.

17.1 Review of Approaches Suggested for Dealing with Sensor Bias

A number of ways of dealing with sensor bias have been proposed. One option is to estimate and eliminate the bias. This approach, which takes advantage of analytical redundancy and requires a large number of temperature and flow sensors to be installed (Wang and Wang, 1999), assumes that the bias is constant and is computationally too demanding for many

Monitoring and Control of Information-Poor Systems: An Approach based on Fuzzy Relational Models, First Edition.
Arthur L. Dexter.
© 2012 John Wiley & Sons, Ltd. Published 2012 by John Wiley & Sons, Ltd.

applications. Another option is to incorporate the effects of sensor bias into the reference models used by the monitoring and control scheme. This approach allows the measurement errors to be taken into account in model-based fuzzy control schemes (see Chapter 12) and avoids false alarms in fault detection systems (see Chapter 13), but it can result in poor control performance and produce highly ambiguous results when diagnosing faults (especially if a constant worst-case bias is assumed; Ngo and Dexter, 1999). A third option is to discount the 'faulty' sensor reading and use other available measurements to estimate the temperature or flow (Lee *et al.*, 1997). The main weakness of this approach is that estimates based on the output of other sensors may exhibit even more bias because the estimation will be sensitive to modelling errors as well as any bias on the other sensors. Alternatively, a fuzzy sensor (see Section 3.2) could be used, whose fuzzy output reflects the uncertainties associated with the raw measurement (Lee and Dexter, 2002). A fourth option is to take the sensor bias into account when designing the monitoring or control scheme that uses the measurements. For example, a detection threshold can be introduced into a fault diagnosis scheme to avoid false alarms or a dead-band can be incorporated into a control algorithm to avoid unnecessary switching between different modes of operation. The main difficulty of this approach is the selection of an appropriate value for the threshold or dead-band. Too high a value reduces the fault or control sensitivity too much; too low a value causes too many false alarms or too frequent mode switching.

17.2 An Example Application

Many large-scale industrial processes involve the flow of air down ducts with large cross-sections. The associated monitoring and control schemes are often based on mass and energy balance equations which assume that the velocity and temperature of the air is the same over all of the cross-section. In practice, this is unlikely to be the case and an estimate of the spatial averages of the velocity and temperature must be used that are based on a finite number of single-point measurements (Lee and Dexter, 2005).

17.2.1 Air Temperature Estimation Using a Single-Point Sensor with Bias Correction

The average temperature $\overline{T}_{\text{true}}$ over the cross-section of the duct weighted by velocity is given by

$$\overline{T}_{\text{true}} = \frac{\int_A VT \, dA}{\int_A V \, dA} = \frac{\int_L VT \, dl}{\int_L V \, dl} \qquad (17.1)$$

The systematic or bias errors are defined as the difference between the mean of the measurements and the true average value:

$$B_T = T_{\text{meas}} - \overline{T}_{\text{true}} \qquad (17.2)$$

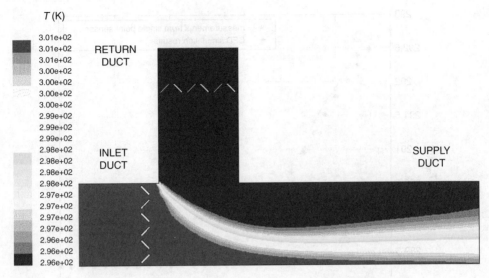

Figure 17.1 Temperature stratification in a symmetric mixing box.

For example, Figure 17.1 shows the results of computer simulation based on computational fluid dynamics (CFD) of the temperature stratification in a part of the duct work of an air-conditioning system (the so-called mixing box) where two streams of air at different temperatures are mixed. It can be seen that there is significant stratification of air temperature in the mixing box. Therefore, there can be a large bias associated with the measurement if a single-point temperature sensor is used to measure the mixed air temperature T_{mix} at the outlet of the mixing box.

Three types of uncertainty cause CFD simulation results to differ from their true values (Carling and Zou, 2001). The first type is physical approximation errors which include physical modelling errors associated with the mathematical models in the form of momentum and energy equations that are used to describe the fluid flow and geometric modelling errors arising from the finite representation of the fluid flow. The second type of uncertainty is numerical error which includes discretization errors, convergence errors and computer round-off errors. It has been shown that the physical approximation errors and numerical errors are negligible for the modelling of a symmetric mixing box (Tan and Dexter, 2006b). The third type of uncertainty is caused by the errors in the boundary conditions which include the values of the inlet airflow rates and air temperatures. The effects of errors in the boundary conditions on the estimated measurement bias can be evaluated by perturbing the inputs of the CFD simulation (Tan and Dexter, 2006b).

Figures 17.2 and 17.3 compare the results of the CFD simulation of the mixing box on a laboratory test rig with the measurements collected from the rig. The duct size of the mixing box is about 0.3 m × 0.3 m. A single-point temperature-sensing probe is mounted 0.3 m downstream of the mixing box in the centre of the supply duct. Measurements from the single-point temperature sensor are collected while traversing the probe across the duct vertically from the lower wall to the upper wall. The experiment is repeated while setting the mixing-box dampers to different positions. It should be noted that the operation of the inlet and return dampers is synchronized so that the supply airflow rate Q_{sup} is approximately constant (i.e. $u_{d_{in}} = 1 - u_{d_{ret}} = u_d$)

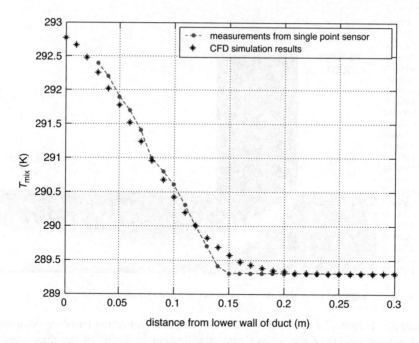

Figure 17.2   CFD simulation results and experimental measurements with $u_{d_{in}} = 0.35$ and $u_{d_{ret}} = 0.65$.

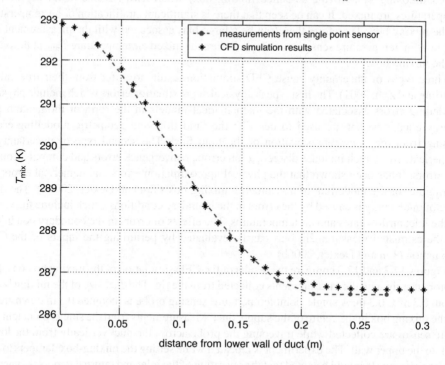

Figure 17.3   CFD simulation results and experimental measurements with $u_{d_{in}} = 0.5$ and $u_{d_{ret}} = 0.5$.

To minimize the errors of the CFD simulation caused by boundary settings, the inlet and return air temperatures are measured using averaging temperature sensors installed in the inlet duct and return duct, respectively. The inlet and return airflow rates are estimated using a measurement of the supply airflow rate and a simple mixing-box model (Tan and Dexter, 2006a). The measured data are obtained from the single-point probe at different vertical positions in the outlet duct. It can be seen that the CFD simulation results are in good agreement with the measurement values when the errors in the boundary settings are negligible. The maximum discrepancy between the simulation and measurements is approximately 0.5 K.

The CFD simulation can therefore be used to investigate how the measurement bias changes with the different operating parameters:

- positions of the mixing-box dampers;
- value of the supply airflow rate;
- temperature difference between the inlet air and return air;
- size of the ducts of the mixing box; and
- horizontal positions of the sensor probes.

The results of different CFD simulation studies have shown that, for a given geometric design, the temperature stratification pattern in the mixing box is most sensitive to the damper positions. Consequently, biases associated with the measurements from a mixed air temperature sensor will change with the control signals for the mixing-box dampers. The simulation results have also shown that the magnitude of the measurement bias is linearly proportional to the temperature difference T_{diff} between the return air and inlet air and that it is insensitive to the value of the supply airflow rate or the size of the duct. The measurement bias is however very sensitive to the vertical and horizontal position of the sensor because of the temperature stratification of the air leaving the mixing box. Here it is assumed that a single-point sensor is installed in the centre of the supply duct $4D/3$ away from the mixing box where D is the height of the square ducts in the mixing box.

An estimate of the measurement bias $\hat{B}_{T_{\text{mix}}}$ is constructed as follows:

$$\hat{B}_{T_{\text{mix}}} = T_{\text{diff}}\, f_{B_T}(u_d) \tag{17.3}$$

where

$$T_{\text{diff}} = T_{\text{in}} - T_{\text{ret}} \tag{17.4}$$

and

$$f_{B_T}(u_d) = \sum_{i=1}^{n+1} p_{B_T}(i) . u_d^{n-i+1} \tag{17.5}$$

The parameters of the polynomial $p_{B_T}(i)$ are obtained by fitting a nth order polynomial function to the measurement biases obtained from the simulation results. As can be seen in Figure 17.4, the fitting errors are very small when a 9th-order polynomial curve is used.

An estimate of the bias is therefore given by

$$\hat{B}_{T_{\text{mix}}} = f_{B_T}(u_d)(T_{\text{in}} - T_{\text{ret}}) \tag{17.6}$$

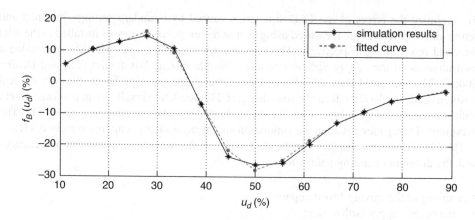

Figure 17.4 Simulation results and fitted curve for the temperature measurement bias.

As it has been found that the measurement bias is insensitive to the value of the supply airflow rate or the size of the duct, Equation (17.6) can be used to estimate the bias on the measurement of the mixed air temperature obtained from a single-point sensor in any symmetric mixing box.

The estimated bias can be used to correct the raw measurements:

$$\hat{T}_{\text{mix}} = T_{\text{mix}} - \hat{B}_{T_{\text{mix}}} \qquad (17.7)$$

The main factors that cause errors in the estimation of the bias are:

- uncertainty in the vertical position of the sensing probe;
- uncertainty in the values of the boundary conditions used in the CFD simulation;
- uncertainty introduced by the assumption that the relationship between the damper angle and the damper control signal is linear;
- errors arising from the polynomial fit; and
- uncertainty in the measurement of T_{in} and T_{ret}.

The overall uncertainty associated with the estimation of the bias can be calculated using the root-sum-square method (see Section 2.2.1) to combine the individual uncertainties (Tan and Dexter, 2006b).

The uncertainty associated with $\hat{T}_{\text{mix}}$ is given by:

$$\delta^2_{\hat{T}_{\text{mix}}} = \delta^2_{T_{\text{mix}}} + \delta^2_{\hat{B}_{\text{mix}}} \qquad (17.8)$$

where $\delta^2_{T_{\text{mix}}}$ is the random error associated with the measurement and $\delta^2_{\hat{B}_{\text{mix}}}$ is the uncertainty associated with the estimation of the bias.

The commissioning of an air-conditioning system in a six-zone building is simulated to examine the effect of using the bias estimator to correct the output of a single-point sensor measuring the mixed air temperature. The measured values of the mixed air temperature are simulated by adding a random noise and a systematic error to the simulated true values. The simulated systematic error is obtained from detailed three-dimensional CFD simulations of

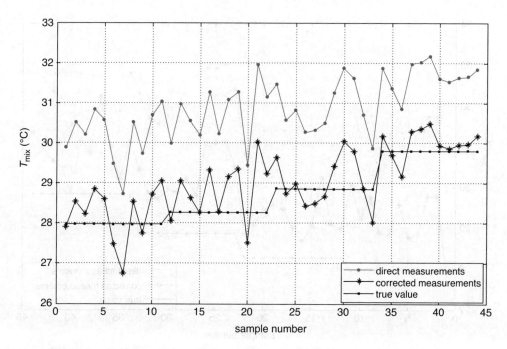

Figure 17.5 Measurements of mixed air temperature before and after correction when the simulated mixing box is symmetric.

the mixing box used in the simulated air-conditioning system. Two cases are considered: one when the geometry of the mixing box is symmetric and the other when the geometry is asymmetric. The simulated measurements, the corrected measurements and the true values for the two cases are plotted in Figures 17.5 and 17.6.

It can be seen from Figure 17.5 that, when the geometry of the mixing box is symmetric, the estimate of the bias associated with the measurement of the mixed air temperature is accurate and measurement bias is successfully eliminated after correction. However, the results presented in Figure 17.6 show that the estimation of the bias is not accurate when the geometry of the mixing box is asymmetric; the corrected measurements are still biased to some extent.

17.2.2   Air Temperature Estimation Based on Mass and Energy Balances

Mass and energy balances can also be used to estimate the mixed air temperature from the inlet airflow rate and air temperature and the return airflow rate and air temperature. In this case the estimate is given by

$$\hat{T}_{mix} = \frac{(Q_{in}T_{in} + Q_{ret}T_{ret})}{Q_{sup}} = \gamma\, T_{in} + (1 - \gamma)T_{ret} \tag{17.9}$$

where $\gamma = Q_{in}/Q_{sup}$ is the ratio of the inlet airflow rate to the supply airflow rate. Unfortunately, it is unusual for the supply airflow rate to be measured and accurate measurement of the inlet airflow rate is extremely difficult (Tan and Dexter, 2006a). Nevertheless, this approach can be

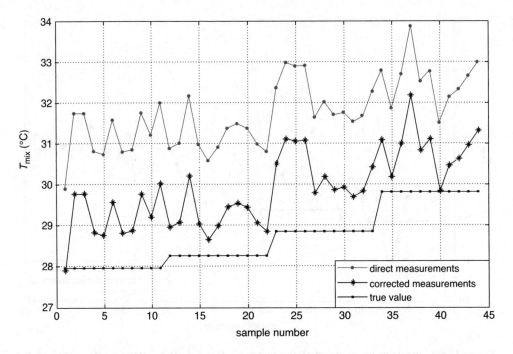

Figure 17.6 Measurements of mixed air temperature before and after correction when the simulated mixing box is asymmetric.

used to estimate the mixed air temperature if reasonably accurate estimates of the airflow rates are available.

17.2.2.1 Estimating the Supply and Extract Airflow Rates Using the Fan Control Signals

Computer simulation of the air circuits in the air-conditioning system (see Figure 17.7) has shown that there are approximately linear relationships between the supply and extract fan control signals and the flow rates through the fans, and that the straight-line relationship

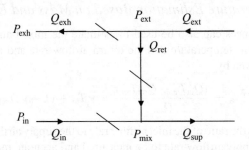

Figure 17.7 Air circuits in the air-conditioning system.

passes through the origin in the case of the extract flow and extract fan control signal (Tan and Dexter, 2006a).

The supply airflow rate Q_{sup} and the extract airflow rate Q_{ext} can therefore be estimated using the equations:

$$Q_{sup} = k_{sup} u_{f_{sup}} + a_{sup} \quad \text{and} \quad Q_{ext} = k_{ret} u_{f_{ret}} \tag{17.10}$$

where the coefficients k_{sup}, a_{sup} and k_{ret} can be determined from experimental data obtained during commissioning.

The results of uncertainty analysis show that the uncertainty associated with the supply airflow rate estimator is within 10% of the true value and the uncertainty associated with the extract airflow rate estimator is about 3% of the true value.

17.2.2.2   Estimating the Inlet Airflow Rate Based on a Mixing-Box Model

The mixing-box model is based on the mass balance constraints and the pressure drop equations in the mixing box. Given the values of the supply Q_{sup} and extract airflow rates Q_{ext}, and assuming that both P_{in} and P_{exh} are equal to the atmospheric pressure and there is no reverse flow in the exhaust duct, the value of the inlet airflow rate Q_{in} is given by:

$$Q_{in} = \frac{\left(-b + \sqrt{b^2 - 4ac}\right)}{2a} \tag{17.11}$$

where

$$\begin{aligned} a &= R_{exh} + R_{in} - R_{ret} \\ b &= 2R_{ret}Q_{sup} - 2R_{exh}(Q_{sup} - Q_{ext}) \\ c &= R_{exh}(Q_{sup} - Q_{ext})^2 - R_{ret}Q_{sup}^2 \end{aligned} \tag{17.12}$$

and R_{in}, R_{ret} and R_{exh} are the pneumatic resistance of the duct and dampers in the inlet, the return and the exhaust sections of the ductwork, respectively. These values can be obtained from design information and empirical relationships.

Monte Carlo simulation can be used to evaluate the effect of the uncertainty associated with the resistances, and the estimates of the supply and extract airflow rates, on the estimation of the inlet airflow rate. For example, if it is assumed that the uncertainty in resistance of the dampers is given by $\delta_R/R = 24\%$, the uncertainties associated with the estimation of the airflow rates are given by $\delta Q_{sup}/Q_{sup} = 10\%$ and $\delta Q_{ext}/Q_{ext} = 5\%$ and 10 000 independent samples of the inputs are generated, the uncertainties associated with the estimated inlet airflow rates for different damper positions are given in Table 17.1.

Table 17.1   Uncertainty in the outputs of the mixing-box model.

u_d	0.3	0.4	0.5	0.6	0.7	0.8	0.9
$\delta Q_{in}/Q_{in}(\%)$	32	27	18.6	13.7	11.2	11	10
γ	0.12	0.24	0.44	0.66	0.81	0.88	0.93
$\delta_\gamma(\%)$	5.4	5.5	5.4	4.4	2.3	1.5	1.0

It should be noted that, although the uncertainty associated with the estimated inlet airflow rate is between 10 and 32% of the true value, the uncertainty associated with γ (the ratio of inlet airflow rate to the supply airflow rate) is always smaller than 5.5%.

The uncertainty associated with the estimate of the mixed air temperature based on energy balance equation is given by:

$$\delta_{T_{mix}}^2 = (T_{sup} - T_{ret})^2 \delta_\gamma^2 + (1 - \gamma)^2 \delta_{T_{ret}}^2 + \gamma^2 \delta_{T_{sup}}^2 \tag{17.13}$$

where the value of δ_γ is obtained from Table 17.1 and $\delta_{T_{ret}}^2$ and $\delta_{T_{sup}}^2$ are the uncertainties associated with the measurements of the return air temperature and supply air temperature.

Figure 17.8 shows the results obtained when this method of estimating the mixed air temperature is used during the simulated commissioning test described in Section 17.3.1. It can be seen that the maximum error associated with the estimates of the mixed air temperature, based on an energy balance and the estimated airflow rates, is about $1°C$.

The estimation errors can be further reduced by using fuzzy data fusion (see Section 3.4) to combine the two independent ways of estimating the mixed air temperature.

17.3 Using Bias Estimation and Fuzzy Data Fusion to Improve Automated Commissioning in Air-Handling Units

The fuzzy FRM-based fault diagnosis described in Section 13.2 can be very sensitive to any bias on the output of the sensors used to monitor the plant (Ngo and Dexter, 1999). Bias estimation and fuzzy data fusion can be used to increase the sensitivity to faults, improve the precision of the diagnosis and reduce the rate of false alarms.

Consider the analysis of the data collected from a simulated commissioning test on the cooling coil subsystem of the air-handling unit. The mixed air temperature (T_{mix}), the supply

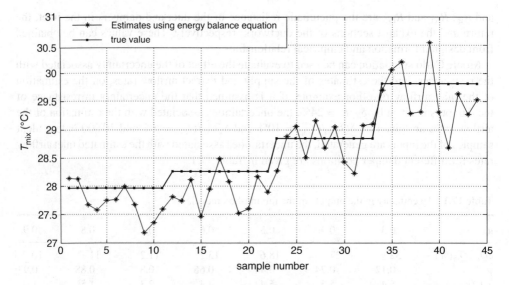

Figure 17.8 Estimates of mixed air temperature based on energy balance equation.

airflow rate (Q_{sup}) and the temperature difference across the cooling coil (T_{diff}) are used for the diagnosis. The results of fault diagnosis based on direct measurements of these quantities are compared to those obtained from estimates generated by defuzzifying the outputs of a fuzzy data fusion scheme. Two cases, corresponding to whether or not the bias associated with the measurement of the mixed air temperature is estimated accurately, are considered.

17.3.1 Diagnosis When the Measurement Bias is Estimated Accurately

Fuzzy data fusion is used to combine (1) measurements of the supply and extract airflow rates with estimates of the flow rates obtained from the fan control signals (see Section 17.2.2.1); and (2) a single-point measurement of the mixed air temperature after it has been corrected for bias (see Section 17.2.1), with an estimate of the mixed air temperature based on energy and mass balances (see Section 17.2.2). The inlet airflow rate is estimated using the mixing-box model described in Section 17.2.2.2. The bias on the measurement of the mixed air temperature is calculated using Equation (17.6) with parameters estimated from data obtained from a CFD simulation of a symmetric mixing box. The simulated air-conditioning system to be commissioned has a similar, but larger, symmetric mixing box.

The direct measurements, the estimates after fuzzy data fusion and the true values observed during a commissioning test are shown in Figure 17.9. It can be seen that the errors associated with the estimates after data fusion are relatively small compared to the errors associated with the direct measurements.

The results of the fault diagnosis using each of the three datasets are compared in Table 17.2. It can be seen that the fault diagnosis scheme generates the correct diagnosis when the true values are used. Ambiguous diagnoses or incorrect diagnoses are generated when the direct measurements are used because of the systematic errors associated with the measurements. The fault detection scheme generates the correct diagnoses when the estimates obtained from fuzzy data fusion are used and the sensor biases have been greatly reduced.

17.3.2 Diagnosis When the Estimate of the Measurement Bias is Inaccurate

The bias on the measurement of the mixed air temperature is again calculated using Equation (17.6) with parameters estimated from data obtained from a CFD simulation of a

Table 17.2 Results of fault diagnosis when the bias estimation model is correct.

| | Actual state of the tested system | | |
Test data	Fault-free	Fouled-coil	Leaky-valve
True values (%)	Bel (c) = 41.2 (correct diagnosis)	Bel (f) = 46.5 (correct diagnosis)	Bel (l) = 84.1 (correct diagnosis)
Direct measurements (%)	Bel (c) = 22.3 Bel (m) = 6.7 (ambiguous diagnosis)	Bel (l) = 7.1 (incorrect diagnosis)	Bel (l) = 87.6 Bel (m) = 7.4 (ambiguous diagnosis)
Estimates obtained from fuzzy fusion (%)	Bel (c) = 53.2 (correct diagnosis)	Bel (f) = 61.3 (correct diagnosis)	Bel (l) = 65.3 (correct diagnosis)

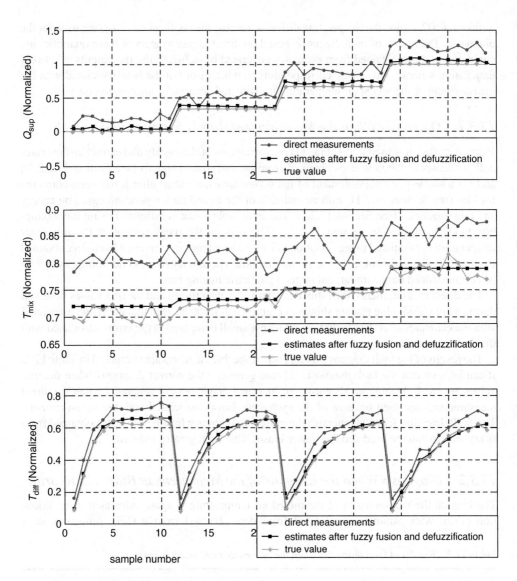

Figure 17.9 Comparison of the true values with direct measurements and estimates obtained after data fusion when the design of the mixing box is symmetric.

symmetric mixing box. In these experiments, however, the mixing box used in the simulated air-conditioning system has an asymmetric design. As before, fuzzy data fusion is used to combine the measurements and estimates of the air flows and temperatures.

The direct measurements, the estimates after data fusion and the true values are shown in Figure 17.10. It can be seen that, although the measurement errors are smaller, the estimates obtained after data fusion are still biased to some extent.

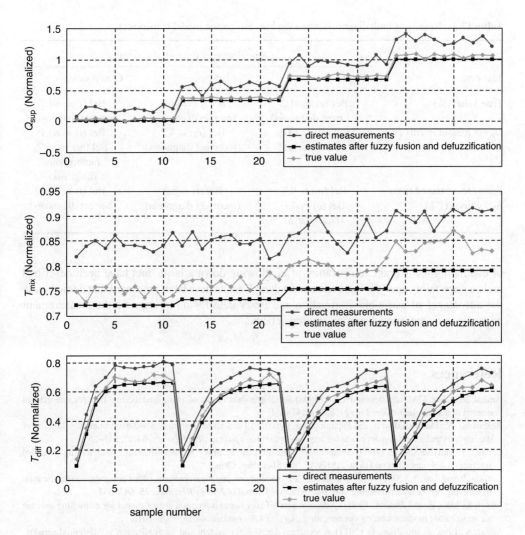

Figure 17.10 Comparison of the true values with direct measurements and estimates obtained after data fusion when the design of the mixing box is asymmetric.

The results of the fault diagnosis are compared in Table 17.3. Note that the fault diagnosis scheme now generates false alarms when no faults are present and an incorrect diagnosis when the coil is fouled.

17.4 Concluding Remarks

The problem of measuring spatially distributed quantities has been considered in this chapter. A brief introduction to the issues associated with measuring spatial averages has been given, and approaches to dealing with sensor bias have been reviewed. As an example, two methods

Table 17.3 Results of fault diagnosis when the bias estimation model is incorrect.

Test data	Actual state of the tested system		
	Fault-free	Fouled-coil	Leaky-valve
True value (%)	Bel (c) = 41.2 (correct diagnosis)	Bel (f) = 46.5 (correct diagnosis)	Bel (l) = 84.1 (correct diagnosis)
Direct measurements (%)	Bel (m) = 13.3 (false alarm)	Bel (m) = 4.2 (incorrect diagnosis)	Bel (l) = 80.1 Bel (m) = 15.7 (ambiguous diagnosis)
Estimates obtained from fuzzy fusion (%)	Bel (m) = 9.7 Bel (c) = 4.1 (false alarm)	Bel (l) = 8.3 (incorrect diagnosis)	Bel (l) = 89.5 (correct diagnosis)

of estimating the average temperature of air flowing down a large duct have been described and an uncertainty analysis has been undertaken. Results have been presented that demonstrate the advantages of using bias estimation and fuzzy data fusion to improve the measurement accuracy in an automated commissioning scheme.

References

Arahal, M., Cirre, C.M. and Berenguel, M. (2008) Serial grey-box model of a stratified thermal tank for hierarchical control of a solar plant. *Solar Energy*, **82**(5), 441–451.

Belessiotis, V., Mathioulakis, E. and Papanicolaou, E. (2010) Theoretical formulation and experimental validation of the input-output modelling approach for large solar thermal systems. *Solar Energy*, **84**(2), 245–255.

Carling, P. and Isakson, P. (1999) Temperature measurement accuracy in an air-handling unit mixing box. 3rd International Symposium on HVAC, ISHVAC '99, Shenzhen, China.

Carling, P. and Zou, Y. (2001) A comparison of CFD-simulations and measurements of the temperature stratification in a mixing box of an air-handling unit. *International Journal of Energy Research*, 25, 643–653.

Jassar, S., Liao, Z. and Zhao, L. (2009) Adaptive neuro-fuzzy based inferential sensor model for estimating average air temperature in space heating systems. *Building and Environment*, **44**(8), 1609–1616.

Jassar, S., Liao, Z. and Zhao, L. (2011) A recurrent neuro-fuzzy system and its application in inferential sensing. *Applied Soft Computing*, **11**(3), 2935–2945.

Kreuzinger, T., Bitzer, M. and Marquardt, W. (2008) State estimation of a stratified storage tank. *Control Engineering Practice*, **16**(3), 308–320.

Lee, P.S. and Dexter, A.L. (2002) Automated commissioning of air-handling units using fuzzy temperature sensors. Proceedings of International Conference on Advanced Building Technology, Hong Kong, 1, 157–164.

Lee, P.S. and Dexter, A. L. (2005) A fuzzy sensor for measuring the mixed air temperature in air-handling unit. *Measurement*, **37**(1), 83–93.

Lee, W-Y, Park, C., House, J.M. and Shin, D.R. (1997) Fault diagnosis and temperature sensor recovery for an air-handling unit. *Transactions of ASHRAE* **103**, 1.

Liao, Z. and Dexter, A.L. (2010) An inferential model-based predictive control scheme for optimising the operation of boilers in of building space heating systems. *IEEE Transactions on Control Systems Technology*, **18**(5), 1092–1102.

Ngo, D. and Dexter, A. L. (1999) A robust model-based approach to diagnosing faults in air-handling units. *ASHRAE Transactions*, **105**(1).

Robinson, K.D. (1999) Mixing effectiveness of AHU combination mixing/filter box with and without filters. *Transactions on ASHRAE*, **1**(Pt. 1).

Tan, H. and Dexter, A.L. (2005) Improving the accuracy of sensors in building automation systems. Proceedings of the 16th IFAC World Congress, Prague, Czech Republic, 2005.

Tan, H. and Dexter, A.L. (2006a) estimating airflow rates in air-handling units from actuator control signals. *Building and Environment*, **41**(10), 1291–1298.

Tan, H. and Dexter, A.L. (2006b) Automated commissioning of a cooling coil using a smart mixed-air temperature sensor. Proceedings of 7th International Conference on System Simulation in Buildings SSB'2006, Paper P14, Liege, Belgium.

Wang, S.W. and Wang, J.B. (1999) Law-based sensor fault diagnosis and validation for building air-conditioning systems. *International Journal of HVAC&R Research*, **5**(4), 353–380.

Wichman, A. and Braun, J.E. (2009) A smart mixed-air temperature. *HVAC & R Research*, **15**(1), 101–115.

Yu, D., Li, H. and Yang, M. (2011) A virtual supply airflow rate meter for rooftop air-conditioning units. *Building and Environment*, **46**(6), 1292–1302.

Index

Monitoring and Control of Information-Poor Systems: An Approach based on Fuzzy Relational Models, First Edition.
Arthur L. Dexter.
© 2012 John Wiley & Sons, Ltd. Published 2012 by John Wiley & Sons, Ltd.